Modelling and Mechanics of Carbon-based Nanostructured Materials

Modelling and Mechanics of Carbon-based Nanostructured Materials

Duangkamon Baowan

Barry J. Cox

Tamsyn A. Hilder

James M. Hill

Ngamta Thamwattana

ELSEVIER

William Andrew
Applied Science Publishers
elsevier.com

William Andrew is an imprint of Elsevier
The Boulevard, Langford Lane, Kidlington, Oxford, OX5 1GB, United Kingdom
50 Hampshire Street, 5th Floor, Cambridge, MA 02139, United States

Notices
Knowledge and best practice in this field are constantly changing. As new research and experience broaden our
understanding, changes in research methods, professional practices, or medical treatment may become necessary.

Practitioners and researchers must always rely on their own experience and knowledge in evaluating and using
any information, methods, compounds, or experiments described herein. In using such information or methods
they should be mindful of their own safety and the safety of others, including parties for whom they have a
professional responsibility.

To the fullest extent of the law, neither the Publisher nor the authors, contributors, or editors, assume any liability
for any injury and/or damage to persons or property as a matter of products liability, negligence or otherwise, or
from any use or operation of any methods, products, instructions, or ideas contained in the material herein.

Library of Congress Cataloging-in-Publication Data
A catalog record for this book is available from the Library of Congress

British Library Cataloguing-in-Publication Data
A catalogue record for this book is available from the British Library

ISBN: 978-0-12-812463-5

For information on all William Andrew publications
visit our website at https://www.elsevier.com/

Working together
to grow libraries in
developing countries

www.elsevier.com • www.bookaid.org

Publisher: Matthew Deans
Acquisition Editor: Simon Holt
Editorial Project Manager: Lindsay Lawrence
Production Project Manager: Priya Kumaraguruparan
Cover Designer: Greg Harris

Typeset by SPi Global, India

Contents

Preface

This book is designed to support third- and fourth-year undergraduate courses in applied mathematical modelling in the currently topical area of nanotechnology. It is purposely designed as a self-contained text, which demonstrates to the student the process of utilising elementary geometry and mechanics combined with some special function theory to formulate simple applied mathematical models in a theoretical physics context, which have been traditionally analysed through computational procedures such as molecular dynamics simulations.

The essential modelling content of the book is the use of the continuum assumption for atomic surface densities and the replacement of double summations with double surface integrals involving atomic potential functions $\Phi(\rho) = -A\rho^{-n}$. For certain surfaces, such integrals can often be evaluated in terms of well-known analytical functions, but generally these integrations are highly nontrivial. For example, even for the electrostatic or Newtonian gravitational potential $\Phi(\rho) = -A\rho^{-1}$, Sir James Jeans states in page 33 of 'The Mathematical Theory of Electricity and Magnetism' that 'An attempt to perform the integration, in even a few simple cases, will speedily convince the student that the form is not one which lends itself to rapid progress'. The mathematical perspective of this book is that many of the integrals can be identified from integral representations of the various hypergeometric functions, such as

$$F(\alpha, \beta; \gamma; z) = \frac{\Gamma(\gamma)}{\Gamma(\beta)\Gamma(\gamma - \beta)} \int_0^1 \frac{t^{\beta-1}(1-t)^{\gamma-\beta-1}}{(1-tz)^\alpha} dt,$$

valid for $\mathrm{Re}(\gamma) > \mathrm{Re}(\beta) > 0$, and Γ denotes the usual gamma function; there are similar results for the Appell hypergeometric functions.

Such formulae are important from two points of view. Firstly, such an identification subsequently leads to the evaluation of the integral in terms of better-known special functions, such as Legendre polynomials, and secondly, through algebraic packages such as MAPLE and MATLAB, the hypergeometric functions can be readily evaluated numerically. Throughout the book such packages are frequently adopted for the numerical evaluation of integrals. In situations where a large numerical landscape is required, these approaches are far more computationally effective than molecular dynamics simulations. We further comment that many of the integrals arising from the hypergeometric function or the Appell hypergeometric function can also be evaluated equally well in terms of elliptic functions. These details are not included in the book; the interested reader should consult the original publications cited in the bibliography.

The first six chapters are intentionally introductory in nature, focussed on the key ideas, and restricted to material considered to be accessible by students in their third year of the traditional Australian undergraduate mathematics degree. The final five chapters deal with the same ideas introduced earlier, but the situations and geometry are more complicated than earlier examples. Many of the exercises and examples throughout the book emanate from the research work of the Nanomechanics Group at the University of Wollongong. In nanotechnology and nanobiotechnology at the moment, there is not a lack of information; rather, there is an abundance of information but very little insight into underlying physical or biological mechanisms. In order to move forward there is a need to condense this information in the form of simple models which properly encapsulate the known behaviour. We believe that at least for some situations, this book demonstrates to the student how simple geometry, mechanics, and mathematical analysis might be exploited to obtain physically meaningful models.

The authors are especially grateful to Professor Quanshui Zheng from the Department of Engineering Mechanics at Tsinghua University, Beijing, whose helpful advice has materially assisted in the research work underlying this text.

D. Baowan, B.J. Cox, T.A. Hilder, J.M. Hill and N. Thamwattana

GEOMETRY AND MECHANICS OF CARBON NANOSTRUCTURES

1.1 BACKGROUND

The advent of nanoscience and nanotechnology has generated considerable advances in many industries such as composite materials, electronics, and medicine. The prefix 'nano' means small and derives from the Greek word for dwarf, namely nanos. In a scientific context, 'nano' means one billionth, and a nanometer (nm) means one billionth of a meter or one millionth of a millimeter. One angstrom (Å) is ten nanometers; for convenience angstrom units are used throughout the book. In this book, the term 'nanoscience' refers to the study of the structures and properties of materials on the nanometer scale (nanomaterials), and the term nanotechnology refers to the synthesis, control, engineering, and manipulation of the nanomaterials. By nanometer scale, we mean a length scale at the level of several atoms and molecules.

The unique physical properties observed at the nanoscale are often counter-intuitive, sometimes surprising researchers and thus driving numerous investigations into their special properties and their potential applications. Typically, existing research has been conducted through experimental studies and molecular dynamics simulations. However, mathematical modelling often facilitates device development and provides a quicker route to applications of the technology. As stated by Ferrari (2005) 'Novel mathematical models are needed, in order to secure the full import of nanotechnology into oncology'. Mathematical models are not only important for nanomedicine but in all areas of nanotechnology, where it is important to fully comprehend often subtle or complex phenomena and perhaps save on time-consuming experimental studies. Applied mathematics is an important tool which may provide insight, assess feasibility and deliver overall guidelines for subsequent experimental and molecular dynamics studies. This book provides an introduction to the use of mathematics and mechanics in nanoscience and nanotechnology. We refer the reader to Dresselhaus et al. (1996), Saito et al. (1998), and Harris (2002) for more physically based accounts of the various carbon structures.

Until recently only two types of all-carbon crystalline structures were known, namely the naturally occurring allotropes diamond and graphite. The breakthrough discovery of carbon nanotubes and fullerenes have revolutionised carbon science from experiments on clusters formed by laser vaporisation of graphite. The discovery of fullerenes and nanotubes has generated considerable research into their properties and potential applications, and many more carbon nanostructures have also been discovered. For example, the nanopeapod, which is a carbon nanotube with many fullerenes encapsulated within its interior, is one such novel nanostructure. Other carbon nanostructures which have received attention are nanotori, nano-onions, and nanobundles, which are illustrated in Fig. 1.1.

Modelling and Mechanics of Carbon-based Nanostructured Materials. http://dx.doi.org/10.1016/B978-0-12-812463-5.00001-7

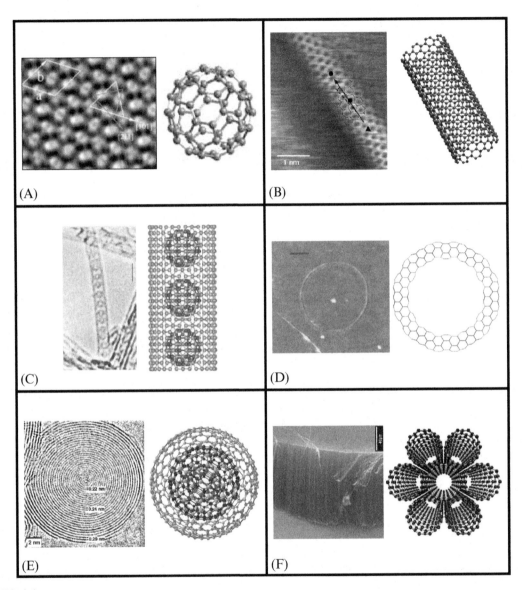

FIG. 1.1

Various carbon nanostructures. (A) C_{60} fullerene, (B) carbon nanotube, (C) nanopeapod, (D) nanotori, (E) nano-onion, and (F) nanobundle.

Briefly, a nanotorus or fullerene 'crop circle' is a circular formation of a carbon nanotube which joins together seamlessly; a nano-onion, like the name suggests, is a collection of concentric fullerenes of varying radii encapsulated within each other; and a nanobundle is a stable circular arrangement of several nanotubes. Our focus is the use of geometry and mechanics for modelling; in contrast to quantum mechanical, electrical, or optical applications.

As a consequence of their unique mechanical and electronic properties, carbon nanostructures are being investigated for their use in numerous applications such as:

- Nanoreinforced composites: producing more durable, lightweight materials
- Nanocapsules: used for drug delivery, toxin storage, and magnetic resonance imaging scans
- Shock absorbers, or man-made molecular springs
- Nanoscale instruments: nanoprobes for scanning force microscopy, nanobearings with super low friction, molecular oscillators with frequencies as high as several gigahertz, and nanosensors
- Nanostraws: to penetrate living cells and inject molecules without damaging the cellular structure
- Computer memory: to further miniaturise the computer industry
- Superconductors: increasing the current carrying capacity

and there are many more potential applications of nanotechnology other than those listed here.

The vast amount of potentially important applications of carbon nanostructures and their unique physical, mechanical, and electronic properties has stimulated many investigations. Modelling provides a critical basis for providing a predictive capacity and for understanding and designing nanostructures and their applications. This book focuses on the existing mathematics and mechanics, which has proved useful for the modelling of nanostructures. The following section provides an introduction to the topic and discusses various carbon nanostructures. In particular, the section outlines: graphene, including details on carbon–carbon bonding; carbon nanotubes and the rolled up model which is used to describe their structure; and fullerenes and cones. In addition, Section 1.2 includes a discussion on Euler's theorem and how it may be applied to these carbon nanostructures. Following this, in Section 1.3 the basic equations and the theory which are used to determine the interaction between molecular structures is outlined. Finally, in Section 1.4 the organisation of the book is outlined, and relevant exercises are given at the end of the chapter.

1.2 CARBON NANOSTRUCTURES

1.2.1 GRAPHENE AND C-C BONDING

As a result of the chemical structure of carbon, a carbon atom can bond with itself and other atoms to form an endlessly varied combination of chains and rings. As a result of this ability to form several distinct types of valence bonds, carbon can assume many distinct structural forms. For example, a planar sheet forms graphene or a tetrahedral structure forms diamond. Carbon is the sixth element in the periodic table, and a free carbon atom has six electrons which occupy $1s^2$, $2s^2$, and $2p^2$ atomic orbitals, which are illustrated in Fig. 1.2. In other words, carbon has two electrons which occupy the $1s$ orbital which is the orbital closest to the nucleus, two in the $2s$ orbital, and the remaining two electrons occupy two separate $2p$ orbitals, since the p orbitals have the same energy and the electrons would prefer to be separate. The $2s$ and $2p$ orbitals have similar energy levels. In practice, it is not possible to know exactly where the electrons are, since they appear smeared into orbitals. Despite this, the orbital notation is commonly used.

When bonding with other atoms, the electronic structure of a single carbon atom may be hybridised to adapt to various structural arrangements. The energy difference between the upper $2p$ orbital and the lower $2s$ orbital is small as compared to the binding of chemical bonds. Therefore, these electrons

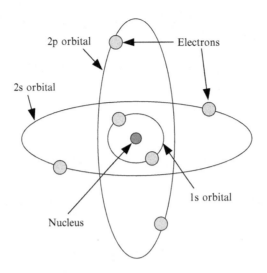

FIG. 1.2

The electron structure of a single carbon atom. Two electrons are in the $1s$ orbital, two in the $2s$ orbital, and two in the $2p$.

can readily mix with each other to enhance the binding energy of the carbon atom with neighbouring atoms, and this process is called hybridisation. For example, in order to form covalent bonds in graphene (nanotubes and fullerenes), one of the $2s$ electrons is promoted to a $2p$ orbital, and the orbitals are then hybridised to form what are called sp^2 bonds. In sp^2 bonding, each carbon atom forms three bonds with three other neighbouring atoms. Similarly in diamond, the bonds are hybridised to form sp^3 bonds so that each carbon atom forms four bonds to give a tetrahedral bonding structure. Generally, the bonds which are in-plane and form strong covalent bonds are referred to as σ bonds, and the bonds which are perpendicular to the plane and form weak van der Waals bonds are referred to as π bonds.

The most commonly occurring form of carbon is graphite, which is formed from many layers of graphene sheets. Graphene is a planar sheet consisting of a tessellation of hexagonal rings of carbon atoms, or a honeycomb lattice as shown in Fig. 1.3, all with the hybridised sp^2 bonds. Graphite, originally found useful for marking sheep in the 16th century, is best known for its use in pencils, where it is commonly referred to as 'lead'. In graphite, the graphene layers are stacked on top of each other so that for example in a pencil, the sheets slide past each other and onto the page. The graphene layers slide past each other because of the relatively weak van der Waals forces which exist between layers, not as a result of the strong covalent bonds which exist between each atom in the planar sheet.

The sp^2 carbon–carbon bond in the basal plane of graphite is the strongest of all chemical bonds, but the interplanar forces (or van der Waals forces) are relatively weak. In fact, the sp^2 bonding in graphene is stronger than the sp^3 bonding in diamond, as demonstrated by the carbon–carbon bond lengths (1.42 Å and 1.54 Å, respectively); and the shorter the bond length, the stronger the bond. For a more detailed account of the molecular bonding of carbon we refer the reader to Companion (1979).

Immediately after the discovery of carbon nanotubes and fullerenes, there was less interest in graphene. However, graphene sheets have recently shown considerable promise for applications in

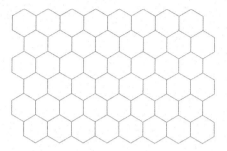

FIG. 1.3

Planar sheet consisting of a tessellation of hexagonal carbon rings (graphene).

electronics. For example, it was found that graphene exhibits a much higher thermal conductivity than carbon nanotubes, which is vital in electronic applications to dissipate heat. Since graphene is basically an unrolled carbon nanotube, many of the applications investigated for nanotubes may also apply to graphene. As mentioned in Section 1.1, there are now many forms of carbon structures and two of these, the carbon nanotube and fullerene, are examined in some detail in Sections 1.2.2 and 1.2.3, respectively. Interestingly, fullerenes and nanotubes exhibit structures remarkably similar to those found in nature. For example, C_{60} and other symmetrical fullerenes have a similar structure to icosahedral viruses, and carbon nanotubes have a similar structure to certain bacteriophages (bacteriolytic viruses).

1.2.2 CARBON NANOTUBES AND THE ROLLED-UP MODEL

The discovery by Iijima (1991) that carbon nanotubes could be grown without a catalyst has generated considerable research into numerous potential applications, ranging from prospective devices in biology to electronics. His paper outlined the experimental identification of multiwalled carbon nanotubes and had a significant impact on subsequent scientific research.

This is often considered to be the first occurrence of carbon nanotubes, but in fact they were observed much earlier. The first evidence, using transmission electron microscopy, of the tubular nature of some nanosized carbon filaments appeared in 1952 by Radushkevich and Lukyanovich in the *Russian Journal of Physical Chemistry*. Figures from this paper clearly show carbon filaments with a continuous inner cavity and tubes that formed as a result. The nanotubes are now thought to be multiwalled carbon nanotubes with 15–20 layers. However, this paper received very little attention from the world scientific community, perhaps because of the Cold War and the general difficulties of obtaining Russian scientific documents. However, certainly Radushkevich and Lukyanovich should be credited with the discovery that carbon filaments could be hollow and have a nanometer-size diameter or, in other words, for the actual discovery of carbon nanotubes. Of course, Iijima was the first to fully appreciate the nature and importance of these structures.

Laboratory methods to synthesise single-walled carbon nanotubes were discovered in 1993 by Bethune and colleagues at the IBM Almaden Research Center in San Jose, California, and also independently by Sumio Iijima at the NEC Laboratories in Japan. An image found in a paper by Oberlin as early as 1976 shows a nanotube resembling a single-walled carbon nanotube, although this is not explicitly claimed by the authors.

Carbon fibres are the macroscopic analogue of carbon nanotubes. Such fibres have been known for some time, and they were first used by Thomas A. Edison as a filament for an early model of the electric light bulb. Since then, carbon fibres have been used in numerous applications, such as for sporting equipment, boat hulls, and aeronautics. Carbon nanotubes have the potential to vastly improve technologies which already employ carbon fibres, in addition to providing an avenue for many new and exciting applications. Carbon nanotubes have many fascinating and unique mechanical and electronic properties, including but not limited to: high strength and flexibility, low density, completely reversible deformation and their capacity to be either metallic or semiconducting depending upon their geometric structure.

Carbon nanotubes may be thought of as one or several graphene sheets rolled up into a seamless hollow cylinder, forming either a single-wall (having only one atomic layer) or a multiwall carbon nanotube (having two or more walls). Experimental studies confirm that a carbon nanotube consists of a hexagonal lattice sheet, just like graphene, which is rolled into a seamless cylinder. Fig. 1.4 illustrates single-walled and multiwalled carbon nanotubes.

Single-walled carbon nanotubes frequently occur in tightly packed bundles so as to minimise their energy. Laboratory samples of single-walled carbon nanotubes tend to be more uniform than multiwalled carbon nanotubes, and they have a smaller range in diameters and fewer obvious defects. In other words, single-walled carbon nanotubes have more perfect structures than multiwalled carbon nanotubes. Several nanotube bundles can be bound together to form a nanorope. Generally, the diameter of single-walled carbon nanotubes is approximately 7–100 Å, with the majority of observed single-wall nanotubes having a diameter less than 20 Å. The aspect ratio (length/diameter) of nanotubes is extremely large, and it can be as high as 10^4–10^5, and therefore nanotubes are often considered to be one-dimensional structures. Multiwalled carbon nanotubes are concentric with a separation distance which is typically of the order of 3.4 Å, which is close to the interlayer separation distance of graphite.

The structure of a carbon nanotube is described by its chiral vector **C**, which is obtained by unrolling the carbon nanotube onto a planar sheet and connecting two crystallographically equivalent sites. The

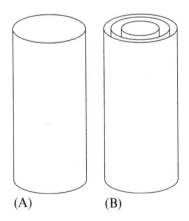

(A)   (B)

FIG. 1.4

Illustration of (A) single-wall and (B) multiwall carbon nanotubes.

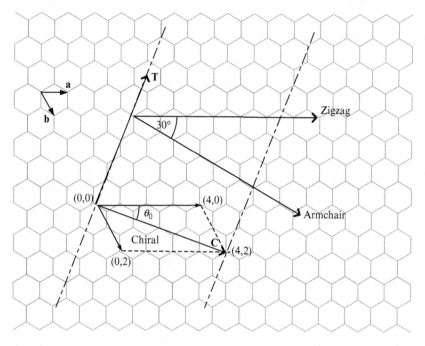

FIG. 1.5

Conventional rolled up model for carbon nanotubes.

chiral vector is defined by $\mathbf{C} = n\mathbf{a} + m\mathbf{b}$ or alternatively written simply as (n, m), where n and m are integers, and $\mathbf{a}$ and $\mathbf{b}$ are the basis vectors for one hexagonal unit cell on a graphene sheet. The magnitude of both $\mathbf{a}$ and $\mathbf{b}$ is 2.46 Å, or the width of one hexagonal unit, as shown in Fig. 1.5. Note that due to the hexagonal symmetry of the honeycomb lattice, it is only necessary to consider $0 \leq m \leq n$; by convention this is assumed to be the case. In the rolled-up model of carbon nanotubes, the chiral vector $\mathbf{C}$ is always perpendicular to the nanotube axis, and its magnitude is equal to the nanotube's circumference. The indices (n, m) are commonly referred to as the nanotube's chirality and are sometimes referred to as the helicity.

Two special carbon nanotubes are zigzag $(n, 0)$, and armchair (n, n); these two types are normally referred to as achiral. The general carbon nanotube is referred to as chiral (n, m). See Fig. 1.5 for illustrations of the three types of tubes. With reference to Fig. 1.5, the zigzag line refers to the line at which a zigzag tube will be formed, and the zigzag structure ($\bigwedge\!\bigvee\!\bigwedge$) is visible along this line. Similarly, the armchair line refers to the line at which an armchair tube would be formed, and an armchair structure ($\bigvee\!\underline{}\!\bigwedge$) is visible along this line. Interestingly, carbon nanotubes can be either metallic or semiconducting, and this property is completely dependent upon their geometric structure, namely their diameter and chirality. When $(n - m)$ is a multiple of 3 the tubes are metallic; otherwise, they are semiconducting. For example, an armchair (n, n) tube is always metallic, while a zigzag $(n, 0)$ tube is metallic only if n is a multiple of 3.

A carbon nanotube is also defined by the chiral angle θ_0 shown in Fig. 1.5, which is the angle the chiral vector $\mathbf{C}$ makes with the zigzag line, and which is defined by

$$\theta_0 = \cos^{-1}\left[\frac{2n + m}{2\sqrt{(n^2 + nm + m^2)}}\right].$$

We note that $\theta_0 = 0$ for zigzag tubes, while $\theta_0 = \pi/6$ for armchair tubes and generally $0 < \theta_0 < \pi/6$ for chiral tubes. The radius of a carbon nanotube, written in terms of (n, m), is given by

$$r = \frac{\sigma}{2\pi}\sqrt{3(n^2 + nm + m^2)}, \tag{1.1}$$

where σ is the length of one carbon–carbon bond, which is approximately $1.42\,\text{Å}$ for a carbon nanotube. For example, using Eq. (1.1) the radius of a $(10, 10)$ carbon nanotube is $6.78\,\text{Å}$. In fact, the length of the carbon–carbon bond σ in a nanotube is believed to be $1.44\,\text{Å}$, and as such the radius of a $(10, 10)$ nanotube would be $6.875\,\text{Å}$. Despite this discrepancy the length of the carbon–carbon bond for graphene ($1.42\,\text{Å}$) is typically assumed for carbon nanotubes.

WORKED EXAMPLE 1.1

Prove the equation for the carbon nanotube radius (Eq. 1.1) for the rolled up model.

Solution

On noting that the magnitude of the chiral vector $\mathbf{C}$ is equal to the circumference of the nanotube $|\mathbf{C}| = 2\pi r$, we obtain

$$r = \frac{|\mathbf{C}|}{2\pi} = \frac{\sqrt{\mathbf{C}.\mathbf{C}}}{2\pi} = \frac{\sqrt{(n\mathbf{a} + m\mathbf{b}).(n\mathbf{a} + m\mathbf{b})}}{2\pi}$$

$$= \frac{\sqrt{n^2\mathbf{a}.\mathbf{a} + m^2\mathbf{b}.\mathbf{b} + 2nm\mathbf{a}.\mathbf{b}}}{2\pi}$$

$$= \frac{|\mathbf{a}|}{2\pi}\sqrt{n^2 + nm + m^2},$$

where the magnitude of $\mathbf{a}$ is $2.46\,\text{Å}$ or the width of one hexagon ring, which is also equivalent to $\sqrt{3}\sigma$. We note that the vectors $\mathbf{a}$ and $\mathbf{b}$ are not orthogonal and their inner products yield

$$\mathbf{a}.\mathbf{a} = \mathbf{b}.\mathbf{b} = |\mathbf{a}|^2, \quad \mathbf{a}.\mathbf{b} = |\mathbf{a}|^2/2.$$

There are many other formulae and vectors which exist to describe a carbon nanotube, and these are derived from the chiral vector $\mathbf{C}$. One important definition is the length of the unit cell. The translational vector $\mathbf{T}$ of a unit cell is parallel to the nanotube axis; therefore it is perpendicular to the chiral vector $\mathbf{C}$, as illustrated in Fig. 1.5, and it is given by

$$\mathbf{T} = t_1\mathbf{a} + t_2\mathbf{b},$$

where t_1 and t_2 are integers, which are defined by

$$t_1 = (2m + n)/d_R, \quad t_2 = -(2n + m)/d_R,$$

and d_R is the highest common divisor of $(2n + m, 2m + n)$, or more specifically

$$d_R = \begin{cases} d & \text{if } n - m \text{ not a multiple of } 3d \\ 3d & \text{if } n - m \text{ a multiple of } 3d, \end{cases}$$

where d is the highest common divisor of (n, m). For example, for a $(5, 5)$ armchair tube $d_R = 3d = 15$, while for a $(9, 0)$ zigzag tube $d_R = d = 9$ and for a $(7, 4)$ chiral tube $d_R = 3d = 3$.

The above relation for d and d_R comes from one of Euclid's Laws (7th Book, Proposition 2). The law is commonly referred to as the Euclidean algorithm or the greatest common divisor, $gcd(a, b)$. The Euclidean algorithm is one of the oldest algorithms known, since it appeared in Euclid's *Elements* in 200 BC. Euclid originally formulated the problem geometrically as the greatest common measure for two line lengths.

The length of **T** is thus given by

$$T = \frac{\sqrt{3}|\mathbf{C}|}{d_R}. \tag{1.2}$$

We note that the length of **T** is greatly reduced when (n, m) has a common divisor or when $(n - m)$ is a multiple of $3d$. The number of hexagons per unit cell N; therefore the number of carbon atoms n_C per unit cell are given by

$$N = \frac{2(n^2 + nm + m^2)}{d_R}, \quad n_C = 2N. \tag{1.3}$$

WORKED EXAMPLE 1.2

Prove the formulae given in Eq. (1.3).

Solution

Note that the number of hexagons per unit cell is equal to the area of the unit cell divided by the area of one hexagonal unit.

Fig. 1.6 illustrates the cross product of the vectors **a** and **b**, where the cross product is perpendicular to both vectors and its magnitude is equal to the area between them. The cross product of any two vectors may be determined from the determinant $det(\mathbf{a}\,\mathbf{b})$ or simply $\mathbf{a} \times \mathbf{b} = ab \sin\theta\hat{n}$. The magnitude of the cross product is thus $|\mathbf{a} \times \mathbf{b}| = |\mathbf{a}||\mathbf{b}| \sin\theta$.

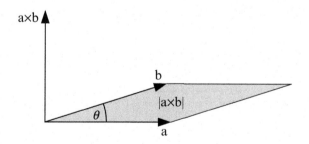

FIG. 1.6

The cross product between two vectors **a** and **b**.

Therefore the area of the unit cell can be determined by finding the magnitude of the cross product of the vectors $\mathbf{C}$ and $\mathbf{T}$, thus

$$|\mathbf{C} \times \mathbf{T}| = |\mathbf{C}||\mathbf{T}| \sin(\pi/2)$$
$$= \frac{\sqrt{3}(|\mathbf{C}|)^2}{d_R} = \frac{\sqrt{3}|\mathbf{a}|^2(n^2 + nm + m^2)}{d_R},$$

where $|\mathbf{a}| = \sqrt{3}\sigma$. Note that the vectors $\mathbf{C}$ and $\mathbf{T}$ are always perpendicular to one another so that the angle between them is always 90°, i.e. $\pi/2$.

Similarly, the area of one hexagonal unit is the magnitude of the cross product of the unit vectors $\mathbf{a}$ and $\mathbf{b}$, thus

$$|\mathbf{a} \times \mathbf{b}| = |\mathbf{a}||\mathbf{b}| \sin(\pi/3) = \frac{\sqrt{3}|\mathbf{a}|^2}{2}.$$

Note that the angle between the vectors $\mathbf{a}$ and $\mathbf{b}$ is always 60° or $\pi/3$ and that $|\mathbf{a} \times \mathbf{a}| = |\mathbf{b} \times \mathbf{b}| = 0$. The number of hexagons per unit cell is thus given by

$$N = \frac{\text{Area of unit cell}}{\text{Area of one hexagon unit}}$$
$$= \frac{|\mathbf{C} \times \mathbf{T}|}{|\mathbf{a} \times \mathbf{b}|}$$
$$= \frac{2(n^2 + nm + m^2)}{d_R}.$$

It then follows that the number of carbon atoms per unit cell is $n_C = 2N$.

As an example, Fig. 1.7 illustrates a unit cell for a $(5,5)$ carbon nanotube. Further details for this nanotube are given in Table 1.1, where the structural parameters of a selected number of carbon nanotubes are given. For more details on the structural parameters of carbon nanotubes, we refer the reader to Dresselhaus et al. (1995) and Saito et al. (1998).

Generally, carbon nanotubes are assumed to be formed from perfect hexagonal rings of carbon. However, in reality, there are often defects occurring along the nanotube wall. Nanotubes with diameters greater than 20 Å tend to show many more structural defects and are inclined to become unstable due to atmospheric pressure. On the other hand, the very small diameter nanotubes, although they tend to be more perfect with fewer structural defects tend to be intrinsically unstable giving rise to a definite window of diameters which are seen to occur in practice. The Stone–Wales transformation, which is

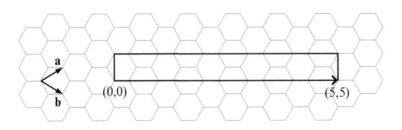

FIG. 1.7

Unit cell for a $(5, 5)$ armchair nanotube.

Table 1.1 General Structural Parameters for a Selection of Carbon Nanotubes

C	d	d_R	θ_0	r	$|C|$	T	T	N
(5,5)	5	15	$\pi/6$	3.39 Å	21.3 Å	(1,-1)	2.46 Å	10
(9,0)	9	9	0	3.52 Å	22.1 Å	(1,-2)	4.26 Å	18
(7,4)	1	3	0.3674	3.77 Å	23.7 Å	(5,-6)	13.69 Å	62
(n,n)	n	3n	$\pi/6$	$3n\sigma/(2\pi)$	$3n\sigma$	(1,-1)	a_0	2n
(n,0)	n	n	0	$\sqrt{3}n\sigma/(2\pi)$	$\sqrt{3}n\sigma$	(1,-2)	$\sqrt{3}a_0$	2n

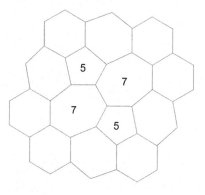

FIG. 1.8

Stone–Wales transformation, pentagon–heptagon pair (5-7-7-5).

illustrated in Fig. 1.8 and discussed in more detail in Section 1.2.5, is one such defect in which excess strain energy is accomodated by a rotation of the carbon–carbon bonds so that two adjacent hexagons form a pentagon–heptagon pair (5-7-7-5).

It is possible for carbon nanotubes to be capped or uncapped. Naturally occurring carbon nanotubes are grown with caps, and these may then be removed by oxidation because the pentagonal rings which form the caps are more readily removed by oxidation. The caps of carbon nanotubes are formed as hemispheres from fullerenes, and fullerenes are discussed in the next section. Different sized fullerenes are capable of capping various sizes of carbon nanotube, and a few examples of this are illustrated in Fig. 1.9. The C_{60} fullerene, discussed in the following section, is capable of capping both a (5,5) tube and a (9,0) tube. All nanotubes larger than the (5,5) and (9,0) nanotubes can be capped, and the number of possible caps increases rapidly with increasing diameter. In practice, caps can have a variety of shapes, and more complex caps are often observed. For example, caps are frequently conical in shape, formed from the introduction of fewer pentagonal rings than the six required to form a hemisphere, as discussed in Section 1.2.4.

Fig. 1.9 also illustrates the various chiralities which are possible for a carbon nanotube. From Fig. 1.9, the armchair, zigzag, and chiral types are ascertained by examination of the end of the nanotube, and Table 1.2 provides a summary of the classification of carbon nanotubes.

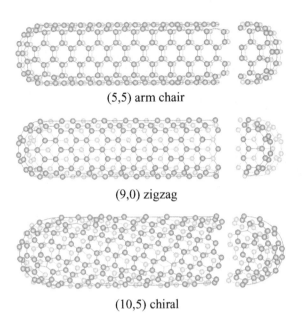

(5,5) arm chair

(9,0) zigzag

(10,5) chiral

FIG. 1.9

Different sized and chirality carbon nanotubes with various caps.

Source: Mechanics of Carbon Nanotubes, Appl. Mech. Rev. 2002, 55(6): 495–533. doi:10.1115/1.1490129

Table 1.2 Classification of Carbon Nanotubes

Type	θ_0	C	Shape of Cross-Section
Armchair	$\pi/6$	(n, n)	*cis*-type _/
Zigzag	0	$(n, 0)$	*trans*-type /\\/\\
Chiral	$0 < \theta_0 < \pi/6$	(n, m)	Both *cis* and *trans*

Carbon nanotubes can also be divided into symmetry groups. For example, armchair and zigzag tubes are in a symmorphic group. This refers to the fact that rotations can be created by a simple point group representation. On the other hand, chiral tubes exhibit no mirror planes and they are therefore nonsymmorphic. Their symmetry operations involve both translations and rotations; for further details on these symmetry groups, we refer the reader to Dresselhaus et al. (1996).

1.2.3 GOLDBERG FULLERENES

Fullerenes were discovered in 1985 by a research group at Rice University led by Robert F. Curl (1933-), Sir Harold W. Kroto (1939-), and Richard E. Smalley (1943–2005). For their discovery all three received the Nobel Prize for Chemistry in 1996. Their discovery, which subsequently has generated considerable research and development, was in fact purely accidental. These researchers were conducting experiments on clusters formed by laser vapourisation of graphite for various interesting and quite unrelated reasons. Kroto believed that studying these clusters might provide some insight into the processes occurring on the surface of stars, and Smalley's main interest was for applications to semiconductors. In the soot resulting from their experiments they discovered a closed carbon cluster consisting of precisely 60 carbon atoms with unique stable and symmetrical characteristics. These structures had been proposed in earlier work, however their discovery, for which they won the Nobel Prize, was for the realisation that carbon itself could form truncated icosahedral molecules and larger geodesic structures.

The first fullerene to be observed was the C_{60} fullerene or buckyball, named after the designer and architect Richard Buckminster Fuller (1895–1983), who first proposed geodesic dome structures. A geodesic dome is an almost spherical structure formed from a network of great circles lying on approximately the surface of a sphere. This fullerene was so named due to the direct association with geodesic domes. Since then many more different fullerenes have been observed, such as the C_{70} fullerene which is not spherical. Fullerenes can be thought of as hollow, closed cages of carbon atoms, where the carbon atoms are approximately located on either the surface of a sphere, such as the C_{60} fullerene, or on the surface of a spheroid, such as the C_{70} fullerene. The term spheroid refers to an ellipsoid which has two of the axes equal. In some cases the fullerenes can be shaped like capsules resembling short cylinders with two hemispherical caps, such as the C_{80} fullerene. Three fullerene molecules are illustrated in Fig. 1.10, including the famous buckyball or C_{60} fullerene. Fullerenes consist of only hexagonal or pentagonal faces, and all involve precisely 12 pentagons but any number of hexagons. This is an immediate consequence of Euler's theorem as will be described in Section 1.2.5.

The 60 atoms making up the C_{60} fullerene are now known to be located at the vertices of a truncated icosahedron. This is also true for other fullerenes, and therefore a rule for generating icosahedral fullerenes is formulated using equilateral triangles as the basic building blocks. An icosahedral fullerene consists of exactly 20 equilateral triangles, each specified by a pair of integers (n, m). The 20 equilateral triangles which make up a fullerene consist of 10 equilateral triangles in what is referred to as the belt area and five equilateral triangles in the cap regions, as illustrated in Fig. 1.11. Note that with reference to Fig. 1.11, the vertices of each equilateral triangle contain a number between 1 and 12, and that these numbers refer to the vertices on the icosahedral fullerene. For example, points **A** and **B** and the remaining vertices with the number 2 give rise to the same vertex in the resulting icosahedral fullerene.

C_{60} C_{70} C_{80}

FIG. 1.10

Fullerenes C_{60}, C_{70}, and C_{80}.

Source: Mechanics of Carbon Nanotubes, Appl. Mech. Rev. 2002, 55(6): 495–533. doi:10.1115/1.1490129

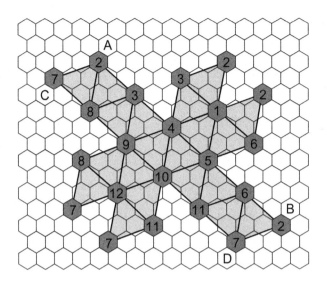

FIG. 1.11

Projection construction for an icosahedral fullerene.

Source: Phys Rev B 45 13834 (1992).

Different fullerenes have a different number of hexagonal rings of carbon atoms per equilateral triangle building block. The number of carbon atoms N in a fullerene C_N is given by

$$N = 20(n^2 + nm + m^2), \tag{1.4}$$

and the diameter of the icosahedron is given by

$$d = \frac{5\sqrt{3}a_{C-C}}{\pi}(n^2 + nm + m^2)^{1/2}, \tag{1.5}$$

where a_{C-C} denotes the nearest neighbour carbon–carbon distance on the fullerene. In a C_{60} fullerene, the bonding is also sp^2 with an average carbon–carbon distance of 1.44 Å, as compared to graphene which has a carbon–carbon distance of 1.42 Å. In general, there are both single and double bonds in the fullerene and as such the bond angles are not necessarily the same, although this is generally assumed to be the case.

The lattice of equilateral triangles is formed by assuming an origin on a graphene sheet and attaching a pentagon at this origin. The lattice point (n, m) is then placed on the sheet and another pentagonal ring is placed at this point. The third pentagon is placed equilaterally with respect to the other two, thus forming an equilateral triangle. The lattice point (n, m) which is used to generate the equilateral triangle is the same as that used in Eqs (1.4) and (1.5). Worked Example 1.3 illustrates the construction of one equilateral triangle, using as an example the $(1, 1)$ fullerene.

WORKED EXAMPLE 1.3

In this example we construct one equilateral triangle unit for a $(1, 1)$ fullerene. This equilateral triangle will then form the basis for the resulting icosahedral fullerene, which consists of 20 identical equilateral triangles and will be similar in appearance to Fig. 1.11.

Solution

The first step involves locating an origin on the graphene sheet and placing a pentagonal ring next to this origin, as shown in Fig. 1.12.

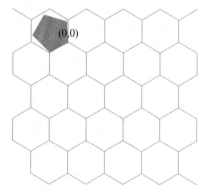

FIG. 1.12

Step 1: Select an origin $(0, 0)$ and place a pentagon at the origin on a graphene sheet.

In the next step, we move to the selected lattice point, in this case $(1, 1)$ and place another pentagonal ring next to it, as shown in Fig. 1.13.

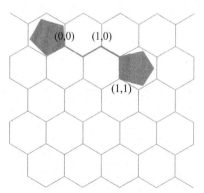

FIG. 1.13

Step 2: Choose a lattice point (n, m) and place another pentagon next to it. In this example, we have chosen a lattice point $(1, 1)$.

The third step involves placing a third pentagonal ring equilaterally from the other two, as shown in Fig. 1.14, thus generating the triangular building block which generates the fullerene. On using Eq. (1.4), it is clear that this is the building lattice for a C_{60} fullerene, since $N = 20(1^2 + 1 + 1^2) = 60$.

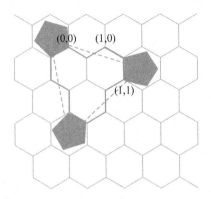

FIG. 1.14

Step 3: Locate a third pentagon equilaterally to the other two.

The vector between two pentagons in the triangle unit is given by $\mathbf{d_{nm}} = n\mathbf{a} + m\mathbf{b}$, so that the length of the edge of the equilateral triangle is simply

$$d_{nm} = |\mathbf{a}|(n^2 + nm + m^2)^{1/2}, \tag{1.6}$$

where $|\mathbf{a}|$ is the width of one hexagon on a graphene sheet, 2.46 Å. It is simple to show that the area of each equilateral triangle is given by

$$A_{nm} = \frac{\sqrt{3}a_0^2}{4}(n^2 + nm + m^2). \tag{1.7}$$

WORKED EXAMPLE 1.4

Show that the equation for the radius of a fullerene is given by Eq. (1.5).

Solution

Since the belt, shown in Fig. 1.11, consists of 10 triangles, the length around the entire sphere (approximately the circumference) consists of five edges, with a total length of **AB**, as shown in Fig. 1.11. Using the equation for the length of an arc, we obtain

$$\text{Length of arc} = 2\pi r,$$

$$5d_{nm} = \pi d,$$

$$d = \frac{5\sqrt{3}a_{C-C}}{\pi}(n^2 + nm + m^2)^{1/2},$$

where d_{nm} is the length of the edge of the equilateral triangle which is given by Eq. (1.6).

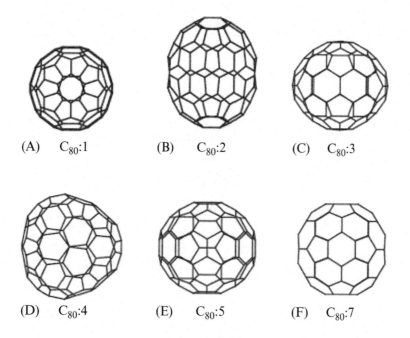

FIG. 1.15

Six of the seven isomers for the C_{80} fullerene. (A) C_{80}:1. (B) C_{80}:2. (C) C_{80}:3. (D) C_{80}:4. (E) C_{80}:5. (F) C_{80}:7.

    Every fullerene C_N belongs to a symmetry class which is either the symmetry I or I_h. The Goldberg fullerene has the symmetry I_h, which is further divided into two types: type 1 when $n = m$ and type 2 when $n = 0$ or $m = 0$. Fullerenes of I symmetry are for $n \neq m$. The fullerene C_{60} is the smallest of the I_h type 1 fullerene, whereas C_{80} is the smallest for type 2, because C_{20} does not generally exist in the sense that it is unstable. Note that molecules with I_h symmetry, of which C_{60} is the most common, have the highest degree of symmetry of any known molecule. The symmetry of a molecule is useful for determining certain material properties, such as their vibrational modes or their electronic states. For further details on the symmetry properties of fullerenes, we refer the reader to Dresselhaus et al. (1996).

    The particular structural configuration of fullerenes is not necessarily unique. For a given number of carbon atoms, there may be more than one way to arrange the atoms into a fullerene, and each possibility is termed an isomer. For example, a C_{80} fullerene has seven different structural arrangements or isomers, and six of these isomers are illustrated in Fig. 1.15,[1] and we refer the reader to Dresselhaus et al. (1996) for further details.

[1]G. Sun, M. Kertesz Chemical Physics Letters 328 (2000) 387–395.

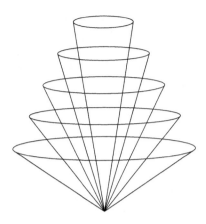

FIG. 1.16

Five possible nanocones with vertex angles 19.20°, 38.90°, 60.00°, 83.60°, and 112.90°.

1.2.4 CONES

Many studies utilise molecular dynamics simulation to calculate the energy of carbon nano systems and these investigations emphasise systems involving C_{60} fullerenes and carbon nanotubes, but very little of the existing literature deals with carbon nanocones. Carbon nanocones have received less attention primarily because only a small amount tend to occur in the production process, and most research on nanocones deals with their electronic structure.

There are five possible ways to construct carbon nanocones, depending on the number of pentagons which are needed to close the vertex. It is believed that the different number of pentagons in carbon nanocones is the key to the puzzle of nucleation in atomic construction. The catalytic chemical vapour deposition method can be used to synthesise carbon nanocones inside carbon nanotubes, and the resulting structures have different physical and electronic properties from that of the original carbon structure. Carbon nanocones are ideal candidates for nanoprobes in scanning tunnelling microscopy, and their electronic structure is dependent on the position of the pentagons. Carbon nanocones also have a nonlinear mechanical behaviour for both the original shape and the inverse carbon nanocone, which is obtained from the original cone by inversion.

Carbon nanocones were originally discovered by Ge and Sattler (1994) and subsequently synthesised by Krishnan et al. (1997). Typically, carbon nanocones are observed together with carbon nanotubes and nanotube bundles during the synthesis process, and carbon nanocones tend to be found at the cap of carbon nanotubes. There are five possible structures for nanocones, as indicated in Fig. 1.16 and shown in Fig. 1.17 by the transmission electron microscope images, since the cone angle depends on the number of pentagons needed to close the structure. Cones are formed from hexagons on a honeycombed lattice by adding fewer pentagons than the six which are needed by Euler's theorem to form the closed structure of a semifullerene. As discussed in Section 1.2.5, a closed-cage fullerene may be generated from a hexagonal lattice, provided that there are precisely 12 pentagons. The carbon nanotube cap, which is half a fullerene or semifullerene, contains six pentagons, and therefore carbon nanocones must have a number of pentagons which is less than six.

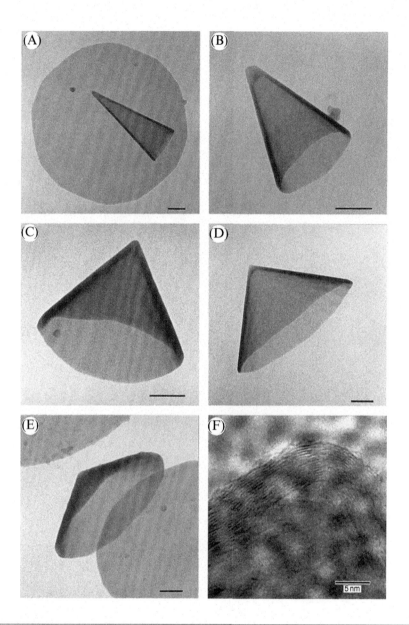

FIG. 1.17

Examples of the five types of nanocones with vertex angles (A) 19.20°, (B) 38.90°, (C) 60.00°, (D) 83.60°, and (E) 112.90°, where (F) is a higher magnification image of a cone tip Krishnan et al. (1997).

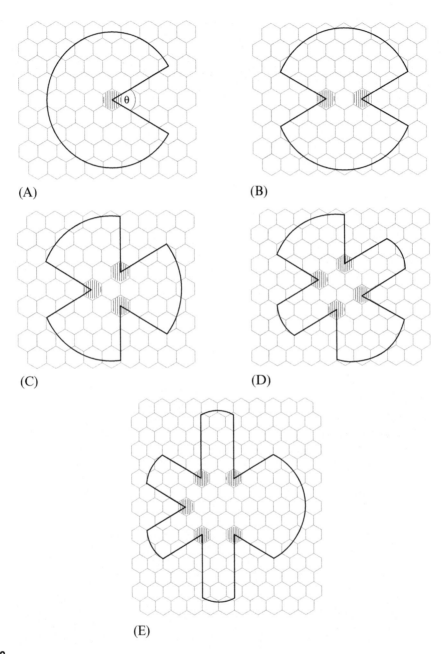

FIG. 1.18

Construction of cones by creation of disclinations (*triangular wedges*) in the graphene lattice. The part of the graphene plane bounded by the *thick lines* is folded into a cone; (A)–(E) display the graphene cutouts used to construct the cones with one, two, three, four, and five pentagons, respectively (*shaded hexagons* are replaced by pentagons when forming cones).

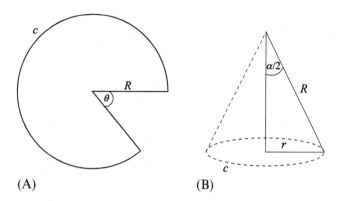

(A)   (B)

FIG. 1.19

(A) Graphene sheet (B) forming the carbon nanocone.

Table 1.3 Relation of Number of Pentagons N_p and Open Angle α for Carbon Nanocones	
Number of Pentagons (N_p)	**Angle of Cone (α)**
1	1.9702
2	1.4595
3	$\pi/6$
4	0.6797
5	0.3349

The disclination number of pentagons on the graphene sheet gives the change with θ in the form (see Fig. 1.18A)

$$\theta = \frac{\pi N_p}{3},$$

where N_p is the number of the pentagons which ranges from 1 to 5. From the diagram of the cone shown in Fig. 1.19, it is clear that $\sin(\alpha/2) = r/R$ and $c = 2\pi r = 2\pi(1 - N_p/6)R$. Therefore the relation of the cone angle and the number of pentagons is obtained as

$$\sin(\alpha/2) = 1 - \frac{N_p}{6}.$$

There are five possible values of the angle α depending on the number of pentagons, and these are shown in Table 1.3. Note that for $N_p = 0$ we have a graphene sheet and for $N_p = 6$ a cap of a carbon nanotube or half of a spherical fullerene is obtained. Hence these five values of α as shown in Table 1.3 give rise to carbon nanocones. The mean atomic surface density of the carbon nanocones is assumed to be the mean atomic surface density of a graphene sheet, $0.3812\,\text{Å}^{-2}$, due to the fact that the carbon nanocones are envisaged to be formed from a graphene sheet.

A right cone is one for which the vertex is directly above the centre of the base. However, when used without qualification, the term 'cone' often means right cone. A right cone is the surface in

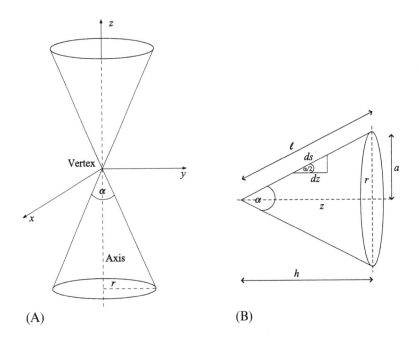

FIG. 1.20

(A) Geometry for a circular cone and (B) diagram for surface integrations.

three-dimensional space generated by a line that revolves about a fixed axis in such a way that the line passes through a fixed point on the axis and always makes the same angle with the axis. The fixed point is called the vertex of the cone. The surface shown in Fig. 1.20A is a double right cone which consists of two parts called nappes that intersect at the vertex. The quadratic equation in Cartesian coordinates (x, y, z) for a double elliptical cone is given by

$$\frac{x^2}{a^2} + \frac{y^2}{b^2} = \frac{z^2}{c^2}, \tag{1.8}$$

where a, b, and c are constants. The case $b = a$ gives rise to a double circular cone. Alternatively, in cylindrical coordinates (r, θ, z) with the vertex at the origin, the equation for the right cone becomes $r = z \tan(\alpha/2)$, where α is the vertex cone angle. Surface integrals of the cone are required in order to calculate the Lennard-Jones potential energy using the continuum approach, and one such evaluation to determine the area of the cone is outlined in Worked Example 1.5.

WORKED EXAMPLE 1.5

Find the surface area of a cone of height h and vertex angle α.

Solution

The surface area of a single cone is given by

$$\text{Area} = \int_0^\ell \int_0^{2\pi} r \, d\theta \, ds.$$

From Fig. 1.20B, it can be seen that $ds = dz/\cos(\alpha/2)$ and $r = z\tan(\alpha/2)$, and therefore the surface integral takes the form

$$\text{Area} = \frac{\tan(\alpha/2)}{\cos(\alpha/2)} \int_0^h \int_0^{2\pi} z\,d\theta\,dz = \frac{a\ell}{h^2} \int_0^h \int_0^{2\pi} z\,d\theta\,dz = \pi a\ell,$$

where $\ell = \sqrt{a^2 + h^2}$.

1.2.5 EULER'S THEOREM AND THE ISOLATED PENTAGON

One of Euler's theorems relates to regular polyhedra and can be used to describe the geometry of carbon structures, such as fullerenes and capped carbon nanotubes. Euler's theorem is embodied in the equation

$$v - e + f = 2(1 - g), \tag{1.9}$$

where v, e, and f are the number of vertices, edges, and faces, respectively, and g is the genus of the particular structure. A surface with genus g requires $2g$ suitable cuts, which all begin and end at the same point, to unfold it into a $4g$-gon. For example, a sphere has a genus of 0 and a torus has a genus of 1. Worked Example 1.6 illustrates how Euler's theorem can be used to predict the structure of a fullerene.

WORKED EXAMPLE 1.6

Determine the number of pentagons in a fullerene using Euler's theorem (Eq. 1.9).

Solution

Assuming that the fullerene can be approximated by a sphere and that the genus of a sphere is zero, Euler's theorem for a sphere becomes

$$v - e + f = 2.$$

Assuming that there exist pentagons (p) and hexagons (h), we obtain the following equations

$$f = p + h, \quad v = \frac{5p + 6h}{3}, \quad e = \frac{5p + 6h}{2},$$

since every vertex is connected to three edges, and every edge forms two faces. Substituting these into Euler's theorem for a sphere, we obtain the result

$$\frac{5p + 6h}{3} - \frac{5p + 6h}{2} + p + h = 2$$
$$10p + 12h - 15p - 18h + 6p + 6h = 12,$$

from which we may deduce $p = 12$, and thus for a fullerene, no matter the size, there can be any number of hexagons, but always precisely 12 pentagons. This remarkable result says that in a hexagonal–pentagonal structure, we need precisely six pentagons to close either end.

A number of famous mathematicians have proved Euler's theorem (Eq. 1.9). For example, Augustin-Louis Cauchy (1789–1857) proved the formula using networks as illustrated in Worked Example 1.7, and Adrien-Marie Legendre (1752–1833) proved the formula using the concept of projecting a polyhedron onto the surface of a sphere of unit radius. For more details of various derivations of Euler's theorem and for a detailed account of the mathematics of polyhedra, we refer the reader to Coxeter (1969) and Cromwell (1996).

WORKED EXAMPLE 1.7

The following example proves Euler's theorem making use of the proof first devised by the French mathematician Cauchy, famous for a number of contributions, including results in complex variables, and for his formulation of the basic equations underlying continuum mechanics.

Solution

To start, we have a cube as shown in Fig. 1.21(i). Looking down on the cube, we imagine that we disassemble the cube so that it lies flat, as shown in Fig. 1.21(ii) and (iii).

It is possible to create a table which outlines the number of vertices (v), edges (e), and faces (f) of Fig. 1.21(iii), as shown in Table 1.4. Note that we must also include the exterior face which consists of the area outside the network. With reference to Fig. 1.21(iii), each face that is not a triangle is divided so that it becomes a triangle, as shown in Fig. 1.21(iv), and we again record the number of vertices, edges, and faces.

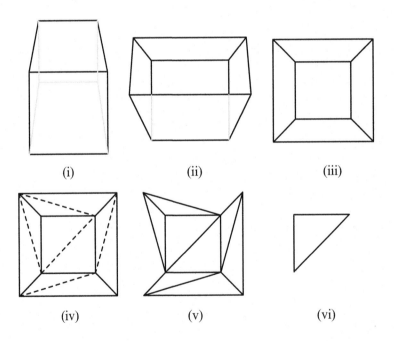

(i)   (ii)   (iii)

(iv)   (v)   (vi)

FIG. 1.21

The typical steps used to prove Euler's theorem (Eq. 1.9) based on Cauchy's proof.

Table 1.4 The Number of Vertices (v), Edges (e), and Faces (f) at Various Points in Cauchy's Proof

Step Number	v	e	f	$v - e + f$
(iii)	8	12	6	2
(iv)	8	17	11	2
(v)	8	15	9	2
(vi)	3	3	2	2

Next, we remove the faces which share only one edge with the exterior face, and in Fig. 1.21(v) two of these faces have been removed. Once there are no more faces with only one edge with the exterior face we remove all faces which share two edges with the exterior face, until finally we obtain Fig. 1.21(vi). We note that throughout all these steps Euler's formula (1.9) remains unchanged, as illustrated in Table 1.4.

Another useful tool is the so-called isolated pentagon rule, which relates to the idea that a pentagonal ring prefers to be completely surrounded by hexagonal forms of carbon rings so as to minimise its strain energy. Interestingly, the C_{60} fullerene is the smallest fullerene that abides by the isolated pentagon rule, so that it has minimum strain energy, and optimal stability arises due to the pentagons being completely surrounded by hexagons. As a result, the strain energy is uniformly distributed over the fullerene surface, and thus it is the most stable fullerene, as well as the form of carbon predominantly found in soot. Other fullerenes which also obey the isolated pentagon rule are C_{70}, which happens to be the next most abundant, and also C_{76}, C_{78}, C_{82}, and C_{84}.

Using Euler's theorem, the smallest fullerene possible would be C_{20}, but as with carbon nanotubes of small radius, this small structure tends not to be stable. It consists of all pentagons and no hexagons; therefore all the pentagons are fused to one another, thus creating a strain related instability.

Carbon nanotubes also follow these geometric rules, provided that they are capped at both ends. In addition, Euler's theorem (Eq. 1.9) may be used to explain the previously mentioned Stone–Wales transformation, which is a topological defect in a nanotube structure. It occurs as a consequence of excess strain in the structure and is a natural release in the form of a rotation of the carbon–carbon bond, which produces two pentagons and two heptagons coupled in pairs. With reference to Fig. 1.22, the highlighted edges of the hexagon lattice are rotated through 90°, and the lattice reassembles to form the 5-7-7-5 structure. To maintain Euler's theorem, in Fig. 1.22A we remove five edges and two vertices and in Fig. 1.22B we have added five edges and two vertices.

The Stone–Wales transformation is named after Professor Anthony J. Stone and David J. Wales. Their paper appeared in 1986 shortly after the discovery of fullerenes by Kroto and others. In it Stone and Wales proposed the possibility of many stable structures of C_{60} fullerenes other than the icosahedral structure, and these are related to each other by the transformation of two atoms or the now so-called Stone–Wales transformation.

Alternatively, let us assume that there exist topological defects on a graphene sheet, which is subsequently rolled up to form a carbon nanotube. These defects may be any numbered carbon ring n_i where i refers to the number of sides on the ring. From Euler's theorem, in order to close the hexagonal

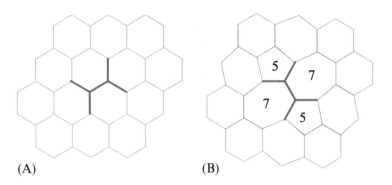

(A)   (B)

FIG. 1.22

Stone–Wales transformation. (A) Original hexagonal lattice with edges to be rotated (removed) highlighted. (B) Stone–Wales transformation with rotated (added) edges highlighted.

network on a graphene sheet we have

$$3n_3 + 2n_4 + n_5 - n_7 - 2n_8 - 3n_9 = 12. \tag{1.10}$$

For example, in a Stone–Wales transformation only hexagons (n_6), pentagons (n_5), and heptagons (n_7) are present so that this equation simplifies to become

$$n_5 - n_7 = 12,$$

so that there is a balance between the pentagonal and heptagonal pairs, as previously illustrated above. We note that the hexagons are not present in these equations since they do not contribute to the curvature of the system. The pentagons contribute to the curvature by creating a convex surface, such as in fullerenes, while heptagons create concave surfaces, such as for tori or nanotube elbows.

Table 1.5 outlines various constants which are used throughout this book, in particular the carbon–carbon bond length in graphene, various radii of fullerenes and nanotubes and the mass of both an atom and C_{60} fullerene.

1.3 INTERACTION BETWEEN MOLECULAR STRUCTURES
1.3.1 INTERACTION ENERGY

For two separate molecular structures (i.e. nonbonded), the interaction energy E can be evaluated using either a discrete atom–atom formulation or approximating by a continuous approach. Thus the nonbonded interaction energy may be obtained as a summation of the interaction energy between each atom pair, namely

$$E = \sum_i \sum_j \Phi(\rho_{ij}), \tag{1.11}$$

Table 1.5 Various Constants Used Throughout Book	
Carbon–carbon bond length	$\sigma = 1.421\,\text{Å}$
Radius of $(5, 5)$	$3.392\,\text{Å}$
Radius of $(8, 8)$	$5.428\,\text{Å}$
Radius of $(10, 10)$	$6.784\,\text{Å}$
Radius of $C_{60}{}^a$	$3.55\,\text{Å}$
Radius of C_{240}	$7.12\,\text{Å}$
Radius of C_{540}	$10.5\,\text{Å}$
Radius of C_{960}	$13.8\,\text{Å}$
Radius of C_{1500}	$17.5225\,\text{Å}$
Mass of a single carbon atom	$M = 1.993 \times 10^{-26}\,\text{kg}$
Mass of a single C_{60} fullerene	$M = 1.196 \times 10^{-24}\,\text{kg}$

^{a}Radii of fullerenes taken from Lu and Yang (1994) apart from C_{1500}, which is taken from Dunlap and Zope (2006).

where $\Phi(\rho_{ij})$ is the potential function for atoms i and j located a distance ρ_{ij} apart on two distinct molecular structures, and it is assumed that each atom on the two molecules has a well-defined coordinate position. The continuum approximation assumes that the atoms are uniformly distributed over the entire surface of the molecule, and the double summation in Eq. (1.11) is replaced by a double integral over the surface of each molecule, thus

$$E = \eta_1 \eta_2 \iint \Phi(\rho)\, dS_1\, dS_2, \tag{1.12}$$

where η_1 and η_2 represent the mean surface density of atoms on the two interacting molecules, and ρ represents the distance between the two typical surface elements dS_1 and dS_2 located on the two interacting molecules. Note that the mean atomic surface density is determined by dividing the number of atoms which make up the molecule by the surface area of the molecule; an example is given in Worked Example 1.8. Table 1.6 outlines the surface density of carbon atoms on various carbon nanostructures which are examined in this book. The continuum approximation is rather like taking the average or mean behaviour and in the limit of a large number of atoms, the discrete model approaches the continuum approximation.

WORKED EXAMPLE 1.8

Determine the surface density of carbon atoms on a sheet of graphene or a carbon nanotube.

Solution

A graphene sheet consists of a tessellation of hexagonal rings, so that each atom contributes one-third of the total number of atoms in the ring because each atom is also bonded to two other hexagonal rings. Since there are six carbon atoms per hexagon, the surface density of carbon atoms is thus

$$\eta = \frac{1/3 \times 6}{SA} = \frac{2}{SA},$$

Table 1.6 Typical Mean Surface Densities Used Throughout This Book

	Mean Surface Density (Å^{-2})
Graphene	0.3812
Carbon nanotube[a]	0.3812
C_{60}^{b} fullerene	0.3789
C_{240}	0.3767
C_{540}	0.3898
C_{960}	0.4011
C_{1500}	0.3888

[a] *Mean surface density of fullerene C_N radius b can be deduced from $N/(4\pi b^2)$, where N is the number of carbon atoms in the molecule.*
[b] *Mean surface density of a carbon nanotube can be deduced from $4\sqrt{3}/(9\sigma^2)$, where the carbon–carbon bond length $\sigma = 1.421$ Å.*

where *SA* is the surface area of one hexagonal ring. The surface area of one hexagon ring is $3\sqrt{3}\sigma^2/2$, so that the surface density of carbon atoms becomes

$$\eta = \frac{4}{3\sqrt{3}\sigma^2},$$

and multiplying both the numerator and denominator by $\sqrt{3}$, we obtain

$$\eta = \frac{4\sqrt{3}}{9\sigma^2},$$

where the units of the atomic surface density are thus atoms/Å^2. For a sheet of graphene with $\sigma = 1.42$ Å the atomic surface density is $\eta = 0.382$ atoms/Å^2.

The continuum approach is an important approximation, and Girifalco et al. (2000) state that,

> From a physical point of view the discrete atom–atom model is not necessarily preferable to the continuum model. The discrete model assumes that each atom is the centre of a spherically symmetric electron distribution while the continuum model assumes that the electron distribution is uniform over the surface. Both of these assumptions are incorrect and a case can even be made that the continuum model is closer to reality than a set of discrete Lennard-Jones centres.

One such example is a C_{60} fullerene, in which the molecule rotates freely at high temperatures so that the continuum distribution averages out the effect. Qian et al. (2003) suggest that the continuum approach is more accurate for the case where the 'C nuclei do not lie exactly in the centre of the electron distribution, as is the case for carbon nanotubes'. However, one of the constraints of the continuum approach is that the shape of the molecule must be reasonably well defined in order to evaluate the integral analytically, and therefore the continuum approach is mostly applicable to highly symmetrical structures, such as cylinders, spheres, and cones. We believe the importance of analytical evaluation, which is usually in terms of hypergeometric functions, lies in the fact that such functions may be readily evaluated using many of the standard mathematical packages, such as MAPLE and MATHEMATICA. Hodak and Girifalco (2001) point out that the continuum approach ignores the effect of chirality and that nanotubes are only characterised by their diameters. The continuum or continuous approximation has been successfully applied to a number of systems, including C_{60}–nanotube, C_{60}–C_{60}, and nanotube–nanotube. For the graphite-based and C_{60}-based potentials, Girifalco et al. (2000) state that calculations using the continuum and discrete approximations give similar results, such that the difference between equilibrium distances for the atom–atom interactions is less than 2%.

In the interest of modelling irregularly shaped molecules, such as drugs, an alternative hybrid discrete-continuum approximation can also be used to determine the interaction energy. The hybrid model, which is represented by elements of both Eqs (1.11) and (1.12), applies when a symmetrical molecule is interacting with a molecule comprising asymmetrically located atoms and is given by

$$E = \sum_i \eta \int \Phi(\rho_i)\,dS, \tag{1.13}$$

where η is the surface density of atoms on the symmetrical molecule, ρ_i is the distance between a typical surface element dS on the continuously modelled molecule and atom i in the molecule which is modelled as discrete. Again $\Phi(\rho_i)$ is the potential function, and the energy is obtained by summing over all atoms in the drug or the molecule, which is represented discretely. For further details on the hybrid model given by Eq. (1.13) and a comparison of the approaches given by Eqs (1.11) and (1.12), we refer the reader to Hilder and Hill (2007).

1.3.2 INTERACTION FORCE

The van der Waals force refers to the attractive and repulsive forces between molecules, or the intermolecular force, and arises in nature. For example, the gecko climbs walls with use of the van der Waals force between the wall and the gecko's setae, or hair-like structures, on their feet. The van der Waals interaction force between two typical atoms on two nonbonded molecules is given by

$$F_{vdW} = -\nabla E, \tag{1.14}$$

where the interaction energy E is given by either Eqs (1.11), (1.12), or (1.13), and the symbol ∇ refers to the vector gradient. The gradient in Cartesian coordinates is given by

$$\nabla E(x, y, z) = \frac{\partial E}{\partial x}\hat{\mathbf{i}} + \frac{\partial E}{\partial y}\hat{\mathbf{j}} + \frac{\partial E}{\partial z}\hat{\mathbf{k}},$$

where $(\hat{\mathbf{i}}, \hat{\mathbf{j}}, \hat{\mathbf{k}})$ denote unit vectors in the (x, y, z) directions, respectively. So for example, a resultant axial force in the z-direction is obtained by differentiating the integrated interaction energy with respect to Z, which may be defined as the distance between the centres of two molecules, and Eq. (1.14) simplifies to become

$$F_Z = -\frac{\partial E}{\partial Z}. \tag{1.15}$$

1.3.3 LENNARD-JONES POTENTIAL

Sir John Lennard-Jones (1894–1954), a mathematician knighted in 1946, is regarded as the father of modern computational chemistry due to his contributions in theoretical chemistry. In particular, Sir Lennard-Jones had interests in quantum mechanics and intermolecular forces which led to his most notable contribution, the atomic interaction potential proposed in 1924 and which now bears his name.

The Lennard-Jones potential is a simple mathematical model which describes the interaction between two nonbonded atoms and is given by

$$\Phi(\rho) = -A\rho^{-m} + B\rho^{-n}, \tag{1.16}$$

where A and B are positive constants referred to as the attractive and repulsive constants, respectively, and ρ is the distance between the atoms. In many cases, the values $m = 6$ and $n = 12$ are adopted, and this is commonly referred to as the 6–12 potential. For hydrogen bonding interactions, a 10–12 potential is used; it should be noted that there are a number of other empirically motivated potentials in the literature, such as the Morse potential. Alternatively, the Lennard-Jones potential can be written in the form

$$\Phi(\rho) = 4\epsilon \left[-\left(\frac{\sigma}{\rho}\right)^6 + \left(\frac{\sigma}{\rho}\right)^{12} \right], \tag{1.17}$$

where σ is the distance at which the intermolecular potential between the two atoms is zero and ϵ denotes the energy well depth, $\epsilon = A^2/(4B)$. From Eq. (1.17), the equilibrium distance ρ_0 for two atoms is given by $\rho_0 = 2^{1/6}\sigma = (2B/A)^{1/6}$, which is shown in Fig. 1.23.

In this book, we refer to the van der Waals force as that which is derived from the above Lennard-Jones potential (Eq. 1.17) by means of Eq. (1.14). With this interpretation, the van der Waals force is a short-range force, and therefore when using the Lennard-Jones potential to represent the interaction between molecular structures, it may only be necessary to include the nearest neighbour interactions. For example, if we are investigating the behaviour of a molecule near the open end of a carbon nanotube, we can for this purpose consider the tube to be semiinfinite in length. We further comment that in the chemical literature, the van der Waals force sometimes refers simply to the force derived from the attractive component of Eq. (1.17).

WORKED EXAMPLE 1.9

Determine the equilibrium distance ρ_0 for two atoms, as depicted in Fig. 1.23.

Solution

The equilibrium distance between two atoms occurs when the Lennard-Jones potential energy is a minimum, as shown in Fig. 1.23. Thus we have

$$\frac{d\Phi}{d\rho} = 0.$$

Differentiating Eq. (1.17) we obtain

$$\frac{d\Phi}{d\rho} = 4\epsilon \left[\frac{6\sigma^6}{\rho^7} - \frac{12\sigma^{12}}{\rho^{13}} \right].$$

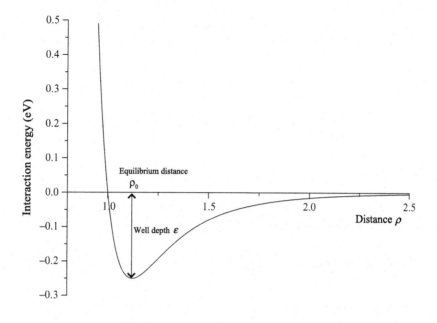

FIG. 1.23

Lennard-Jones potential.

Thus letting this equal zero we obtain the equation

$$\frac{6\sigma^6}{\rho_0^7} = \frac{12\sigma^{12}}{\rho_0^{13}}$$

$$\rho_0^6 = 2\sigma^6,$$

to deduce that $\rho_0 = 2^{1/6}\sigma$. Alternatively, in terms of the attractive and repulsive constants A and B, respectively, the equilibrium distance is $\rho_0 = (2B/A)^{1/6}$, where $\sigma = (B/A)^{1/6}$.

The Lennard-Jones potential is a semiempirical potential in that the attractive term originates from an examination of the mechanics of a quantum oscillator, while the repulsive term is based upon empirical considerations (Lennard-Jones, 1931). The Lennard-Jones potential is believed to apply between nonbonded (i.e. atoms on separate molecules) and nonpolar (i.e. nonelectrostatic) atomic interactions and it has been applied to a number of molecular configurations of carbon nanostructures. For example, two identical parallel carbon nanotubes, between two C_{60} fullerenes, and between a carbon nanotube and C_{60} (both inside and outside the tube). Numerical values of the Lennard-Jones constants for carbon–carbon atoms and atoms in graphene–graphene, C_{60}–C_{60}, and C_{60}–graphene are shown in Table 1.7, and these parameter values are used throughout this book.

Table 1.7 Lennard-Jones Constants for Graphitic Systems

	A (eV×Å^6)	B (eV×Å^{12})	ρ_0 (Å)
Carbon–carbon	19.97	34.81×10^3	3.89
Graphene–graphene	15.2	24.1×10^3	3.83
C_{60}–C_{60}	20.0	34.8×10^3	3.89
C_{60}–graphene	17.4	29.0×10^3	3.86

WORKED EXAMPLE 1.10

In this example, we work through the code necessary to plot the 6–12 Lennard-Jones potential shown in Fig. 1.23. This example uses the algebraic package MAPLE but any similar package may be used, such as Mathematica or MatLab.

Solution

```
> restart; with(plots):
```

Define the Lennard-Jones potential given by Eq. (1.16) for $m = 6$ and $n = 12$.

```
> E := -A/r^6 + B/r^(12):
```

Define the Lennard-Jones constants; note that we let both A and B equal unity.

```
> A := 1: B := 1:
```

We plot the Lennard-Jones potential against the distance ρ.

```
> plot(E,r=0.95..2);
```

1.4 BOOK OVERVIEW

In the following chapter, we give some mathematical preliminaries for various standard functions used throughout the book, such as the gamma and beta functions, the hypergeometric function, and the associated Legendre functions. In the chapter thereafter, we detail a number of specific scenarios relating to the interaction energy of linear objects, such as a single atom with a plane, and of spherical objects, such as a plane with a sphere. Chapter 4 details the interaction of nested carbon nanostructures, in particular carbon nano-onions, double-walled carbon nanotubes, and nanobundles. Chapter 5 describes the important concepts of acceptance condition and suction energy, then details a number of specific scenarios with regard to these terms, such as the encapsulation of a fullerene into a nanotube, double-walled carbon nanotubes, and nanobundles. Following this in Chapter 6, these concepts are further utilised and applied to nano-oscillators, in particular the C_{60}-nanotube, double-walled nanotube, and nanobundle oscillators.

The first six chapters are intentionally introductory in nature; they are intended to be appropriate for third-year students. The remaining chapters involve developments of the ideas previously introduced in Chapters 1–6, but the situations and geometry are more complicated than the earlier examples.

Chapter 7 uses the concepts previously introduced in Chapters 5 and 6 to investigate more complicated geometries, in particular nanopeapods and ellipsoids. Following this in Chapter 8, the concepts detailed in Chapter 5 are utilised to investigate the encapsulation of drug molecules into nanotubes constructed from various materials. In Chapter 9, a new model for the geometric structure of a carbon nanotube is formulated. Following this in Chapters 10 and 11, this new geometric model is used in conjunction with least squares and calculus of variations, respectively, to investigate the joining of various carbon nanostructures. A compilation of key works can be found in the bibliography, and the solutions and hints for the exercises in all chapters are given at the end of the book.

EXERCISES

1.1. Explain how a pencil works. What is meant by sp^2 bonding?

1.2. What is meant when we refer to the chirality of a carbon nanotube? Explain the difference between zigzag, armchair, and chiral tubes.

1.3. Prove that the chiral angle θ_0 is equal to

$$\theta_0 = \cos^{-1}\left[\frac{2n + m}{2(n^2 + nm + m^2)^{1/2}}\right].$$

1.4. If the translational vector is given by $\mathbf{T} = t_1\mathbf{a} + t_2\mathbf{b}$, obtain the expressions for t_1 and t_2.

1.5. Find d_R for a (i) $(10, 10)$ tube, (ii) $(6, 0)$ tube, and a (iii) $(9, 3)$ tube.

1.6. Prove that the length of the translational vector $\mathbf{T}$ of a 1D unit cell of a carbon nanotube is equal to $\sqrt{3}|C|/d_R$. Determine the lengths the unit cell in (i) $(10, 10)$ and (ii) $(9, 6)$ tubes.

1.7. Draw the unit cell for the $(9, 6)$ tube above.

1.8. Create code using MAPLE, or a similar package, to determine the greatest common divisor, i.e. d or d_R. Note: do not use built in functions such as $gcd(a, b)$.

1.9. Draw the equilateral triangle building block for a $(3, 1)$ fullerene. What type of fullerene is this?

1.10. Prove that the area of a general equilateral triangle building block is given by Eq. (1.7).

1.11. Prove that the number of carbon atoms in a fullerene is given by Eq. (1.4). Remember that each fullerene consists of 20 equilateral triangles.

1.12. Prove Euler's theorem (Eq. 1.9) using Legendre's proof.

1.13. Prove the Eq. (1.10) for the closing of graphene into a carbon nanotube.

1.14. Determine Euler's theorem for a nanotori, where there may be hexagon (n_6), pentagon (n_5), and heptagon (n_7) rings.

1.15. Explain the isolated pentagon rule.

1.16. When is it possible to use the continuum approximation to determine the energy between interacting molecular structures? Explain some advantages of this model.

1.17. Create a plot of the 10–12 Lennard-Jones potential.

MATHEMATICAL PRELIMINARIES

2.1 INTRODUCTION

Throughout this book, we will be evaluating a number of integrals which have a long history of investigation in mathematical analysis. This is to our advantage because the enormous amount of existing theory on integrals makes the task of analysing and evaluating them much easier. In this chapter, we introduce some special functions, their definitions, and some important identities.

Although we deal with a number of special functions in this chapter, the most important section concerns the hypergeometric functions. Hypergeometric functions represent a very wide class of functions, and special cases give rise to well-known elementary and other special functions. Throughout this book, we will be making use of various transformations of hypergeometric functions, which allows us to express solutions perhaps in a more elegant and economical form that might have certain computational advantages. While this chapter deals with a variety of special functions, the central theme is that of hypergeometric functions, which act as a crossroad to almost every other section in this chapter.

2.2 DIRAC DELTA FUNCTION: $\delta(x)$

The Dirac delta function is a mathematical construct which is called a generalised function or a distribution and was originally introduced by the British theoretical physicist Paul Dirac. The function $\delta(x)$ has the value zero everywhere except at $x = 0$, where its value is infinitely large and is such that its total integral is 1. This function is very useful as an approximation for a tall narrow spike function, namely an impulse. For example, to calculate the dynamics of a baseball being hit by a bat, approximating the force of the bat hitting the baseball by a delta function is a useful device. The delta function not only enables the equations to be simplified, but it also allows the motion of the baseball to be calculated by only considering the total impulse of the bat against the ball, rather than requiring the details of how the bat transferred energy to the ball. However, we note that the Dirac delta function is not strictly a function, although for many purposes it can be manipulated as such and can be formally defined as a generalised function or as a distribution that is also a measure.

Modelling and Mechanics of Carbon-based Nanostructured Materials. http://dx.doi.org/10.1016/B978-0-12-812463-5.00002-9

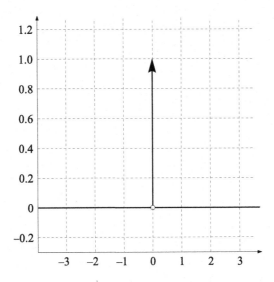

FIG. 2.1

Plot of the Dirac delta function.

The Dirac delta function can be thought of as a function on the real line, which is zero everywhere except at the origin where it becomes very large, as shown in Fig. 2.1, such that

$$\delta(x) \to \infty, \quad x = 0$$
$$= 0, \quad x \neq 0,$$

and it is constrained to satisfy the identity

$$\int_{-\infty}^{\infty} \delta(x)dx = 1. \tag{2.1}$$

This definition gives an intuitive grasp of the Dirac delta function but should not be taken too seriously because no normal function has the above properties. Moreover, there exist descriptions of the delta function which differ from the above conceptualisation. Formally, we define a delta function as a limit of a delta convergent sequence. For example, one such delta convergent sequence is the Lorentz–Cauchy functions, given by

$$F_\epsilon(x) = \frac{1}{\pi} \left(\frac{\epsilon}{x^2 + \epsilon^2} \right), \tag{2.2}$$

which satisfies the property that

$$\int_{-\infty}^{\infty} F_\epsilon(x)\, dx = 1,$$

for all $\epsilon > 0$ and is said to converge to the delta function in the limit $\epsilon \to \infty$; that is

$$\lim_{\epsilon \to 0} F_\epsilon(x) \to \delta(x).$$

WORKED EXAMPLE 2.1

Another delta convergent sequence is the Gaussian functions, defined by

$$F_\alpha(x) = \frac{1}{2\sqrt{\pi \alpha}} e^{-x^2/4\alpha}.$$

Show that the Gaussian functions satisfy the properties of a delta convergent sequence as $\alpha \to 0$. Specifically, that $F_\alpha(x) \to 0$ for $x \neq 0$ and that $\int_{-\infty}^{\infty} F_\alpha(x)dx = 1$.

Solution

When $x \neq 0$, then the exponential term dominates the behaviour of the function; that is

$$\lim_{\alpha \to 0} F_\alpha(x) \to \frac{1}{2\sqrt{\pi \alpha}} e^{-\infty} \to 0.$$

However, at $x = 0$ the exponential term is identically 1 regardless of the value of α and therefore

$$\lim_{\alpha \to 0} F_\alpha(x) \to \frac{1}{2\sqrt{\pi \alpha}} e^0 \to \infty.$$

To evaluate the integral, we consider the integral

$$\int_{-\infty}^{\infty} \int_{-\infty}^{\infty} e^{-(x^2+y^2)} dx\, dy = \int_{-\infty}^{\infty} e^{-x^2} dx \int_{-\infty}^{\infty} e^{-y^2} dy$$

$$= \left(\int_{-\infty}^{\infty} e^{-x^2} dx \right)^2.$$

Now we take the LHS and make the substitution from Cartesian to polar coordinates

$$\int_{-\infty}^{\infty} \int_{-\infty}^{\infty} e^{-(x^2+y^2)} dx\, dy = \int_0^{2\pi} \int_0^{\infty} re^{-r^2} dr\, d\theta$$

$$= 2\pi \int_0^{\infty} re^{-r^2} dr$$

$$= -\pi \left[e^{-r^2} \right]_0^{\infty}$$

$$= \pi.$$

Therefore we may deduce

$$\frac{1}{\sqrt{\pi}} \int_{-\infty}^{\infty} e^{-x^2} dx = 1.$$

Now we make the substitution $x = z/2\sqrt{\alpha}$, which gives

$$\frac{1}{2\sqrt{\pi \alpha}} \int_{-\infty}^{\infty} e^{-z^2/4\alpha} dz = 1.$$

The delta function can be viewed as the derivative of another generalised function known as the Heaviside step function $H(x)$, namely

$$\frac{d}{dx} H(x) = \delta(x),$$

where the Heaviside step function $H(x)$ is defined in the following section (see Section 2.3). Moreover, another definition of the delta function is as a generalised function which has the fundamental property

$$\int_{-\infty}^{\infty} f(x)\delta(x-a)dx = f(a)$$

for every $f(x)$, which is a suitable test function. A suitable test function is one which is differentiable any number of times and for which the function and all its derivatives vanish outside some finite interval, so that the above equation can be written as

$$\int_{a-\epsilon}^{a+\epsilon} f(x)\delta(x-a)dx = f(a)$$

for $\epsilon > 0$.

WORKED EXAMPLE 2.2

Using a delta convergent sequence, show that

$$\int_{-\infty}^{\infty} f(x)\delta(x-a)\,dx = f(a).$$

Solution

By rearrangement of the Lorentz–Cauchy functions (2.2), the integral above is given by

$$\int_{-\infty}^{\infty} f(x)\delta(x-a)\,dx = \lim_{\epsilon \to 0} \frac{1}{\pi} \int_{-\infty}^{\infty} \frac{f(x)\epsilon\,dx}{(x-a)^2 + \epsilon^2},$$

which on making the substitution $x = a + \epsilon \tan \theta$ gives

$$\int_{-\infty}^{\infty} f(x)\delta(x-a)\,dx$$

$$= \lim_{\epsilon \to 0} \frac{1}{\pi} \int_{-\pi/2}^{\pi/2} f(a + \epsilon \tan \theta)\,d\theta,$$

$$= \lim_{\epsilon \to 0} \frac{1}{\pi} \left\{ \int_{-\pi/2}^{-b} f(a + \epsilon \tan \theta)\,d\theta + \int_{-b}^{b} f(a + \epsilon \tan \theta)\,d\theta + \int_{b}^{\pi/2} f(a + \epsilon \tan \theta)\,d\theta \right\}.$$

Since

$$\left| \frac{1}{\pi} \int_{-\pi/2}^{\pi/2} f(a + \epsilon \tan \theta) \, d\theta - f(a) \right| \leq \frac{1}{\pi} \int_{-b}^{b} |f(a + \epsilon \tan \theta) - f(a)| \, d\theta + 2 \left(\frac{\pi}{2} - b \right) \| f \|_\infty,$$

and

$$|a + \epsilon \tan \theta - a| < \epsilon \tan b \to 0,$$

as $\epsilon \to 0$, and by the continuity of f, we have

$$|f(a + \epsilon \tan \theta) - f(a)| \to 0,$$

which is independent of θ, therefore

$$\int_{-b}^{b} |f(a + \epsilon \tan \theta) \, d\theta - f(a)| \, d\theta \to 0,$$

and

$$\left| \frac{1}{\pi} \int_{-\pi/2}^{\pi/2} f(a + \epsilon \tan \theta) \, d\theta - f(a) \right| \leq 2 \left(\frac{\pi}{2} - b \right) \| f \|_\infty.$$

It can be seen that LHS does not depend on b, then let $b \to (\pi/2)^-$, which gives

$$\frac{1}{\pi} \int_{-\pi/2}^{\pi/2} f(a + \epsilon \tan \theta) \, d\theta = f(a) = \int_{-\infty}^{\infty} f(x) \delta(x - a) \, dx.$$

2.3 HEAVISIDE FUNCTION: $H(x)$

The Heaviside step function $H(x)$, also called the unit step function, is a discontinuous function, whose value is zero for negative arguments $x < 0$ and one for positive arguments $x > 0$, as illustrated in Fig. 2.2. The function is commonly used in the mathematics of control theory and signal processing to represent a signal that switches on at a specified time and stays switched on indefinitely. In this book, the Heaviside function is utilised to represent the oscillatory behaviour of nanostructures, such as in Chapter 6.

The Heaviside function might be viewed as an antiderivative of the Dirac delta function, outlined in Section 2.2, and can be written as

$$H(x) = \int_{-\infty}^{x} \delta(t) \, dt.$$

Moreover, this function is related to the boxcar function by

$$\prod_{a,b}(x) = c[H(x - a) - H(x - b)],$$

which is equal to c for $a \leq X \leq b$ and zero otherwise.

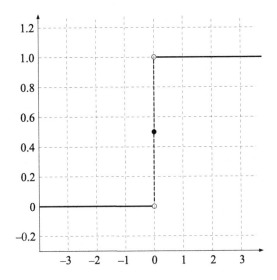

FIG. 2.2

Heaviside step function.

2.4 GAMMA FUNCTION: $\Gamma(z)$

The gamma function is in many ways the fundamental building block of special functions. For $\Re(z) > 0$, the gamma function $\Gamma(z)$ can be defined as the definite integral

$$\Gamma(z) = \int_0^\infty t^{z-1} e^{-t}\, dt, \tag{2.3}$$

which is known as the Euler integral. Integrating $\Gamma(z+1)$ by parts yields

$$\Gamma(z+1) = \left[-t^z e^{-t}\right]_0^\infty + z \int_0^\infty t^{z-1} e^{-t}\, dt,$$

and we observe that the first term is zero and that the second is $z\Gamma(z)$. This allows us to write

$$\boxed{\Gamma(z+1) = z\Gamma(z),} \tag{2.4}$$

which is the basic recurrence relationship for the gamma function. It can be seen from Eq. (2.4), and the fact that $\Gamma(1) = 1$, that when z is a positive integer n, the gamma function corresponds to the factorial operator given by the expression

$$\boxed{\Gamma(n+1) = n!} \tag{2.5}$$

WORKED EXAMPLE 2.3

The integral given by Eq. (2.3) has singularities at the negative integers. In this example, we replace the nonanalytical Euler integral form of the gamma function with a function and an integral which is well-defined for all values of z.

Solution

From Eq. (2.3) we have

$$\Gamma(z) = \int_0^\infty t^{z-1} e^{-t} \, dt$$

$$= \int_0^1 t^{z-1} e^{-t} \, dt + \int_1^\infty t^{z-1} e^{-t} \, dt$$

$$= P(z) + \int_1^\infty t^{z-1} e^{-t} \, dt.$$

Expanding e^{-t} as a power series and integrating term by term gives

$$P(z) = \sum_{n=0}^\infty \frac{(-1)^n}{n!\,(z+n)},$$

and therefore

$$\Gamma(z) = \sum_{n=0}^\infty \frac{(-1)^n}{n!\,(z+n)} + \int_1^\infty t^{z-1} e^{-t} \, dt. \tag{2.6}$$

The final integral term in Eq. (2.6) is analytic for all z, and we can see from the sum that the gamma function is always analytic except at the points $z = 0, -1, -2, \ldots$ and at $z = -n$, where the gamma function has simple poles with residues of $(-1)^n/n!$

In addition to $\Gamma(1) = 1$, it is often important to evaluate the gamma function for half integer arguments. These can be calculated using the recurrence formula (2.4) and the special value

$$\Gamma(1/2) = \sqrt{\pi},$$

which can be derived from the Euler integral (2.3), as shown in Worked Example 2.4.

WORKED EXAMPLE 2.4

In this example, we show that $\Gamma(1/2) = \sqrt{\pi}$, using the Euler integral given by Eq. (2.3).

Solution

Substituting $z = 1/2$ into Eq. (2.3) and taking the square obtains

$$\Gamma(1/2)^2 = \int_0^\infty \frac{e^{-x}}{\sqrt{x}} \, dx \int_0^\infty \frac{e^{-y}}{\sqrt{y}} \, dy.$$

On making the substitution $x = u^2$ and $y = v^2$, we obtain

$$\Gamma(1/2)^2 = 4 \int_0^\infty \int_0^\infty e^{-(u^2+v^2)} \, du \, dv$$

$$= 4 \int_0^\infty \int_0^{\pi/2} e^{-r^2} r \, d\theta \, dr,$$

where we have made the substitution $u = r\cos\theta$ and $v = r\sin\theta$ so that $du \, dv = r \, dr \, d\theta$. Upon integration, we obtain

$$\Gamma(1/2)^2 = 4\frac{\pi}{2}\left[-\frac{e^{-r^2}}{2}\right]_0^\infty$$

$$= \pi,$$

so that $\Gamma(1/2) = \sqrt{\pi}$.

The gamma function is analytic everywhere in the complex plane except for the points $z = 0$, $-1, -2, \ldots$ and it satisfies the following functional identity

$$\boxed{\Gamma(z)\Gamma(1-z) = \frac{\pi}{\sin(\pi z)},} \tag{2.7}$$

which relates the gamma function to the trigonometric functions. The gamma function also satisfies the well-known duplication formula

$$\boxed{\Gamma(2z) = 2^{2z-1}\pi^{-1/2}\Gamma(z)\Gamma(z+1/2),} \tag{2.8}$$

which is due to Legendre. The duplication formula can be extended to the general multiplication formulae, thus

$$\prod_{l=0}^{m-1} \Gamma(z+l/m) = (2\pi)^{(m-1)/2} m^{1/2-mz} \Gamma(mz), \tag{2.9}$$

where $m = 2, 3, 4, \ldots$, and where $\prod$ denotes the product notation.

Functions of integer arguments, which are defined in terms of factorials, can now be generalised to complex arguments by substitution of the gamma functions for the factorial terms. For example, the binomial coefficient is defined for integer arguments by

$$\binom{m}{n} = \frac{m!}{n! \, (m-n)!}, \tag{2.10}$$

and can be extended to noninteger and even complex arguments by simply defining it as

$$\binom{p}{q} = \frac{\Gamma(p+1)}{\Gamma(q+1)\Gamma(p-q+1)},$$

where p and q can be any complex number, provided p, q, and $p - q$ are not negative integers.

Another useful function is the product of n successive terms from some starting value a, which is known as the Pochhammer symbol

$$(a)_n = a(a+1)(a+2)\ldots(a+n-1).$$

If the argument a is an integer, then we can write

$$(a)_n = \frac{(a+n-1)!}{(a-1)!},$$

and this can be extended to noninteger and complex values of a by the definition

$$(a)_n = \frac{\Gamma(a+n)}{\Gamma(a)}. \qquad (2.11)$$

We also note that when $a = 1$, then $(1)_n = n!$

WORKED EXAMPLE 2.5

In this example, we wish to derive an expression for $(-n)_m$, where $n = 1, 2, 3, \ldots$ for $n \geqslant m$, in terms of simple factorial terms.

Solution

From the definition of the Pochhammer symbol, we have

$$(-n)_m = (-n)(1-n)(2-n)\ldots(m-n-1)$$
$$= (-1)^m n(n-1)(n-2)\ldots(n-m-1)$$
$$= (-1)^m \frac{n!}{(n-m)!},$$

and therefore

$$(-n)_m = \begin{cases} \dfrac{(-1)^m n!}{(n-m)!}, & m \leqslant n, \\ 0, & m > n, \end{cases} \qquad (2.12)$$

WORKED EXAMPLE 2.6

In this example, we give an independent definition for the gamma function.

Solution

We write the factorial operator as

$$x! = \frac{(x+n)!}{(x+1)_n}$$
$$= \frac{n!\,(n+1)_x}{(x+1)_n}$$
$$= \frac{n!\,n^x}{(x+1)_n} \cdot \frac{(n+1)_x}{n^x}.$$

We may show that

$$\lim_{n \to \infty} \frac{(n+1)_x}{n^x} = 1,$$

which allows us to write

$$x! = \lim_{n \to \infty} \frac{n! \, n^x}{(x+1)_n}.$$

We may now replace $x + 1$ by any complex number $z \neq 0, -1, -2, \ldots$, and define

$$\Gamma(z) = \lim_{n \to \infty} \frac{n! \, n^{z-1}}{(z)_n}, \tag{2.13}$$

which can be shown to be equivalent to the integral definition (Eq. 2.3).

There are several useful series expressions for $\log \Gamma(z)$, and some of these are given by Binet's formula; thus

$$\log \Gamma(z) = \left(z - \frac{1}{2}\right) \log z - z + \frac{1}{2} \log(2\pi) + \frac{1}{2} \sum_{m=1}^{\infty} \frac{m \zeta(m+1, z+1)}{(m+1)(m+2)},$$

where $\zeta(z, \alpha)$ is the Hurwitz zeta function, named after the German mathematician Adolf Hurwitz (1859–1919). The Hurwitz zeta function is defined by the integral expression

$$\zeta(z, \alpha) = \frac{1}{\Gamma(z)} \int_0^{\infty} \frac{t^{z-1} e^{-\alpha t}}{1 - e^{-t}} \, dt, \tag{2.14}$$

and the standard zeta function arises from this expression when $\alpha = 1$, namely

$$\zeta(z) = \frac{1}{\Gamma(z)} \int_0^{\infty} \frac{t^{z-1}}{e^t - 1} \, dt, \tag{2.15}$$

which is illustrated in an exercise at the end of this chapter. Another useful series representation for $\log \Gamma(z)$ is Burnside's formula, which is given by

$$\log \Gamma(z) = \left(z - \frac{1}{2}\right) \log\left(z - \frac{1}{2}\right) - z - \frac{1}{2} + \frac{1}{2} \log(2\pi) - \sum_{m=1}^{\infty} \frac{\zeta(2m, z)}{2m(2m+1)} 2^{-2m},$$

for $\Re(z) > -1/2$. These give rise to an asymptotic expansion for the logarithm of the gamma function due to Stirling

$$\log \Gamma(z) = \left(z - \frac{1}{2}\right) \log z - z + \frac{1}{2} \log(2\pi) + \sum_{m=1}^{n} \frac{B_{2m}}{2m(2m-1)} z^{1-2m} + O(z^{-1-2m}), \tag{2.16}$$

where B_{2m} are the Bernoulli numbers which are defined by the expansion

$$\frac{t}{e^t - 1} = \sum_{n=0}^{\infty} \frac{B_n t^n}{n!}, \quad 0 < |t| < 2\pi,$$

where $B_0 = 1$ and all B_n for n odd are zero except $B_1 = -1/2$ (see the exercises at the end of this chapter).

Eq. (2.16) can be used to give an asymptotic expansion for the gamma function, which is

$$\Gamma(z) = z^{z-1/2} e^{-z} \sqrt{2\pi} \left\{ 1 + \frac{1}{12z} + \frac{1}{288z^2} - \frac{139}{51840z^3} - \frac{571}{2488320z^4} + O(z^{-5}) \right\},$$

and this leads to the approximation

$$\Gamma(z) \approx \sqrt{\frac{2\pi}{z}} \left(\frac{z}{e}\right)^z,$$

for large $|z|$, which for $n \to \infty$ can also be expressed in terms of the factorial operator as

$$n! \approx \sqrt{2\pi n} \left(\frac{n}{e}\right)^n.$$

2.5 BETA FUNCTION: B(x, y)

The beta function is defined by the integral

$$B(x, y) = \int_0^1 t^{x-1} (1-t)^{y-1} \, dt, \tag{2.17}$$

for $\Re(x) > 0$ and $\Re(y) > 0$, and can be extended to the complex plane for both variables. The beta function is given in terms of gamma functions by

$$B(x, y) = \frac{\Gamma(x)\Gamma(y)}{\Gamma(x+y)}. \tag{2.18}$$

It is clear from this definition that the beta function is symmetric; that is

$$B(x, y) = B(y, x).$$

By substituting $t = \sin^2 \theta$ into Eq. (2.17), we may derive

$$B(x, y) = 2 \int_0^{\pi/2} \sin^{(2x-1)} \theta \, \cos^{(2y-1)} \theta \, d\theta,$$

and therefore

$$\int_0^{\pi/2} \sin^p \theta \, \cos^q \theta \, d\theta = \frac{1}{2} B \left(\frac{p+1}{2}, \frac{q+1}{2} \right). \tag{2.19}$$

The beta function can also be thought of as a reciprocal binomial term. We see from Eq. (2.18) that

$$\frac{1}{B(x, y)} = y \binom{x+y-1}{x-1} = x \binom{x+y-1}{y-1},$$

or alternatively,

$$\binom{x}{y} = \frac{1}{yB(x-y+1,y)},$$

where, as usual, the binomial coefficient is defined by

$$\binom{n}{k} = \frac{n!}{(n-k)!k!}.$$

WORKED EXAMPLE 2.7

Show that

$$\frac{1}{B(x,y)} = y\binom{x+y-1}{x-1} = x\binom{x+y-1}{y-1}.$$

Solution

From the definitions given by Eqs (2.18) and (2.5), we obtain

$$B(x,y) = \frac{\Gamma(x)\Gamma(y)}{\Gamma(x+y)} = \frac{(x-1)!(y-1)!}{(x+y-1)!},$$

so that

$$\frac{1}{B(x,y)} = \frac{(x+y-1)!}{(x-1)!(y-1)!}.$$

On noticing that

$$\binom{x+y-1}{x-1} = \frac{(x+y-1)!}{y!(x-1)!}, \quad \text{and} \quad \binom{x+y-1}{y-1} = \frac{(x+y-1)!}{x!(y-1)!},$$

and using the definition $y = y!/(y-1)!$, we obtain

$$\frac{1}{B(x,y)} = \frac{y(x+y-1)!}{y!(x-1)!} = y\binom{x+y-1}{x-1}$$

and similarly for x.

2.6 HYPERGEOMETRIC FUNCTION: $F(a, b; c; z)$
2.6.1 HYPERGEOMETRIC SERIES

The hypergeometric series may be defined in terms of the general form of a power series

$$\sum_{n=0}^{\infty} \alpha_n z^n,$$

such that the ratio of successive coefficients can be expressed as a rational function of n; thus

$$\frac{\alpha_{n+1}}{\alpha_n} = \frac{A(n)}{B(n)},$$

for some general polynomials $A(n)$ and $B(n)$. One special case of this is the geometric series, for which the ratio of coefficients is a constant. Another special case is the exponential series, given by

$$e^z = 1 + z + \frac{z^2}{2!} + \cdots + \frac{z^n}{n!} + \cdots,$$

for which

$$\frac{\alpha_{n+1}}{\alpha_n} = \frac{1}{n+1}.$$

In general, the exponential function is used as the benchmark for comparisons and the general hypergeometric series assumes the form

$$\sum_{n=0}^{\infty} \beta_n \frac{z^n}{n!}.$$

The convergence of such series were first investigated by Johann Carl Friedrich Gauss (1777–1855). In particular, he considered the case for which the ratio of the coefficients could be expressed as

$$\frac{\beta_{n+1}}{\beta_n} = \frac{(n+a)(n+b)}{n+c},$$

which leads to the series representation of the standard hypergeometric function given by

$$F(a, b; c; z) = \sum_{n=0}^{\infty} \frac{(a)_n (b)_n}{(c)_n} \frac{z^n}{n!}, \tag{2.20}$$

where $(x)_n$ denotes the Pochhammer symbol defined by Eq. (2.11). Evidently from Eq. (2.20), we can make the following observations:

- If either a or b is a negative integer, then the series terminates after a finite number of terms;
- If c is a negative integer, then the series becomes undefined after a finite number of terms;
- The ratio of successive terms approaches z in the limit as $n \to \infty$; and thus
- The series is absolutely convergent for $|z| < 1$.

2.6.2 RELATIONSHIPS TO OTHER FUNCTIONS

Several of the elementary functions can be given in terms of hypergeometric functions, including:

$$(1 + z)^a = F(-a, b; b; -z),$$
$$\arcsin(z) = z F(1/2, 1/2; 3/2; z^2),$$
$$\arctan(z) = z F(1/2, 1; 3/2; -z^2),$$
$$\log(1 + z) = z F(1, 1; 2; -z).$$

WORKED EXAMPLE 2.8

Show that $(1 + z)^n = F(-n, b; b; -z)$ for some positive integer n.

Solution

We expand $F(-n, b; b; -z)$ using the series definition (Eq. 2.20), where we assume n is a positive integer giving

$$F(-n, b; b; -z) = \sum_{m=0}^{\infty} \frac{(-n)_m (b)_m}{(b)_m m!} (-z)^m,$$

using Eq. (2.12) to substitute for $(-n)_m$ yields

$$F(-n, b; b; -z) = \sum_{m=0}^{n} \frac{(-1)^m n!}{(n-m)! m!} (-z)^m$$

$$= \sum_{m=0}^{n} \binom{n}{m} z^m$$

$$= (1 + z)^n.$$

As noted above, when a or b is a negative integer $-n$, then the series for $F(a, b; c; z)$ terminates and becomes a polynomial of degree n. Many of the classical families of orthogonal polynomials, which are discussed later in this chapter, can be expressed as terminating hypergeometric series, such as

$$T_n(x) = F(-n, n; 1/2; (1-x)/2), \tag{2.21}$$

$$U_n(x) = (n+1)F(-n, n+2; 3/2; (1-x)/2), \tag{2.22}$$

$$P_n(x) = F(-n, n+1; 1; (1-x)/2), \tag{2.23}$$

where $T_n(x)$ and $U_n(x)$ denote the Chebyshev polynomials of the first and second kind detailed in Section 2.9, and $P_n(x)$ denotes the Legendre polynomials detailed in Section 2.8.

The hypergeometric function is also related to elliptic integrals, such as

$$F(1/2, 1/2; 1; k^2) = 2K(k)/\pi,$$

$$F(1/2, -1/2; 1; k^2) = 2E(k)/\pi,$$

where $K(k)$ and $E(k)$ are the complete elliptic integrals of the first and second kinds, respectively, which are defined in Section 2.10.

2.6.3 HYPERGEOMETRIC DIFFERENTIAL EQUATION

The hypergeometric functions, as defined by Eq. (2.20), represent an extremely wide class of special functions occurring in physics and engineering. Frequently, identification of such functions is facilitated through the linear second order differential equation, for which Eq. (2.20) is one of the solutions. Differentiation of the hypergeometric function is straightforward, and the nth derivative is given by

$$\frac{d^n}{dz^n} F(a, b; c; z) = \frac{(a)_n (b)_n}{(c)_n} F(a+n, b+n; c+n; z).$$

The hypergeometric function is a solution of a homogeneous linear second order differential equation, with regular singularities at $z = 0$, 1 and ∞, which is called the hypergeometric equation and is given by

$$z(1-z)\frac{d^2u}{dz^2} + [c - (a+b+1)z]\frac{du}{dz} - abu = 0, \tag{2.24}$$

where a, b, and c are constants which are independent of z. Now, any second order linear ordinary differential equation with at most three regular singular points can be transformed into the hypergeometric equation, so that $u_1 = F(a,b;c;z)$ is a solution of Eq. (2.24) that is regular at $z = 0$. For the other solutions of the differential equation, we must consider the following three cases.

2.6.3.1 *None of the numbers c, a − b, or c − a − b are integers*
In this case, there are two linearly independent solutions of Eq. (2.24) in the vicinity of the three singular points, which are given by

$$
\begin{aligned}
u_1^{(0)} &= F(a,b;c;z), \\
&= (1-z)^{c-a-b}F(c-a,c-b;c;z), \\
&= (1-z)^{-a}F(a,c-b;c;z/(z-1)), \\
&= (1-z)^{-b}F(c-a,b;c;z/(z-1)), \\
u_2^{(0)} &= z^{1-c}F(a-c+1,b-c+1;2-c;z), \\
&= z^{1-c}(1-z)^{c-a-b}F(1-a,1-b;2-c;z), \\
&= z^{1-c}(1-z)^{c-a-1}F(a-c+1,1-b;2-c;z/(z-1)), \\
&= z^{1-c}(1-z)^{c-b-1}F(1-a,b-c+1;2-c;z/(z-1)), \\
u_1^{(1)} &= F(a,b;a+b-c+1;1-z), \\
&= z^{1-c}F(1+b-c,a-c+1;a+b-c+1;1-z), \\
&= z^{-a}F(a,a-c+1;a+b-c+1;1-1/z), \\
&= z^{-b}F(b,b-c+1;a+b-c+1;1-1/z), \\
u_2^{(1)} &= (1-z)^{c-a-b}F(c-a,c-b;c-a-b+1;1-z), \\
&= z^{1-c}(1-z)^{c-a-b}F(1-a,1-b;c-a-b+1;1-z), \\
&= z^{a-c}(1-z)^{c-a-b}F(c-a,1-a;c-a-b+1;1-1/z), \\
&= z^{b-c}(1-z)^{c-a-b}F(c-b,1-b;c-a-b+1;1-1/z), \\
\end{aligned}
$$

$$
\begin{aligned}
u_1^{(\infty)} &= z^{-a}F(a,a-c+1;a-b+1;1/z), \\
&= z^{b-c}(z-1)^{c-a-b}F(1-b,c-b;a-b+1;1/z), \\
&= (z-1)^{-a}F(a,c-b;a-b+1;1/(1-z)), \\
&= (z-1)^{-b}F(c-a,b;b-a+1;1/(1-z)), \\
\end{aligned}
$$

$$u_2^{(\infty)} = z^{-b}F(b-c+1,b;b-a+1;1/z),$$
$$= z^{a-c}(z-1)^{c-a-b}F(1-a,c-a;b-a+1;1/z),$$
$$= z^{1-c}(z-1)^{c-a-1}F(a-c+1,1-b;a-b+1;1/(1-z)),$$
$$= z^{1-c}(z-1)^{c-b-1}F(1-a,b-c+1;b-a+1;1/(1-z)),$$

which are the 24 solutions to the hypergeometric equations due to Kummer. From this list of solutions, the following two important linear transformations are apparent:

$$F(a,b;c;z) = (1-z)^{c-a-b}F(c-a,c-b;c;z), \tag{2.25}$$

$$F(a,b;c;z) = (1-z)^{-a}F(a,c-b;c;z/(z-1)). \tag{2.26}$$

2.6.3.2 One of the numbers a, b, c − a, or c − b are integers
In this case, the hypergeometric series terminates or can be made to terminate by transforming the hypergeometric function. Therefore the solution to the hypergeometric equation is called degenerate and may be written in the form

$$u = z^{\lambda}(1-z)^{\mu}p_n(z),$$

where λ and μ depend on the values of a, b, c, and $p_n(z)$ is a polynomial in z of degree n. For more information on how to transform the solution to a degenerate case, we refer to the reader to Andrews et al. (1999).

2.6.3.3 One of the numbers c − a − b or c are integers
In this case, the solutions around $z = 0$, $u_1^{(0)}$ and $u_2^{(0)}$ can be extended into the neighbourhoods of $z = 1$ and $z = \infty$ by use of a series involving logarithmic and digamma functions. The digamma function $\psi(z)$ is the logarithmic derivative of the gamma function and is given by

$$\psi(z) = \frac{d}{dz}\ln\Gamma(z) = \frac{\Gamma'(z)}{\Gamma(z)}.$$

For the details of these solutions, we refer the reader to Erdélyi et al. (1953) or Magnus et al. (1966).

2.6.4 CONTIGUOUS RELATIONS
The six hypergeometric functions $F(a \pm 1,b;c;z)$, $F(a,b \pm 1;c;z)$, $F(a,b;c \pm 1;z)$ are said to be contiguous to the function $F(a,b;c;z)$ and a relationship exists between this function and any two contiguous functions, which is called a contiguous relation. There are 15 such relations which are due to Gauss:

$$[c - 2a - (b-a)z]F + a(1-z)F_{a+1} - (c-a)F_{a-1} = 0,$$
$$(b-a)F + aF_{a+1} - bF_{b+1} = 0,$$
$$(c-a-b)F + a(1-z)F_{a+1} - (c-b)F_{b-1} = 0,$$
$$c[a - (c-b)z]F - ac(1-z)F_{a+1} + (c-a)(c-b)zF_{c+1} = 0,$$
$$(c-a-1)F + aF_{a+1} - (c-1)F_{c-1} = 0,$$

$$(c - a - b)F - (c - a)F_{a-1} + b(1 - z)F_{b+1} = 0,$$
$$(b - a)(1 - z)F - (c - a)F_{a-1} + (c - b)F_{b-1} = 0,$$
$$c(1 - z)F - cF_{a-1} + (c - b)zF_{c+1} = 0,$$
$$[a - 1 - (c - b - 1)z]F + (c - a)F_{a-1} - (c - 1)(1 - z)F_{c-1} = 0,$$
$$[c - 2b + (b - a)z]F + b(1 - b)F_{b+1} - (c - b)F_{b-1} = 0,$$
$$c[b - (c - a)z]F - bc(1 - z)F_{b+1} + (c - a)(c - b)zF_{c+1} = 0,$$
$$(c - b - 1)F + bF_{b+1} - (c - 1)F_{c-1} = 0,$$
$$c(1 - z)F - cF_{b-1} + (c - 1)F_{c+1} = 0,$$
$$[b - 1 - (c - a - 1)z]F + (c - b)F_{b-1} - (c - 1)(1 - z)F_{c-1} = 0,$$
$$c[c - 1 - (2c - a - b - 1)z]F +$$
$$(c - 1)(c - b)zF_{c+1} - c(c - 1)(1 - z)F_{c-1} = 0,$$

where $F = F(a, b; c; z)$, $F_{a\pm1} = F(a \pm 1, b; c; z)$, $F_{b\pm1} = F(a, b \pm 1; c; z)$, and $F_{c\pm1} = F(a, b; c \pm 1; z)$.

2.6.5 QUADRATIC TRANSFORMATIONS

The Kummer list of 24 solutions to the hypergeometric equation and the contiguous relations can be thought of as linear transformations of the hypergeometric function, as indicated in Eqs (2.25) and (2.26). For a more complete list of linear transformations, we refer the reader to Andrews et al. (1999). If a, b, and c are unrestricted, no higher order transformations exist. However, if and only if the six numbers

$$\pm(1 - c), \quad \pm(a - b), \quad \pm(a + b - c),$$

are such that one of them equals 1/2 or two of them are equal, then there exists higher order transformations, which are known as quadratic transformations. Of these the fundamental formulae are:

$$F(a, b; 2b; 4z/(1 + z)^2) = (1 + z)^{2a}F(a, a - b + 1/2; b + 1/2; z^2),$$
$$F(a, b; a - b + 1; z) = (1 - z)^{-a}F(a/2, a/2 - b + 1/2; b - a + 1; -4z(1 - z)^2),$$
$$F(a, a + 1/2; b; z) = 2^{2a}[1 + (1 - z)^{1/2}]^{-2a}$$
$$\times F(2a, 2a - b + 1; b; [1 - (1 - z)^{1/2}]/[1 + (1 - z)^{1/2}]),$$
$$F(2a, 2b; a + b + 1/2; z) = F(a, b; a + b + 1/2; 4z(1 - z)),$$

which are due to Gauss and Kummer. Some quadratic transformations which are particularly useful in this book are:

$$F(a, b; 2b; z) = (1 - z)^{-a/2}F\left(a, 2b - a; b + 1/2; -\frac{(1 - \sqrt{1 - z})^2}{4\sqrt{1 - z}}\right), \tag{2.27}$$

$$F(a, b; 2b; z) = \left[\frac{1 + (1 - z)^{1/2}}{2}\right]^{-2a}$$
$$\times F\left(a, a - b + 1/2; b + 1/2; \left[\frac{1 - (1 - z)^{1/2}}{1 + (1 - z)^{1/2}}\right]^2\right). \tag{2.28}$$

For a more complete list of quadratic and cubic transformations, we refer the reader to Andrews et al. (1999).

2.6.6 INTEGRAL FORMS OF THE HYPERGEOMETRIC FUNCTION

In subsequent chapters, we will evaluate a number of surface integrals arising from cylinders, spheres, and cones. Our interest in hypergeometric functions stems from the fact that many of these surface integrals involve the hypergeometric function. There are several integrals which define or involve the hypergeometric functions, and here we will include only the most useful ones. Euler's integral formula is given by

$$F(a, b; c; z) = \frac{\Gamma(c)}{\Gamma(b)\Gamma(c - b)} \int_0^1 t^{b-1}(1 - t)^{c-b-1}(1 - tz)^{-a} \, dt, \tag{2.29}$$

provided that $\Re(c) > \Re(b) > 0$ and $|\arg(1 - z)| < \pi$. There are also the closely related formulae

$$F(a, b; c; 1 - z) = \frac{\Gamma(c)}{\Gamma(b)\Gamma(c - b)} \int_0^\infty s^{b-1}(1 + s)^{a-c}(1 + sz)^{-a} \, ds, \tag{2.30}$$

when $\Re(c) > \Re(b) > 0$ and $|\arg z| < \pi$, and

$$F(a, b; c; 1/z) = \frac{\Gamma(c)}{\Gamma(b)\Gamma(c - b)} \int_1^\infty s^{a-c}(s - 1)^{c-b-1}(s - 1/z)^{-a} \, ds,$$

when $1 + \Re(a) > \Re(c) > \Re(b)$ and $|\arg(z - 1)| < \pi$. These integrals can be converted to trigonometric, hyperbolic or exponential functions by substitution of variables. It is interesting to note that in none of these integrals is the symmetry in a and b readily apparent, which is inherent in the relation $F(a, b; c; z) = F(b, a; c; z)$. However, for this purpose Erdélyi et al. (1953) has proved the following double integral:

$$F(a, b; c; z) = \frac{[\Gamma(c)]^2}{\Gamma(a)\Gamma(c - a)\Gamma(b)\Gamma(c - b)}$$
$$\times \int_0^1 \int_0^1 t^{b-1}\tau^{a-1}(1 - t)^{c-b-1}(1 - \tau)^{c-a-1}(1 - t\tau z)^{-c} \, dt \, d\tau,$$

in which the symmetry is obvious. More integral forms can be found in the bibliography cited at the end of the book.

2.7 APPELL'S HYPERGEOMETRIC FUNCTION: $F_1(a; b, b'; c; x, y)$

Appell's hypergeometric equation is a formal extension of the hypergeometric equation for two variables. The most commonly used form of Appell's hypergeometric function in this book is given by

$$F_1(a; b, b'; c; x, y) = \sum_{m=0}^\infty \sum_{n=0}^\infty \frac{(a)_{m+n}(b)_m(b')_n}{m!n!(c)_{m+n}} x^m y^n, \tag{2.31}$$

which is convergent when $|x| < 1$ and $|y| < 1$. The integral form of Appell's hypergeometric function is given by

$$F_1(a; b, b'; c; x, y) = \frac{\Gamma(c)}{\Gamma(a)\Gamma(c-a)} \int_0^1 t^{a-1}(1-t)^{c-a-1}(1-tx)^{-b}(1-ty)^{-b'} dt.$$ (2.32)

In fact there are four Appell's functions, and the remaining three, as defined by their expansions, are

$$F_2(a; b, b'; c, c'; x, y) = \sum_{m=0}^{\infty}\sum_{n=0}^{\infty} \frac{(a)_{m+n}(b)_m(b')_n}{m!n!(c)_m(c')_n} x^m y^n,$$

$$F_3(a, a'; b, b'; c; x, y) = \sum_{m=0}^{\infty}\sum_{n=0}^{\infty} \frac{(a)_m(a')_n(b)_m(b')_n}{m!n!(c)_{m+n}} x^m y^n,$$

$$F_4(a; b; c, c'; x, y) = \sum_{m=0}^{\infty}\sum_{n=0}^{\infty} \frac{(a)_{m+n}(b)_{m+n}}{m!n!(c)_m(c')_n} x^m y^n.$$

For the most part, this book uses the form given by Eq. (2.31); however, in Chapter 4, we also use the series expansion for $F_2(a; b, b'; c, c'; x, y)$.

There are various transformations of Appell's hypergeometric function, which enable this more complicated hypergeometric function to be simplified in terms of ordinary hypergeometric functions. Some of which are particularly useful in this book are

$$F_1(a; b, b'; c; x, 0) = F(a, b; c; x)$$
$$F_1(a; b, b'; c; 0, y) = F(a, b'; c; y).$$

It is also possible to transform Appell's functions into other Appell's functions, and two such transformations are

$$F_1(a; b, b'; c; x, y) = (1-y)^{-a} F_1\left(a; b, c-b-b'; c; \frac{y-x}{y-1}, \frac{y}{y-1}\right)$$

$$F_1(a; b, b; c; x, y) = (1-y)^{-a} F_1\left(a; b, c-2b; c; \frac{y-x}{y-1}, \frac{y}{y-1}\right)$$

Another useful transformation which enables Appell's hypergeometric function to be expressed in terms of an infinite sum of ordinary hypergeometric functions is

$$F_1(a; b, b'; c; x, y) = \sum_{m=0}^{\infty} \frac{(a)_m(b)_m}{m!(c)_m} F(a+m, b'; c+m; y)x^m,$$ (2.33)

where $(\alpha)_m$ is the Pochhammer symbol. This representation turns out to be useful for computational purposes using mathematical packages (such as MAPLE) which have implemented the usual hypergeometric function, but not Appell's hypergeometric function. For further details on Appell's hypergeometric functions and a complete list of transformations, we refer the reader to Colavecchia et al. (2001), Burchnall and Chaundy (1940), and Bailey (1972).

2.8 ASSOCIATED LEGENDRE FUNCTIONS: $P_\nu^\mu(z)$ AND $Q_\nu^\mu(z)$

The Legendre functions are solutions to the differential equation

$$(1-z^2)\frac{d^2w}{dz^2} - 2z\frac{dw}{dz} + [\nu(\nu+1) - \mu^2(1-z^2)^{-1}]w = 0, \tag{2.34}$$

with no restrictions on z, ν, and μ. When we make the substitution $w = (z^2 - 1)^{\mu/2}u$, the Legendre equation becomes

$$(1-z^2)\frac{d^2u}{dz^2} - 2(\mu+1)z\frac{du}{dz} + (\nu-\mu)(\nu+\mu+1)u = 0, \tag{2.35}$$

and with the subsequent substitution of the independent variable $\zeta = (1-z)/2$, we arrive at

$$\zeta(1-\zeta)\frac{d^2u}{d\zeta^2} + (\mu+1)(1-2\zeta)\frac{du}{d\zeta} + (\nu-\mu)(\nu+\mu+1)u = 0, \tag{2.36}$$

where we note that Eq. (2.36) is a special case of the hypergeometric equation (2.24) with $a = \mu - \nu$, $b = \mu + \nu + 1$, and $c = \mu + 1$. From the previous section, it follows that

$$w = P_\nu^\mu(z) = \frac{1}{\Gamma(1-\mu)}\left(\frac{z+1}{z-1}\right)^{\mu/2} F(-\nu, \nu+1; 1-\mu; (1-z)/2), \tag{2.37}$$

with $|1-z| < 2$ is a solution of the Legendre equation (2.34). It is known as the associated Legendre function of the first kind of degree ν and order μ.

If instead we substitute $\zeta = z^2$ in Eq. (2.35), we yield

$$4\zeta(1-\zeta)\frac{d^2u}{d\zeta^2} + [2 - (4\mu+6)\zeta]\frac{du}{d\zeta} + (\nu-\mu)(\nu+\mu+1)u = 0,$$

which is another special case of Eq. (2.24) with $a = (\mu+\nu+1)/2$, $b = (\mu-\nu)/2$, and $c = 1/2$, and therefore Eq. (2.34) has a solution given by

$$w = Q_\nu^\mu(z) = e^{\mu i\pi}2^{-\nu-1}\pi^{1/2}\frac{\Gamma(\nu+\mu+1)}{\Gamma(\nu+3/2)}z^{-\nu-\mu-1}(z^2-1)^{\mu/2}$$
$$\times F((\nu+\mu)/2+1, (\nu+\mu+1)/2; \nu+3/2; z^{-2}), \tag{2.38}$$

with $|z| > 1$. This is known as the associated Legendre function of the second kind of degree ν and order μ.

When $\mu = 0$, it is customary to write $P_\nu^0(z) = P_\nu(z)$ and $Q_\nu^0(z) = Q_\nu(z)$, where these are simply called Legendre functions. Furthermore, when ν is a nonnegative integer n, then $P_n(z)$ corresponds to the usual Legendre polynomial of degree n. We also comment that by replacing μ by $-\mu$, z by $-z$, and ν by $-\nu - 1$, then Eq. (2.34) remains unchanged. Therefore

$$P_\nu^{\pm\mu}(\pm z), \quad Q_\nu^{\pm\mu}(\pm z), \quad P_{-1-\nu}^{\pm\mu}(\pm z), \quad Q_{-1-\nu}^{\pm\mu}(\pm z),$$

are also solutions of Eq. (2.34).

Using the various transformation formulae available for the hypergeometric function and applying them to Eqs (2.37) and (2.38), it is possible to derive a large number of transformations between hypergeometric functions and the associated Legendre functions, which are of the form

$$P_\nu^\mu(z) = A_1 F(a_1, b_1; c_1; \zeta) + A_2 F(a_2, b_2; c_2; \zeta),$$

$$e^{-i\mu\pi} Q_\nu^\mu(z) = A_3 F(a_3, b_3; c_3; \zeta) + A_4 F(a_4, b_4; c_4; \zeta),$$

where ζ is a function of z. For an extensive list of such transformations, we refer the reader to Erdélyi et al. (1953).

2.9 CHEBYSHEV POLYNOMIALS: $T_n(X)$ AND $U_n(X)$

The Chebyshev polynomials of the first kind $T_n(x)$ are associated with the weight function $1/\sqrt{1-x^2}$ and are normalised so that $T_n(1) = 1$. The recurrence relation for these polynomials is given by

$$T_{n+1}(x) = 2xT_n(x) - T_{n-1}(x), \tag{2.39}$$

and they can be given explicitly by the equations

$$T_n(\cos\theta) = \cos n\theta,$$

$$T_n(x) = \frac{n}{2} \sum_{m=0}^{\lfloor n/2 \rfloor} (-1)^m \frac{(n-m-1)!}{m!(n-2m)!} (2x)^{n-2m}$$

$$= \sum_{m=0}^{\lfloor n/2 \rfloor} \binom{n}{2m} x^{n-2m} (x^2 - 1)^m.$$

The orthogonality relationship is given by

$$\int_{-1}^1 \frac{T_m(x)T_n(x)}{\sqrt{1-x^2}} \, dx = \begin{cases} 0, & m \neq n, \\ \pi/2, & m = n \neq 0, \\ \pi, & m = n = 0. \end{cases}$$

The Chebyshev polynomials of the second kind $U_n(x)$ are associated with the weight function $\sqrt{1-x^2}$ and normalised such that $U_n(1) = n + 1$. The recurrence relation for these polynomials is given by

$$U_{n+1}(x) = 2xU_n(x) - U_{n-1}, \tag{2.40}$$

and they may be given explicitly by the equations

$$U_n(\cos\theta) = \frac{\sin(n+1)\theta}{\sin\theta},$$

$$U_n(x) = \sum_{m=0}^{\lfloor n/2 \rfloor} (-1)^m \binom{n-m}{m} (2x)^{n-2m} \tag{2.41}$$

$$= \sum_{m=0}^{\lfloor n/2 \rfloor} \binom{n+1}{2m+1} x^{n-2m}(x^2-1)^m.$$

The orthogonal relation for these polynomials is

$$\int_{-1}^{1} U_m(x)U_n(x)\sqrt{1-x^2}\,dx = \begin{cases} 0, & m \neq n, \\ \pi/8, & m = n. \end{cases}$$

The first few polynomials of each kind are given by

$$
\begin{aligned}
T_0(x) &= 1, & U_0(x) &= 1, \\
T_1(x) &= x, & U_1(x) &= 2x, \\
T_2(x) &= 2x^2 - 1, & U_2(x) &= 4x^2 - 1, \\
T_3(x) &= 4x^3 - 3x, & U_3(x) &= 8x^3 - 4x, \\
T_4(x) &= 8x^4 - 8x^2 + 1, & U_4(x) &= 16x^4 - 12x^2 + 1, \\
T_5(x) &= 16x^5 - 20x^3 + 5x, & U_5(x) &= 32x^5 - 32x^3 + 6x,
\end{aligned}
$$

which can be extended indefinitely by use of Eqs (2.39) and (2.40).

2.10 ELLIPTIC INTEGRALS: $F(\phi, k)$ AND $E(\phi, k)$

Two fundamental forms of the elliptic integrals are

$$F(\phi, k) = \int_0^{\phi} \frac{d\varphi}{\sqrt{1 - k^2 \sin^2 \varphi}},$$

$$E(\phi, k) = \int_0^{\phi} \sqrt{1 - k^2 \sin^2 \varphi}\, d\varphi,$$

where we have adopted the Legendre forms and $0 < \phi \leq \pi/2$, is called the argument of the elliptic function and $0 < k < 1$ is referred to as the modulus. However, in special cases these ranges may be extended to a larger range or the complex plane. When $\phi = \pi/2$, then the elliptic function is said to be 'complete' and the argument is usually dropped. The notation we will use is

$$K(k) = F(\pi/2, k), \quad E(k) = E(\pi/2, k),$$

where we note that only in the case of the elliptic function of the first kind is the letter denoting the function changed for the complete case. We also define the complimentary modulus k' which is given in terms of the usual modulus by

$$k' = \sqrt{1 - k^2}.$$

The complete elliptic integrals of the first and second kind are related by the expression

$$E(k)K(k') + E(k')K(k) - K(k)K(k') = \pi/2,$$

which is called Legendre's relation. The elliptic integrals also reduce to greatly simplified expressions for particular value of the argument and modulus. Some of these are

$$F(0, k) = E(0, k) = 0,$$
$$F(\varphi, 0) = E(\varphi, 0) = \varphi,$$
$$F(\varphi, 1) = \sin \varphi, \quad E(\varphi, 1) = \ln(\tan \varphi + \sec \varphi).$$

EXERCISES

2.1. Show that $\Gamma(n + 1/2) = \sqrt{\pi}(2n - 1)!!/2^n$, where $(2n + 1)!! = (2n - 1)(2n - 3)\ldots 3.1$.

2.2. Sketch the equation given by $y = H(x - 2) - H(x - 4)$.

2.3. The standard Riemann zeta function $\zeta(z)$ is defined by

$$\zeta(z) = \sum_{n=1}^{\infty} \frac{1}{n^z}, \quad \Re(z) > 1.$$

Make the substitution $t = nx$ to deduce

$$\frac{1}{n^z} = \frac{1}{\Gamma(z)} \int_0^{\infty} x^{z-1} e^{-nx} \, dx,$$

and derive the integral expression

$$\zeta(z) = \frac{1}{\Gamma(z)} \int_0^{\infty} \frac{x^{z-1}}{e^x - 1} \, dx, \quad \Re(z) > 1.$$

2.4. The Hurwitz zeta function $\zeta(z, \alpha)$ is defined by

$$\zeta(z, \alpha) = \sum_{n=0}^{\infty} \frac{1}{(\alpha + n)^z}, \quad \Re(z) > 1,$$

and $\alpha \neq 0, -1, -2, -3, \ldots$ Observe that when $\alpha = 1$ we have

$$\zeta(z, 1) = \sum_{n=0}^{\infty} \frac{1}{(n + 1)^z} = \sum_{m=1}^{\infty} \frac{1}{m^z} = \zeta(z),$$

and show that

$$\zeta(z, \alpha) = \frac{1}{\Gamma(z)} \int_0^{\infty} \frac{x^{z-1} e^{-\alpha x}}{1 - e^{-x}} \, dx, \quad \Re(z) > 1.$$

2.5. By expanding as a power series, deduce from

$$\frac{t}{e^t - 1} = \sum_{n=0}^{\infty} \frac{B_n t^n}{n!}, \quad 0 < |t| < 2\pi,$$

that $B_0 = 1$ and $B_n = 0$ for n odd, and that

$$llB_1 = -1/2, B_2 = 1/6,$$
$$B_4 = -1/30, B_6 = 1/42,$$
$$B_8 = 1/30, B_{10} = -5/66,$$
$$B_{12} = -691/2730, B_{14} = -7/6.$$

2.6. Show that $B(x, y)B(x + y, z) = B(y, z)B(y + z, x) = B(x, z)B(x + z, y)$.

2.7. Show that $\log(1 + z) = zF(1, 1; 2; -z)$.

2.8. Show that

$$F(a, b; c; z) = \frac{\Gamma(c)}{\Gamma(a)\Gamma(b)} \sum_{n=0}^{\infty} \frac{\Gamma(a + n)\Gamma(b + n)}{\Gamma(c + n)\Gamma(1 + n)} z^n.$$

2.9. Prove that the differential equation (2.24) is satisfied by $F(a, b; c; z)$.

EVALUATION OF LENNARD-JONES POTENTIAL FIELDS

3

3.1 INTRODUCTION

For much of this book, we will be concerned with evaluating the interaction energy between atoms, molecules, and nanostructures, assuming that quantum effects arising from the atomic length scale may be neglected. As mentioned in Chapter 1, there are a number of different potential functions which may be used to model the molecular interaction. However, we will primarily concern ourselves with the Lennard-Jones potential function $\Phi(\rho)$, which accounts for the interaction of two nonbonded atoms and may be written in the form

$$\Phi(\rho) = -A\rho^{-6} + B\rho^{-12}, \tag{3.1}$$

where ρ is the distance between the interacting atoms, and A and B are positive constants, which are empirically determined and correspond to the constants of attraction and repulsion, respectively. We refer the reader to Hunter (2001) for the origin and analysis of the van der Waals force.

As discussed in Chapter 1, when calculating the van der Waals interaction between molecules containing a number of atoms, the pairwise interactions may be summed up to derive a total interaction E, which is given by Eq. (1.11). To refresh the reader, this is written as

$$E = \sum_i \sum_j \Phi\left(\rho_{ij}\right),$$

where the indices i and j vary over all the atoms in each molecule, and ρ_{ij} denotes the distance between two typical atoms i and j. Obviously, this formulation requires that we know the precise location of every atom in both molecules, and we are required to perform ij individual calculations of the potential function $\Phi\left(\rho_{ij}\right)$ to calculate the total interaction energy for the two molecules. These two considerations can be inconvenient or unnecessarily intensive computationally for situations in which the exact orientation of the molecules are not specified or when large nanostructures are involved. Therefore for most of this book, we will assume that all the atoms are smeared over ideal lines or surfaces which represent the molecules that we are modelling; the precise atomic locations will be ignored. With this approximation, we are able to replace the explicit summations in Eq. (1.11) with line or surface integrals that allow us to write

$$E = \eta_1 \eta_2 \int_{S_1} \int_{S_2} \Phi(\rho)\, dS_2\, dS_1, \tag{3.2}$$

Modelling and Mechanics of Carbon-based Nanostructured Materials. http://dx.doi.org/10.1016/B978-0-12-812463-5.00003-0

where ρ denotes the distance between typical infinitesimal surface elements dS_1 and dS_2 of the lines or surfaces S_1 and S_2, respectively, and the terms η_1 and η_2 are the atomic densities (i.e. atoms per unit surface area) of the surfaces S_1 and S_2, respectively. We emphasise that when we talk loosely about cylinders and spheres, we have in mind carbon nanotubes and fullerenes, respectively, and that strictly speaking, we should always refer to them as cylindrical and spherical surfaces because we are tacitly assuming that these molecules can be represented by infinitesimally thin surfaces.

From these considerations, we see that the task of calculating the van der Waals interactions will be made considerably easier provided that there are methods to readily evaluate integrals of the form of Eq. (3.2) over various lines and surfaces, which are relevant to problems in nanotechnology. It turns out that many of the molecules and nanostructures that we encounter can be modelled very realistically by the basic geometric objects of points, straight lines, flat planes, spheres and right circular cylinders, and therefore integrals over these objects will be needed frequently throughout the remainder of this book. Accordingly, the purpose of this chapter is to address these integrals in a systematic way to facilitate the evaluation of ideal van der Waals interactions.

When we investigate the form of the Lennard-Jones potential function $\Phi(\rho)$ given in Eq. (3.1) in the context of the integral formulation of Eq. (3.2), we note that the attractive term ρ^{-6} and the repulsive term ρ^{-12} can be separated and integrated independently. We also note that the two terms only vary in terms of the coefficients A and B, and the magnitude of the index, which is applied to the distance variable ρ. In subsequent sections of this chapter, we shall occasionally make use of the fact that the indices of ρ in Eq. (3.1) are both negative even integers, which leads us to express the Lennard-Jones potential function $\Phi(\rho)$ in the form

$$E(\rho) = -AI_3(\rho) + BI_6(\rho), \tag{3.3}$$

where $I_n(\rho) = \rho^{-2n}$, and we will consider evaluating integrals of the form

$$I = \int_{S_1} \int_{S_2} \rho^{-2n}(\rho) \, dS_2 \, dS_1. \tag{3.4}$$

Finally, before leaving this introduction, we must comment on why we seek analytical evaluation of the integrals of the form given in Eq. (3.4). While such integrals can be handled rather easily using numerical methods, it should be remembered that our goal is to elucidate as much as possible from the models that we develop. In some cases, this can involve constructing a broad landscape of data which might be extremely costly or even inaccessible using purely numerical methods, but which may be accessed by evaluating analytical expressions for the potential functions in question. Analytical expressions can also be employed to derive other quantities, like the force experienced by a molecule, which would otherwise only be determined by further numerical evaluation. One can always move from the analytical to the numerical, but the reverse is usually not possible.

3.2 INTERACTION OF LINEAR OBJECTS
3.2.1 INTERACTION OF A POINT WITH A LINE

In this section, we will consider the Lennard-Jones interactions involving linear objects, such as straight lines and planes, as derived from the Lennard-Jones potential function (3.1). We begin by considering

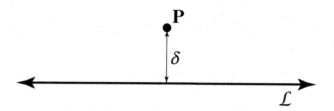

FIG. 3.1

Point **P** interacting with line $\mathcal{L}$ at a distance of δ.

the simplest such case, the interaction between a point and an infinite line. We assume a perpendicular (closest) distance between the point and line, denoted by δ and therefore construct the following geometry in two-dimensional Cartesian coordinates. We define a line parametrically by $\mathcal{L}(p) = (p, 0)$ and the point $\mathbf{P} = (0, \delta)$, as illustrated in Fig. 3.1. We note that the line element is given by dp, and therefore the integral of interest is given by

$$I = \int_{-\infty}^{\infty} \left(p^2 + \delta^2\right)^{-n} dp.$$

We then make a change of variable substitution of $p = \delta \tan \psi$, which transforms the integral to

$$I = \delta^{1-2n} \int_{-\pi/2}^{\pi/2} \cos^{2n-2} \psi \, d\psi, \tag{3.5}$$

which can be evaluated using Eq. (2.19), as illustrated in Worked Example 3.1 to obtain the total interaction E_ℓ; thus

$$E_\ell = \eta \left(-\frac{3\pi A}{8\delta^5} + \frac{63\pi B}{256\delta^{11}}\right), \tag{3.6}$$

where η is the atomic density of the line $\mathcal{L}$.

WORKED EXAMPLE 3.1

Evaluate the integral (3.5) using Eq. (2.19) to obtain Eq. (3.6).

Solution

Since cosine is an even symmetric function, we can rewrite the integral given by Eq. (3.5) as

$$I = 2\delta^{1-2n} \int_{0}^{\pi/2} \cos^{2n-2} \psi \, d\psi.$$

From Eq. (2.19), we see that

$$\int_{0}^{\pi/2} \sin^p \theta \, \cos^q \theta \, d\theta = \frac{1}{2} \mathrm{B}\left(\frac{p+1}{2}, \frac{q+1}{2}\right),$$

where $\mathrm{B}(x, y)$ is the beta function defined in Eq. (2.17), and in this case, $p = 0$ and $q = 2n - 2$. Thus we obtain

$$I = 2\delta^{1-2n}[1/2B(1/2, (2n-2+1)/2)] = \delta^{1-2n}B(n-1/2, 1/2),$$

since $B(x, y) = B(y, x)$.

$$I = \delta^{1-2n}B(n-1/2, 1/2).$$

We may evaluate the beta function given above for $n = 3, 6$ by using Eq. (2.18) to obtain

$$I_3 = \delta^{-5}B(3 - 1/2, 1/2) = \frac{3\pi}{8\delta^5}$$

$$I_6 = \delta^{-11}B(6 - 1/2, 1/2) = \frac{63\pi}{256\delta^{11}},$$

where E_ℓ is given by $E_\ell = \eta(-AI_3 + BI_6)$.

An analysis of Eq. (3.6) shows that when a point is interacting with a line, the point-point six-twelve potential is replaced with a five-eleven potential, and the attractive and repulsive constants A and B are scaled by factors of $3\pi/8$ and $63\pi/256$, respectively. We believe that performing a simple calculation of this kind and deriving the analytical expression (3.6) is far more advantageous than performing numerical calculations or summing up the many pairwise point interactions to estimate the interaction potential.

3.2.2 INTERACTION OF A POINT WITH A PLANE

We now extend the previous section to consider the interaction of a point with an infinite flat plane. This situation is relevant to modelling nanostructures, as it corresponds to the case of an individual atom interacting with a graphene sheet. Again, we assume a perpendicular spacing of δ between the point and the plane; therefore we define a three-dimensional Cartesian coordinate system, in which we locate a plane $\mathcal{P}(p, q) = (p, q, 0)$ and a point $\mathbf{P} = (0, 0, \delta)$, as shown in Fig. 3.2. In this case the area element of the plane is given by $dp\,dq$, and therefore the integral I is given by

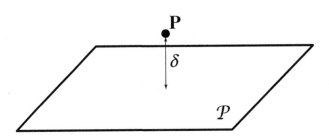

FIG. 3.2

Point **P** interacting with plane $\mathcal{P}$ at a distance of δ.

$$I = \int_{-\infty}^{\infty} \int_{-\infty}^{\infty} \left(p^2 + q^2 + \delta^2 \right)^{-n} dp \, dq. \tag{3.7}$$

We make the substitution $p = \sqrt{q^2 + \delta^2} \tan \psi$ and proceed as in the previous section, which produces

$$I = \mathrm{B} \left(n - 1/2, 1/2 \right) \int_{-\infty}^{\infty} \left(q^2 + \delta^2 \right)^{1/2 - n} dq.$$

On making a further substitution of $q = \delta \tan \phi$, we obtain

$$I = \delta^{2-2n} \mathrm{B} \left(n - 1/2, 1/2 \right) \mathrm{B} \left(n - 1, 1/2 \right)$$
$$= \frac{\pi}{(n-1)\delta^{2n-2}},$$

and therefore the total interaction between a point and a plane E_p is given by

$$\boxed{E_p = \eta \left(-\frac{\pi A}{2\delta^4} + \frac{\pi B}{5\delta^{10}} \right),} \tag{3.8}$$

where η is the atomic density of the plane $\mathcal{P}$. As in the previous section, we see in Eq. (3.8) that when a point is interacting with a plane, the point-point six-twelve potential is replaced with a four-ten potential, and the attractive and repulsive constants A and B are scaled by factors of $\pi/2$ and $\pi/5$, respectively.

3.2.3 INTERACTION OF TWO SKEW LINES

In this section, we consider the van der Waals interaction between two skew lines which are separated by a distance δ and are inclined at an angle θ, as derived from the Lennard-Jones potential function (3.1). The model geometry in this case is a line, which is the x-axis $\mathcal{L}_1(p) = (p, 0, 0)$ and a second line, which is in a plane perpendicular with the xy-plane and at a height δ above it, $\mathcal{L}_2(q) = (q \cos \theta, q \sin \theta, \delta)$, as illustrated in Fig. 3.3. We note that the parametrisation of both lines has been done so that the length elements of each are dp and dq, respectively. Therefore the integral of interest in this case is

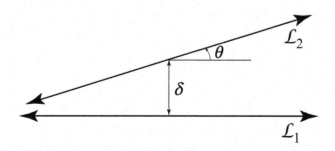

FIG. 3.3

Interacting skew lines $\mathcal{L}_1$ and $\mathcal{L}_2$ at angle θ and at a distance of δ.

$$I = \int_{-\infty}^{\infty} \int_{-\infty}^{\infty} \left[(q\cos\theta - p)^2 + q^2 \sin^2\theta^2 + \delta^2 \right]^{-n} dp\, dq.$$

Firstly, we make the substitution of $p = q\cos\theta + r$, where we assume q is finite and substitute r for the parameter p. This transforms the integral to

$$I = \int_{-\infty}^{\infty} \int_{-\infty}^{\infty} \left(r^2 + q^2 \sin^2\theta^2 + \delta^2 \right)^{-n} dr\, dq.$$

Next, we make the substitution $q = s/\sin\theta$, yielding

$$I = \frac{1}{\sin\theta} \int_{-\infty}^{\infty} \int_{-\infty}^{\infty} \left(r^2 + s^2 + \delta^2 \right)^{-n} dr\, ds.$$

At this point, we note that the integral is now in the same form as Eq. (3.7) but scaled by a factor $1/\sin\theta$. Therefore the interaction can be calculated by exactly the same steps as that used in Section 3.2.2, which gives

$$E_{\ell\ell} = \frac{\eta_1 \eta_2}{\sin\theta} \left(-\frac{\pi A}{2\delta^4} + \frac{\pi B}{5\delta^{10}} \right), \tag{3.9}$$

where η_1 and η_2 are the atomic line densities of $\mathcal{L}_1$ and $\mathcal{L}_2$, respectively. We comment that $E_{\ell\ell} = E_p/\sin\theta$, which raises the question of the case of parallel lines, where $\theta = 0$.

3.2.4 INTERACTION BETWEEN PARALLEL LINES AND PLANES

Up until now, we have been dealing with cases involving infinite lines and planes, where the short distance van der Waals interactions are taking place over a finite local neighbourhood. In these configurations, the effect of the infinite limits of the lines and planes on the evaluation of the Lennard-Jones function $\Phi(\rho)$ tends to zero. However, in the case of infinite parallel lines and planes, this is no longer the case. The lines and planes continue to have short-range interactions for their entire length, and we can see from Eq. (3.9) that it does not make sense to talk of a total interaction energy in this context.

Therefore instead of a total interaction energy, we calculate an interaction energy per unit length, or energy per unit area. For example, a continuous line is a model for a row of atoms, where each atom is spaced by a distance of $1/\eta$. Likewise, a continuous plane is a model for a discrete surface where there is one atom occupying every $1/\eta$ of area. From these considerations and making use of the results thus far, we are able to deduce that for two parallel lines, the interaction per unit length is $E_{\ell||\ell} = \eta_2 E_\ell$, where E_ℓ is the interaction of a point with a line. Substituting from Eq. (3.6), we derive

$$E_{\ell||\ell} = \eta_1 \eta_2 \left(-\frac{3\pi A}{8\delta^5} + \frac{63\pi B}{256\delta^{11}} \right), \tag{3.10}$$

where η_1 and η_2 are the atomic line densities for the two lines and δ is the perpendicular distance between the two lines. Likewise, the interaction energy per unit length of a line parallel with an infinite plane is given by $E_{p||\ell} = \eta_2 E_p$, where E_p is the interaction of a point with an infinite plane. Substituting from Eq. (3.8), we deduce

$$E_{p\|\ell} = \eta_1\eta_2\left(-\frac{\pi A}{2\delta^4} + \frac{\pi B}{5\delta^{10}}\right),$$

(3.11)

where η_1 is the atomic surface density for the plane and η_2 is the atomic line density for the line, which is a perpendicular distance δ from the plane. Finally, the interaction per unit area for two parallel planes is given by $E_{p\|p} = \eta_2 E_p$, and the substitution of Eq. (3.8) yields

$$E_{p\|p} = \eta_1\eta_2\left(-\frac{\pi A}{2\delta^4} + \frac{\pi B}{5\delta^{10}}\right),$$

(3.12)

where in this case, η_1 and η_2 are the atomic surface densities of the two planes with a perpendicular separation distance of δ.

3.3 INTERACTION OF A SPHERICAL SURFACE
3.3.1 SPHERICAL COORDINATE SYSTEM

There are, unfortunately, a number of different conventions that authors have adopted in defining a spherical coordinate geometry. To avoid confusion and to introduce the reader to spherical coordinates, we begin with a description of the convention that we adopt for this book, which is also most commonly adopted in the scientific literature.

We define a spherical coordinate system (r, θ, ϕ) and define r the radial distance, ϕ the azimuthal angle, and θ the polar angle, as illustrated in Fig. 3.4. These are oriented with respect to a Cartesian

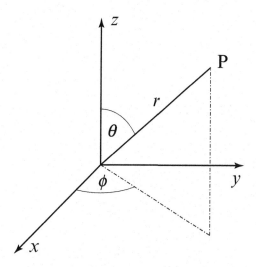

FIG. 3.4

A general point P described by spherical coordinates (r, θ, ϕ).

coordinate system (x, y, z) such that ϕ is measured in the xy-plane beginning from the positive x-axis and proceeding in an anticlockwise direction. The polar angle θ is then measured from the positive direction of the z-axis and proceeds down to the negative direction. Therefore the coordinates have primary values of $r \in [0, \infty)$, $\theta \in (0, \pi]$, and $\phi \in [-\pi, \pi]$, and to map from the spherical to the Cartesian coordinate system we have

$$\boxed{x = r \sin \theta \cos \phi, \quad y = r \sin \theta \sin \phi, \quad z = r \cos \theta,} \tag{3.13}$$

and the inverse mapping is given by

$$r = \sqrt{x^2 + y^2 + z^2}, \quad \theta = \arctan \frac{\sqrt{x^2 + y^2}}{z}, \quad \phi = \arctan \frac{y}{x}.$$

The vector line element for the spherical coordinate system is

$$d\mathbf{s} = dr\, \hat{\mathbf{r}} + r\, d\theta\, \hat{\boldsymbol{\theta}} + r \sin \theta\, d\phi\, \hat{\boldsymbol{\phi}},$$

where $\hat{\mathbf{r}}$, $\hat{\boldsymbol{\theta}}$, and $\hat{\boldsymbol{\phi}}$ denote the unit vectors for the spherical coordinate system. Similarly, the vector area element is given by

$$d\mathbf{S} = r^2 \sin \theta\, d\theta\, d\phi\, \hat{\mathbf{r}}.$$

In the following sections, we will be considering various integrals of the Lennard-Jones potential function $\Phi(\rho)$ over the surface of a sphere of radius a. In such cases, the integral considered will be of the form

$$I = a^2 \int_{-\pi}^{\pi} \int_{0}^{\pi} \frac{\sin \theta}{\rho^{2n}}\, d\theta\, d\phi,$$

where ρ is some function of the geometric parameters of the problem, which will generally be a function of θ and ϕ.

3.3.2 INTERACTION OF POINT WITH A SPHERE

We begin by considering a point given in Cartesian coordinates as $\mathbf{P} = (0, 0, \delta)$, which is at a distance δ from the centre of a sphere of radius a, which is parameterised in spherical coordinates as $\mathcal{S}(\theta, \phi) = (a, \theta, \phi)$, as illustrated in Fig. 3.5. In these coordinates the integral of interest (Eq. 3.4) is given by

$$I = a^2 \int_{-\pi}^{\pi} \int_{0}^{\pi} \frac{\sin \theta}{[a^2 \sin^2 \theta + (a \cos \theta - \delta)^2]^n}\, d\theta\, d\phi.$$

Because the integrand in this case is independent of ϕ, this integration can be effected immediately, and then by reorganising the denominator, we may derive

$$I = 2\pi a^2 \int_{0}^{\pi} \frac{\sin \theta}{[\delta^2 + a^2 - 2\delta a \cos \theta]^n}\, d\theta, \tag{3.14}$$

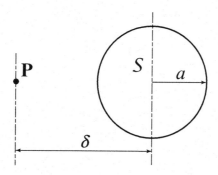

FIG. 3.5

Sphere S of radius a, interacting with point P at a distance of δ from centre of the sphere.

which, by making the substitution $t = \delta^2 + a^2 - 2\delta a \cos\theta$, becomes

$$I = \frac{\pi a}{\delta} \int_{(\delta-a)^2}^{(\delta+a)^2} \frac{dt}{t^n}$$

$$= \frac{\pi a}{\delta(1-n)} \left[\frac{1}{t^{n-1}} \right]_{(\delta-a)^2}^{(\delta+a)^2}$$

$$= \frac{\pi a}{\delta(n-1)} \left[\frac{1}{(\delta-a)^{2(n-1)}} - \frac{1}{(\delta+a)^{2(n-1)}} \right].$$

Therefore the total interaction energy for a point and a sphere is given by

$$E_s = \frac{\pi a \eta}{\delta} \left\{ -\frac{A}{2} \left[\frac{1}{(\delta-a)^4} - \frac{1}{(\delta+a)^4} \right] + \frac{B}{5} \left[\frac{1}{(\delta-a)^{10}} - \frac{1}{(\delta+a)^{10}} \right] \right\}, \qquad (3.15)$$

where η is the atomic surface density of the sphere.

3.3.2.1 An alternative approach

To demonstrate the utility of the hypergeometric function to evaluate these integrals and also to derive an expression which will be of use later, we now repeat the calculation of the integral (3.14) using hypergeometric functions. The method is to express this integral as

$$I = 2\pi a^2 \int_0^\pi \frac{2 \sin(\theta/2) \cos(\theta/2)}{[(\delta-a)^2 + 4\delta a \sin^2(\theta/2)]^n} \, d\theta,$$

by making use of the trigonometric identities $1 - \cos\theta = 2\sin^2(\theta/2)$ and $\sin\theta = 2\sin(\theta/2)\cos(\theta/2)$. We now make the substitution $t = \sin^2(\theta/2)$ and take out a factor of $2(\delta-a)^{-2n}$, which produces

$$I = \frac{4\pi a^2}{(\delta-a)^{2n}} \int_0^1 \left[1 + \frac{4\delta a}{(\delta-a)^2} t \right]^{-n} dt.$$

This integral is in the form of the hypergeometric integral (2.29) with $a = n$, $b = 1$, $c = 2$, and $z = -4\delta a/(\delta - a)^2$. Therefore

$$I = \frac{4\pi a^2}{(\delta - a)^{2n}} F\left(n, 1; 2; -\frac{4\delta a}{(\delta - a)^2}\right).$$

At this point, we note that $c = 2b$, and therefore we may use a quadratic transformation. In this case, we employ Eq. (2.27), which gives

$$I = \frac{4\pi a^2}{(\delta^2 - a^2)^n} F\left(n, 2 - n; 3/2; -\frac{a^2}{\delta^2 - a^2}\right),$$

and because the hypergeometric function is symmetric in a and b, we see from Eq. (2.22) that this may be transformed into a Chebyshev polynomial of the second kind $U_n(x)$, which leads to the expression

$$I = \frac{4\pi a^2}{(n - 1)(\delta^2 - a^2)^n} U_{n-2}\left(\frac{\delta^2 + a^2}{\delta^2 - a^2}\right). \tag{3.16}$$

Therefore the total interaction energy for a point and a sphere is also given by

$$E_s = 4\pi a^2 \eta \left[-\frac{A}{2(\delta^2 - a^2)^3} U_1\left(\frac{\delta^2 + a^2}{\delta^2 - a^2}\right) + \frac{B}{5(\delta^2 - a^2)^6} U_4\left(\frac{\delta^2 + a^2}{\delta^2 - a^2}\right) \right], \tag{3.17}$$

where η is the atomic surface density of the sphere. We comment that Eqs (3.15) and (3.17) are precisely algebraically equivalent. However, the relationship to Chebyshev polynomials (which is not apparent in Eq. 3.15) emerges naturally from the hypergeometric approach, where the relationships between the various transcendental functions and orthogonal polynomials can be readily accessed.

3.3.3 INTERACTION OF LINE WITH A SPHERICAL SURFACE

We now consider the van der Waals interaction between a line and a spherical surface. In this case the method of defining a geometry and determining the distance between typical area elements is not conducive to progress. The approach adopted instead is to take the energy for a point and a sphere E_s given in Section 3.3.2 and integrate this over the line. To do this, we rewrite the integral expression from Eq. (3.16) as

$$I = \frac{4\pi a^2}{(n - 1)(\rho^2 - a^2)^n} U_{n-2}\left(\frac{\rho^2 + a^2}{\rho^2 - a^2}\right),$$

where ρ is the distance between the line element ds and the centre of the sphere.

We now define a problem geometry with a sphere in spherical coordinates $\mathcal{S}(\theta, \phi) = (a, \theta, \phi)$, which is centred on the origin of radius a and line $\mathcal{L}(p) = (p, 0, -\delta)$, which is parallel with the x-axis, and at a distance δ in the negative z-direction, as illustrated in Fig. 3.6. The distance from a typical point on

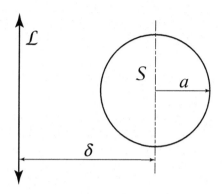

FIG. 3.6

A spherical surface S of radius a, interacting with a line $\mathcal{L}$ at a distance of δ from the centre of the sphere.

$\mathcal{L}(p)$ and the centre of the sphere S is $\rho = \sqrt{\delta^2 + p^2}$. Expanding by using the explicit formula for the Chebyshev polynomials of the second kind (Eq. 2.41) and then making this substitution for ρ yields

$$I = \frac{4\pi a^2}{(n-1)(\rho^2 - a^2)^n} \sum_{m=0}^{\lfloor n/2-1 \rfloor} (-1)^m \binom{n-m-2}{m} 2^{n-2m-2} \left(\frac{\rho^2 + a^2}{\rho^2 - a^2} \right)^{n-2m-2}$$

$$= \frac{2^n \pi a^2}{n-1} \sum_{m=0}^{\lfloor n/2-1 \rfloor} \left(-\frac{1}{4} \right)^m \binom{n-m-2}{m} \frac{(\delta^2 + a^2 + p^2)^{n-2m-2}}{(\delta^2 - a^2 + p^2)^{2n-2m-2}},$$

where the only dependence on p occurs in the last term. Therefore we wish to integrate p from $-\infty$ to ∞ and need only consider the integrals of the form

$$J = \int_{-\infty}^{\infty} \frac{(\delta^2 + a^2 + p^2)^{n-2m-2}}{(\delta^2 - a^2 + p^2)^{2n-2m-2}} \, dp,$$

where in the case of a Lennard-Jones potential, we have $n \in \{3, 6\}$ and $m \in \{0, 1, \ldots, \lfloor n/2 - 1 \rfloor\}$, where $\lfloor x \rfloor$ is the floor function designating the greatest integer less than or equal to x. The first observation we can make is that the integrand is an even function of p and therefore we can split the integral at $p = 0$ and make a change of variable from $p \to -p$ to give

$$J = 2 \int_0^{\infty} \frac{(\delta^2 + a^2 + p^2)^{n-2m-2}}{(\delta^2 - a^2 + p^2)^{2n-2m-2}} \, dp$$

$$= \frac{2(\delta^2 + a^2)^{n-2m-2}}{(\delta^2 - a^2)^{2n-2m-2}} \int_0^{\infty} \frac{[1 + p^2/(\delta^2 + a^2)]^{n-2m-2}}{[1 + p^2/(\delta^2 - a^2)]^{2n-2m-2}} \, dp,$$

and substitute $t = p^2/(\delta^2 + a^2)$, which gives

$$J = \frac{(\delta^2 + a^2)^{n-2m-3/2}}{(\delta^2 - a^2)^{2n-2m-2}} \int_0^{\infty} \frac{(1+t)^{n-2m-2}}{t^{1/2}[1 + t(\delta^2 + a^2)/(\delta^2 - a^2)]^{2n-2m-2}} \, dt,$$

which is now in a fundamental integral form for the hypergeometric function (2.30) with $a = 2n - 2m - 2$, $b = 1/2$, $c = n$, and $1 - z = -2a^2/(\delta^2 - a^2)$. Therefore we may write

$$J = \frac{(\delta^2 + a^2)^{n-2m-3/2}}{(\delta^2 - a^2)^{2n-2m-2}} B(n - 1/2, 1/2) F\left(2n - 2m - 2, 1/2; n; -\frac{2a^2}{\delta^2 - a^2}\right).$$

This is one representation of the integral; however, it is not in the most economical form. It is possible to rearrange the parameters of the hypergeometric function to make the first argument a negative integer, which leads to a terminating series. To achieve this, we use the linear transformation (Eq. 2.25) which produces

$$J = \frac{B(n - 1/2, 1/2)}{(\delta^2 - a^2)^{n-1/2}} F\left(2m - n + 2, n - 1/2; n; -\frac{2a^2}{\delta^2 - a^2}\right).$$

We now expand this hypergeometric function as a series using Eq. (2.20), and we note that if $p \geqslant q > 0$, as shown in Eq. (2.12) $(-p)_q = (-1)^q p!/(p - q)!$, we obtain

$$J = \frac{B(n - 1/2, 1/2)}{(\delta^2 - a^2)^{n-1/2}} \frac{\Gamma(n)}{\Gamma(n - 1/2)}$$

$$\times \sum_{p=0}^{n-2m-2} (-1)^p \frac{(n - 2m - 2)! \Gamma(n + p - 1/2)}{(n - 2m - 2 - p)! \Gamma(n + p)p!} \left(\frac{-2a^2}{\delta^2 - a^2}\right)^p$$

$$= \frac{1}{(\delta^2 - a^2)^{n-1/2}} \sum_{p=0}^{n-2m-2} \binom{n - 2m - 2}{p} B(n + p - 1/2, 1/2) \left(\frac{2a^2}{\delta^2 - a^2}\right)^p.$$

Therefore I is given by

$$I = \frac{2^n \pi a^2}{(n - 1)(\delta^2 - a^2)^{n-1/2}} \sum_{m=0}^{\lfloor n/2-1 \rfloor} \left(-\frac{1}{4}\right)^m \frac{(n - m - 2)!}{m!}$$

$$\times \sum_{p=0}^{n-2m-2} \frac{B(n + p - 1/2, 1/2)}{p!(n - 2m - p - 2)!} \left(\frac{2a^2}{\delta^2 - a^2}\right)^p,$$

which, for the total interaction between the line $\mathcal{L}$ and the spherical surface $\mathcal{S}$, yields

$$\boxed{E_{\ell s} = \pi^2 a^2 \eta_1 \eta_2 \left[-\frac{A(3\delta^2 + 2a^2)}{2(\delta^2 - a^2)^{7/2}} + \frac{B(315\delta^8 + 3360\delta^6 a^2 + 6048\delta^4 a^4 + 2304\delta^2 a^6 + 128a^8)}{320(\delta^2 - a^2)^{19/2}} \right],}$$

(3.18)

where η_1 is the atomic line density of the line $\mathcal{L}$ and η_2 is the atomic surface density of the sphere $\mathcal{S}$.

WORKED EXAMPLE 3.2

If we assume that the radius of the spherical surface a is insignificant compared to the distance δ, then the distance $\rho \approx \delta + a \cos \phi$. Determine an approximation for $E_{\ell s}$ given by Eq. (3.18) and using this value of ρ.

Solution

The approach here is to take the energy for a point and a line E_ℓ given in Eq. (3.6) and integrate this over the sphere. To do this, we rewrite Eq. (3.6) as

$$E_\ell = \eta \left(-\frac{3\pi A}{8} J_3(\rho) + \frac{63\pi B}{256} J_6(\rho) \right),$$

where $J_n(\rho) = \rho^{1-2n}$ and ρ is the perpendicular distance between the typical infinitesimal element on the sphere and the line, which we approximate by $\rho \approx \delta + a \cos \theta$. Therefore we wish to integrate

$$J = a^2 \int_{-\pi}^{\pi} \int_0^\pi \frac{\sin \theta}{(\delta + a \cos \theta)^{2n-1}} \, d\theta \, d\phi$$

$$= 2\pi a^2 \int_0^\pi \frac{\sin \theta}{(\delta + a \cos \theta)^{2n-1}} \, d\theta$$

$$= \frac{2\pi a}{2n-2} \left[\frac{1}{(\delta + a \cos \theta)^{2n-2}} \right]_0^\pi$$

$$= \frac{2\pi a}{2n-2} \left[\frac{1}{(\delta - a)^{2n-2}} - \frac{1}{(\delta + a)^{2n-2}} \right],$$

and therefore the total interaction between a sphere of radius a and line a distance δ from its centre is given by

$$E_{\ell s} \approx \pi^2 a \eta_1 \eta_2 \left\{ -\frac{3A}{16} \left[\frac{1}{(\delta - a)^4} - \frac{1}{(\delta + a)^4} \right] + \frac{63B}{1280} \left[\frac{1}{(\delta - a)^{10}} - \frac{1}{(\delta + a)^{10}} \right] \right\}, \tag{3.19}$$

where η_1 is the atomic line density of the line $\mathcal{L}$ and η_2 is the atomic surface density of the sphere $\mathcal{S}$.

We note here that we should continue with the approximation to eliminate the $(\delta - a)$ terms, since we are assuming that these are approximately equal to δ. To do so, we need the highest order terms for the expressions in square brackets in Eq. (3.19), which can be found as

$$\frac{1}{(\delta - a)^4} - \frac{1}{(\delta + a)^4} = \frac{(\delta + a)^4 - (\delta + a)^4}{(\delta^2 - a^2)^4}$$

$$= \frac{8\delta^3 a + 8\delta a^3}{(\delta^2 - a^2)^4}$$

$$= \frac{8a}{\delta^5} + O(a^3),$$

and likewise,

$$\frac{1}{(\delta - a)^{10}} - \frac{1}{(\delta + a)^{10}} = \frac{20a}{\delta^{11}} + O(a^3).$$

Making these substitutions in Eq. (3.19) leads to

$$E_{\ell s} \approx \pi^2 a^2 \eta_1 \eta_2 \left\{ -\frac{3A}{2\delta^5} + \frac{63B}{64\delta^{11}} \right\}. \tag{3.20}$$

Now, we employ the fact that the atomic surface density for the sphere $\eta_2 = N/4\pi a^2$, where N is the number of atoms comprising the sphere. Making this substitution leads to

$$E_{\ell s} \approx N\pi\eta_1 \left\{ -\frac{3A}{8\delta^5} + \frac{63B}{256\delta^{11}} \right\}, \tag{3.21}$$

which is precisely the formula for the interaction of a point with a line (3.6) scaled by the number of atoms comprising the sphere, as expected.

3.3.4 INTERACTION OF A PLANE WITH A SPHERICAL SURFACE

We will now consider the total interaction for a spherical surface defined by $\mathcal{S}(\theta,\phi) = (a,\theta,\phi)$ in spherical coordinates, and a plane $\mathcal{P}(x,y) = (x,y,-\delta)$ in Cartesian coordinates for $x,y \in (-\infty,\infty)$, as illustrated in Fig. 3.7. Since the plane $\mathcal{P}$ is infinite to the extent that we can take the perpendicular distance from a point on the sphere from the plane as simply $\rho = \delta + a\cos\theta$. Therefore we can calculate the energy exactly by integrating the expression for E_p, given in Eq. (3.8), by using this value for ρ as the perpendicular distance and integrating over the spherical surface. Hence the integral of interest is

$$I = \frac{\pi a^2}{n-1} \int_{-\pi}^{\pi} \int_0^{\pi} \frac{\sin\theta}{(\delta + a\cos\theta)^{2n-2}} \, d\theta \, d\phi.$$

As there is no dependence on ϕ in the integrand, we can perform this integration trivially and obtain

$$I = \frac{2\pi^2 a^2}{n-1} \int_0^{\pi} \frac{\sin\theta}{(\delta + a\cos\theta)^{2n-2}} \, d\theta.$$

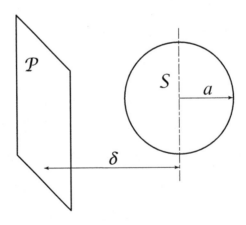

FIG. 3.7

A spherical surface $\mathcal{S}$ of radius a, interacting with a plane $\mathcal{P}$ at a distance of δ from the centre of the sphere.

Making the substitution $t = \delta + a \cos \theta$ yields

$$I = \frac{2\pi^2 a}{n-1} \int_{\delta-a}^{\delta+a} \frac{dt}{t^{2n-2}} = \frac{2\pi^2 a}{(n-1)(2n-3)} \left[\frac{1}{t^{2n-3}} \right]_{\delta+a}^{\delta-a},$$

which gives

$$I = \frac{2\pi^2 a}{(n-1)(2n-3)} \left[\frac{1}{(\delta-a)^{2n-3}} - \frac{1}{(\delta+a)^{2n-3}} \right].$$

Therefore the total interaction energy E_{ps} between the plane $\mathcal{P}$ and the sphere $\mathcal{S}$ is given by

$$E_{ps} = 2\pi^2 a \eta_1 \eta_2 \left\{ -\frac{A}{6} \left[\frac{1}{(\delta-a)^3} - \frac{1}{(\delta+a)^3} \right] + \frac{B}{45} \left[\frac{1}{(\delta-a)^9} - \frac{1}{(\delta+a)^9} \right] \right\}, \tag{3.22}$$

where η_1 and η_2 are the atomic surface densities of the plane and the spherical surface, respectively.

3.3.5 INTERACTION OF TWO CONCENTRIC SPHERICAL SURFACES

In this problem, we have two spherical surfaces $\mathcal{S}_1$ and $\mathcal{S}_2$ with radii of a_1 and a_2, respectively, and with both spheres centred at the origin, as shown in Fig. 3.8. The first sphere has the spherical coordinates $\mathcal{S}_1(\theta, \phi) = (a_1, \theta, \phi)$. We will now employ the formula for the interaction of a point and sphere E_s, given in Eq. (3.17), and replace a by a_2 and δ by the distance from the centre of $\mathcal{S}_2$ to the typical surface element on $\mathcal{S}_1$, which in this case is simply $\delta = a_1$. This expression is then integrated over the surface of $\mathcal{S}_1$, which gives

$$E_{ssc} = 4\pi a_2^2 \eta_2 \left[-\frac{A}{2(a_1^2 - a_2^2)^3} U_1(\gamma) + \frac{B}{5(a_1^2 - a_2^2)^6} U_4(\gamma) \right]$$
$$\times \eta_1 \int_{-\pi}^{\pi} \int_0^{\pi} a_1^2 \sin \phi \, d\theta \, d\phi,$$

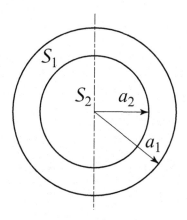

FIG. 3.8

Two interacting concentric spherical surfaces $\mathcal{S}_1$ and $\mathcal{S}_2$ of radii a_1 and a_2, respectively.

where $\gamma = (a_1^2 + a_2^2)/(a_1^2 - a_2^2)$. In this case the integration is trivial and therefore we may immediately write

$$E_{ssc} = 16\pi^2 a_1^2 a_2^2 \eta_1 \eta_2 \left[-\frac{A}{2(a_1^2 - a_2^2)^3} U_1(\gamma) + \frac{B}{5(a_1^2 - a_2^2)^6} U_4(\gamma) \right], \tag{3.23}$$

where η_1 and η_2 are the atomic surface densities for the spherical surfaces S_1 and S_2, respectively.

3.3.6 INTERACTION OF TWO OFFSET SPHERICAL SURFACES

In this problem, we have two spherical surfaces S_1 and S_2 with radii of a_1 and a_2, respectively, and with their centres separated by a distance δ, as illustrated in Fig. 3.9. To calculate the total interaction energy, we define the first sphere in spherical coordinates $S_1(\theta, \phi) = (a_1, \theta, \phi)$ and assume that the second sphere will be centred at the point $(0, 0, \delta)$ in Cartesian coordinates. Therefore the distance ρ from a typical point on S_1 to the centre of S_2 is given by

$$\rho^2 = (a_1 \cos \phi \sin \theta)^2 + (a_1 \sin \phi \sin \theta)^2 + (\delta - a_1 \cos \theta)^2$$
$$= \delta^2 + a_1^2 - 2\delta a_1 \cos \theta,$$

and from Eq. (3.15), the integral we require to evaluate is

$$I = \frac{\pi a_2 \eta_1 \eta_2}{n-1} \int_{-\pi}^{\pi} \int_{0}^{\pi} \frac{a_1^2 \sin \theta}{\rho} \left[\frac{1}{(\rho - a_2)^{2n-2}} - \frac{1}{(\rho + a_2)^{2n-2}} \right] d\theta \, d\phi.$$

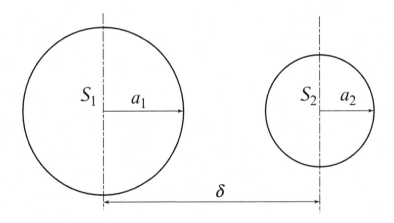

FIG. 3.9

Two interacting spheres S_1 and S_2 of radii a_1 and a_2, respectively, separated by a distance δ.

As there is no dependence on ϕ in the integrand, this can be done immediately and we make the substitution of variable of ρ for θ, which yields

$$
\begin{aligned}
I &= \frac{2\pi^2 a_1 a_2 \eta_1 \eta_2}{\delta(n-1)} \left\{ \int_{\delta - a_1}^{\delta + a_1} \frac{d\rho}{(\rho - a_2)^{2n-2}} - \int_{\delta - a_1}^{\delta + a_1} \frac{d\rho}{(\rho + a_2)^{2n-2}} \right\} \\
&= \frac{2\pi^2 a_1 a_2 \eta_1 \eta_2}{\delta(n-1)(2n-3)} \left\{ \left[\frac{1}{(\rho - a_2)^{2n-3}} \right]_{\delta + a_1}^{\delta - a_1} - \left[\frac{1}{(\rho + a_2)^{2n-3}} \right]_{\delta + a_1}^{\delta - a_1} \right\} \\
&= \frac{2\pi^2 a_1 a_2 \eta_1 \eta_2}{\delta(n-1)(2n-3)} \left[\frac{1}{(\delta - a_1 - a_2)^{2n-3}} - \frac{1}{(\delta + a_1 - a_2)^{2n-3}} \right. \\
&\qquad \left. - \frac{1}{(\delta - a_1 + a_2)^{2n-3}} + \frac{1}{(\delta + a_1 + a_2)^{2n-3}} \right].
\end{aligned}
$$

Therefore the total interaction energy E_{ss} between the two spherical surfaces S_1 and S_2 is given by

$$
\boxed{
\begin{aligned}
E_{ss} &= \frac{2\pi^2 a_1 a_2 \eta_1 \eta_2}{\delta} \\
&\times \left\{ -\frac{A}{6} \left[\frac{1}{(\delta + a_+)^3} + \frac{1}{(\delta - a_+)^3} - \frac{1}{(\delta + a_-)^3} - \frac{1}{(\delta - a_-)^3} \right] \right. \\
&\left. + \frac{B}{45} \left[\frac{1}{(\delta + a_+)^9} + \frac{1}{(\delta - a_+)^9} - \frac{1}{(\delta + a_-)^9} - \frac{1}{(\delta - a_-)^9} \right] \right\},
\end{aligned}
}
\tag{3.24}
$$

where $a_+ = a_1 + a_2$ and $a_- = a_1 - a_2$. We also comment that when the two spheres are equal, such that the radii $a_1 = a_2 = a$, and the atomic surface densities $\eta_1 = \eta_2 = \eta$, then this simplifies to an interaction energy $E_{ss}^\star$ given by

$$
\begin{aligned}
E_{ss}^\star &= \frac{2\pi^2 a^2 \eta^2}{\delta} \left\{ -\frac{A}{6} \left[\frac{1}{(\delta + 2a)^3} + \frac{1}{(\delta - 2a)^3} - \frac{2}{\delta^3} \right] \right. \\
&\left. + \frac{B}{45} \left[\frac{1}{(\delta + 2a)^9} + \frac{1}{(\delta - 2a)^9} - \frac{2}{\delta^9} \right] \right\}.
\end{aligned}
\tag{3.25}
$$

3.4 INTERACTION OF A CYLINDRICAL SURFACE

3.4.1 CYLINDRICAL COORDINATE SYSTEM (r, θ, z)

We employ the usual cylindrical coordinate system (r, θ, z). As usual, θ is measured in an anticlockwise direction from the positive x axis, as shown in Fig. 3.10, and therefore to map from cylindrical to Cartesian coordinates, we have

$$
\boxed{x = r\cos\theta, \quad y = r\sin\theta,}
\tag{3.26}
$$

and the z-coordinate is unchanged. The inverse mapping is given by

$$
r = \sqrt{x^2 + y^2}, \quad \theta = \arctan\frac{y}{x}.
$$

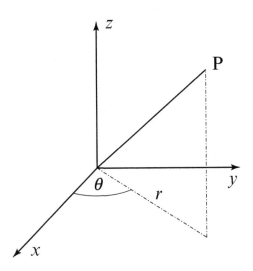

FIG. 3.10

A general point P described by cylindrical coordinates (r, θ, z).

The vector line element in cylindrical coordinates is given by

$$ds = dr\,\hat{\mathbf{r}} + r\,d\theta\,\hat{\boldsymbol{\theta}} + dz\,\hat{\mathbf{z}},$$

where $\hat{\mathbf{r}}$, $\hat{\boldsymbol{\theta}}$, and $\hat{\mathbf{z}}$ denote the unit vectors for the cylindrical coordinate system. Similarly, the vector area element is given by

$$d\mathbf{S} = r\,d\theta\,dz\,\hat{\mathbf{r}}.$$

In this book, we often consider integrals of the Lennard-Jones potential function $\Phi(\rho)$ over the surface of a cylinder of radius b of the form

$$I = b \int_{z_1}^{z_2} \int_{-\pi}^{\pi} \frac{1}{\rho^{2n}}\,d\theta\,dz,$$

where ρ is some function of the geometric parameters of the problem, which will generally also be a function of θ and z.

3.4.2 INTERACTION OF INTERIOR POINT WITH INFINITE CYLINDRICAL SURFACE

In this section, we consider the interaction of an arbitrary point $\mathbf{P}$ with a cylindrical surface $\mathcal{C}$ of radius b and of infinite length. The cylinder is given parametrically by $\mathcal{C}(\theta, z) = (b, \theta, z)$, where $-\pi < \theta \leqslant \pi$ and $-\infty < z < \infty$. Due to the rotational and translational symmetry of the problem, we designate the point $\mathbf{P}$ to be given in Cartesian coordinates by $(\delta, 0, 0)$, where we assume $0 \leqslant \delta < b$, as shown in Fig. 3.11. Therefore the distance from $\mathbf{P}$ to a typical surface element on $\mathcal{C}$ is given by

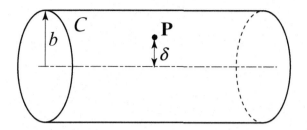

FIG. 3.11

A cylindrical surface C of radii b, interacting with an interior point **P** offset from the cylindrical axis by a distance δ.

$$\rho^2 = (b\cos\theta - \delta)^2 + b^2\sin^2\theta + z^2$$
$$= \delta^2 + b^2 - 2\delta b\cos\theta + z^2$$
$$= (b-\delta)^2 + 4\delta b\sin^2(\theta/2) + z^2.$$

By substitution into the cylindrical surface integral from Section 3.4.1 we need to evaluate the following:

$$I = b\int_{-\infty}^{\infty}\int_{-\pi}^{\pi} \frac{1}{\left[(b-\delta)^2 + 4\delta b\sin^2(\theta/2) + z^2\right]^n}\, d\theta\, dz.$$

We begin by defining $\lambda^2 = (b-\delta)^2 + 4\delta b\sin^2(\theta/2)$, and then we make the substitution $z = \lambda\tan\psi$, which gives

$$I = b\int_{-\pi/2}^{\pi/2}\cos^{2n-2}\psi\, d\psi \int_{-\pi}^{\pi}\frac{1}{\lambda^{2n-1}}\, d\theta$$
$$= b\mathrm{B}(n-1/2, 1/2)\int_{-\pi}^{\pi}\frac{1}{\lambda^{2n-1}}\, d\theta,$$

where we have used Eq. (2.19). Making the further substitution $t = \sin^2(\theta/2)$ yields

$$I = \frac{2b}{(b-\delta)^{2n-1}}\mathrm{B}(n-1/2, 1/2)\int_0^1 t^{-1/2}(1-t)^{-1/2}(1-\mu t)^{1/2-n}\, dt,$$

where $\mu = -4b\delta/(b-\delta)^2$. This integral is now in the Euler form, and from Eq. (2.29), we have

$$I = \frac{2b}{(b-\delta)^{2n-1}}\mathrm{B}(n-1/2, 1/2)\frac{\Gamma(1/2)\Gamma(1/2)}{\Gamma(1)}F(n-1/2, 1/2; 1; \mu)$$
$$= \frac{2\pi b}{(b-\delta)^{2n-1}}\mathrm{B}(n-1/2, 1/2)F(n-1/2, 1/2; 1; \mu).$$

We note that in terms of the usual parameters of the hypergeometric function, we have $c = 2b$, and therefore we may employ the quadratic transformation (Eq. 2.28), which yields

$$I = \frac{2\pi b}{(b-\delta)^{2n-1}} B(n - 1/2, 1/2) \left(\frac{b-\delta}{b}\right)^{2n-1} F\left(n - 1/2, n - 1/2; 1; \delta^2/b^2\right)$$

$$= \frac{2\pi}{b^{2n-2}} B(n - 1/2, 1/2) F\left(n - 1/2, n - 1/2; 1; \delta^2/b^2\right).$$

We now expand the beta function according to Eq. (2.18) and the hypergeometric function using the series representation, given in Eq. (2.20), which gives

$$I = \frac{2\pi}{b^{2n-2}} \frac{\Gamma(n-1/2)\Gamma(1/2)}{\Gamma(n)} \sum_{m=0}^{\infty} \left(\frac{(n-1/2)_m \delta^m}{m! \, b^m}\right)^2.$$

We comment that this series converges, provided that the argument of the hypergeometric function δ^2/b^2 is less than 1; that is, $\delta < b$, or in other words, that the point **P** is inside the cylinder. Replacing the Pochhammer symbol with gamma functions yields

$$I = \frac{2\pi}{b^{2n-2}} \frac{\Gamma(n-1/2)\Gamma(1/2)}{\Gamma(n)} \sum_{m=0}^{\infty} \left(\frac{\Gamma(n+m-1/2)\delta^m}{\Gamma(n-1/2)m! \, b^m}\right)^2$$

$$= \frac{2\pi}{b^{2n-2}} \frac{\Gamma(1/2)}{\Gamma(n)\Gamma(n-1/2)} \sum_{m=0}^{\infty} \left(\frac{\Gamma(n+m-1/2)\delta^m}{m! \, b^m}\right)^2,$$

and from the duplication formula (2.8), we may show that

$$\Gamma(n - 1/2) = \frac{\pi^{1/2}\Gamma(2n-1)}{2^{2n-2}\Gamma(n)},$$

which can be substituted to give

$$I = \frac{2\pi}{b^{2n-2}} \frac{\Gamma(1/2)2^{2n-2}\Gamma(n)}{\Gamma(n)\pi^{1/2}\Gamma(2n-1)} \sum_{m=0}^{\infty} \left(\frac{\pi^{1/2}\Gamma(2n+2m-1)\delta^m}{2^{2n+2m-2}\Gamma(n+m)m! \, b^m}\right)^2$$

$$= \frac{2\pi^2}{(2b)^{2n-2}\Gamma(2n-1)} \sum_{m=0}^{\infty} \left(\frac{\Gamma(2n+2m-1)\delta^m}{\Gamma(n+m)m! \, (4b)^m}\right)^2,$$

where all the gamma functions now have positive integer terms. We may substitute these for the factorial operator, which produces the energy E_{ci} for the interaction between a point and an infinite cylinder; thus

$$\boxed{E_{ci} = \frac{\pi^2 \eta}{192}\left[-\frac{A}{b^4} \sum_{m=0}^{\infty} \left(\frac{(2m+4)!\,\delta^m}{(m+2)!\,m!\,(4b)^m}\right)^2\right.}$$
$$\left. + \frac{B}{9676800\,b^{10}} \sum_{m=0}^{\infty} \left(\frac{(2m+10)!\,\delta^m}{(m+5)!\,m!\,(4b)^m}\right)^2\right].$$

(3.27)

For the special case of the point being on the axis of the tube, $\delta = 0$, and the only term in both series which are nonzero are the cases of $m = 0$, which gives the energy $E_{ci}^\star$ for an axial atom as

$$E_{ci}^\star = 2\pi^2\eta\left(-\frac{3A}{8b^4} + \frac{63B}{256b^{10}}\right),$$

which, as one would expect, is simply $2\pi b E_\ell$, as given in Eq. (3.6).

3.4.3 INTERACTION OF AN EXTERIOR POINT WITH AN INFINITE CYLINDRICAL SURFACE

In this section, we again consider the total interaction of a point $\mathbf{P}$, which is offset from the axis by a distance δ, and a cylindrical surface $\mathcal{C}$ of radius b. However, in this section, we deal with the case where $\delta > b$. In this case the calculation follows along similar lines as in the previous section except that we rearrange the terms so as to pick up a different solution to the hypergeometric equation, which is a solution where the argument is the inverse to that given in the previous section.

We begin with the same cylinder defined in cylindrical coordinates by $\mathcal{C}(\theta, z) = (b, \theta, z)$, where $-\pi < \theta \leqslant \pi$, and $-\infty < z < \infty$. We again define a point $\mathbf{P} = (\delta, 0, 0)$ in Cartesian coordinates except that in this case, we assume $\delta > b$, as illustrated in Fig. 3.12. Following the steps from Section 3.4.2 we derive an expression for the distance ρ from the point $\mathbf{P}$ to an arbitrary area element on the surface of the cylinder $\mathcal{C}$ as

$$\rho^2 = (\delta - b)^2 + 4\delta b \sin^2(\theta/2) + z^2.$$

Similarly as in Section 3.4.2, we may evaluate the integral and derive

$$I = \frac{2\pi b}{(\delta - b)^{2n-1}} B(n - 1/2, 1/2) F\left(n - 1/2, 1/2; 1; -\frac{4\delta b}{(\delta - b)^2}\right),$$

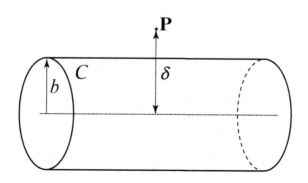

FIG. 3.12

A cylinder $\mathcal{C}$ of radii b, interacting with an exterior point $\mathbf{P}$ offset from the cylindrical axis by a distance δ.

where upon employing the quadratic transformation from Eq. (2.28) yields

$$I = \frac{2\pi b}{\delta^{2n-1}} B(n - 1/2, 1/2) F\left(n - 1/2, n - 1/2; 1; \frac{b^2}{\delta^2}\right).$$

Following the identical steps of writing the hypergeometric function as a series and simplifying all the gamma functions as simple factorial terms, we find that I is given by

$$I = \frac{4\pi^2 b}{(2\delta)^{2n-1}(2n-2)!} \sum_{m=0}^{\infty} \left(\frac{(2n + 2m - 2)! \, b^m}{(n + m - 1)! \, m! \, (4\delta)^m}\right)^2,$$

and therefore the total energy E_{ce}, for an infinite cylinder and an exterior point is given by

$$\boxed{E_{ce} = \frac{\pi^2 b \eta}{192} \left[-\frac{A}{\delta^5} \sum_{m=0}^{\infty} \left(\frac{(2m + 4)! \, b^m}{(m + 2)! \, m! \, (4\delta)^m}\right)^2 \right. \\ \left. + \frac{B}{9676800 \, \delta^{11}} \sum_{m=0}^{\infty} \left(\frac{(2m + 10)! \, b^m}{(m + 5)! \, m! \, (4\delta)^m}\right)^2 \right],}$$

(3.28)

where η is the atomic number density of the cylinder C. We note that the hypergeometric functions derived in this and the preceding section are not the analytic continuations of each other, but they are both solutions of the same hypergeometric equations in the neighbourhood of different singularities.

WORKED EXAMPLE 3.3

Evaluate the interaction between a cylinder and a point in the limit of the radius of the tube b becoming much smaller than the separation distance.

Solution

We can also consider Eq. (3.28) for $b \ll \delta$, and we define the atomic density per unit length for the cylinder to be given by $\eta^\star = 2\pi b \eta$; therefore

$$\eta = \frac{\eta^\star}{2\pi b}.$$

In the limit of $b/\delta \to 0$, the series in Eq. (3.28) is dominated by the first term; therefore substitution yields

$$E_{ce} \approx \left(\frac{\pi^2 b}{192}\right)\left(\frac{\eta^\star}{2\pi b}\right)\left[-\frac{A}{\delta^5}\left(\frac{4!}{2!}\right)^2 + \frac{B}{9676800 \, \delta^{11}}\left(\frac{10!}{5!}\right)^2\right]$$

$$\approx \pi \eta^\star \left(-\frac{3A}{8\delta^5} + \frac{63B}{256\delta^{11}}\right),$$

which is equivalent to the expression for E_ℓ derived in Eq. (3.6). The physical interpretation of this result is that for a point far enough away from the cylinder so that its radius is insignificant as compared to the separation distance will be indistinguishable from a line.

3.4.4 INTERACTION OF A SPHERICAL SURFACE WITH AN INFINITE CYLINDRICAL SURFACE

We now consider a situation of considerable practical interest, which is the interaction between a spherical surface and a cylindrical surface of infinite extent. In performing this calculation, we take the sphere to be of radius a and centred at $(\epsilon, 0, 0)$. The cylinder has a radius b and is centred on the z-axis, such that in Cartesian coordinates the surface has the parametric form $(b \cos \theta, b \sin \theta, z)$, where $\theta \in (-\pi, \pi]$ and $z \in (-\infty, \infty)$. Therefore the distance between the centre of the sphere and a typical surface element on the cylinder is given by

$$
\begin{aligned}
\delta^2 &= (b \cos \theta - \epsilon)^2 + b^2 \sin^2 \theta + z^2 \\
&= \epsilon^2 + b^2 - 2\epsilon b \cos \theta + z^2 \\
&= (b - \epsilon)^2 + 4\epsilon b \sin^2(\theta/2) + z^2.
\end{aligned}
$$

We now take the interaction energy for a point interacting with a sphere E_s, given in Eq. (3.17), which can be written as

$$
E_s = 4\pi a^2 \eta \left(-\frac{A}{2} J_1 + \frac{B}{5} J_4 \right),
$$

$$
J_n = (\delta^2 - a^2)^{-n-2} U_n \left(\frac{\delta^2 + a^2}{\delta^2 - a^2} \right),
$$

where $U_n(x)$ is the Chebyshev polynomial of the second kind, as described in Section 2.9. To evaluate the integration of E_s over the surface of the cylinder, we may express J_n using the explicit formula for Chebyshev polynomials (Eq. 2.41), which yields

$$
\begin{aligned}
J_n &= (\delta^2 - a^2)^{-n-2} \sum_{m=0}^{\lfloor n/2 \rfloor} (-1)^m \binom{n-m}{m} 2^{n-2m} \left(\frac{\delta^2 + a^2}{\delta^2 - a^2} \right)^{n-2m} \\
&= (\delta^2 - a^2)^{-n-2} \sum_{m=0}^{\lfloor n/2 \rfloor} (-1)^m \binom{n-m}{m} 2^{n-2m} \left(1 + \frac{2a^2}{\delta^2 - a^2} \right)^{n-2m} \\
&= (\delta^2 - a^2)^{-n-2} \sum_{m=0}^{\lfloor n/2 \rfloor} (-1)^m \binom{n-m}{m} 2^{n-2m} \sum_{\ell=0}^{n-2m} \binom{n-2m}{\ell} \left(\frac{2a^2}{\delta^2 - a^2} \right)^\ell,
\end{aligned}
$$

which reduces the terms involving δ to be of the form $(\delta^2 - a^2)^{-k}$, where k is a positive integer; in particular, we find

$$
J_1 = \frac{2}{(\delta^2 - a^2)^3} + \frac{4a^2}{(\delta^2 - a^2)^4},
$$

$$
J_4 = \frac{5}{(\delta^2 - a^2)^6} + \frac{80a^2}{(\delta^2 - a^2)^7} + \frac{336a^4}{(\delta^2 - a^2)^8} + \frac{512a^6}{(\delta^2 - a^2)^9} + \frac{256a^8}{(\delta^2 - a^2)^{10}}.
$$

Therefore we may follow similar steps as those employed in Section 3.4.2 to evaluate the integral

$$I = b \int_{-\infty}^{\infty} \int_{-\pi}^{\pi} \frac{d\theta \, dz}{\left[(b - \epsilon)^2 - a^2 + 4\epsilon b \sin^2(\theta/2) + z^2 \right]^n},$$

where in this case, $n \in \{3, 4, 6, 7, 8, 9, 10\}$. As in Section 3.4.2, we may evaluate the integral and derive

$$I = \frac{2\pi b B(n - 1/2, 1/2)}{\left[(b - \epsilon)^2 - a^2 \right]^{n-1/2}} F\left(n - \frac{1}{2}, \frac{1}{2}; 1; -\frac{4\epsilon b}{(b - \epsilon)^2 - a^2} \right).$$

It may be shown that

$$B(n - 1/2, 1/2) = \frac{\pi(2n - 3)!!}{(2n - 2)!!},$$

where !! is the double factorial operator, such that $(2n - 1)!! = (2n - 1)(2n - 3)(2n - 5) \ldots 3.1$ and $(2n)!! = 2n(2n - 2)(2n - 4) \ldots 4.2$. By simplifying and collecting terms, we may express the interaction energy between a sphere and a cylinder by

$$
\boxed{
\begin{aligned}
E_{cs} &= 8\pi^3 a^2 b \eta_1 \eta_2 \left[-\frac{A}{2} \left(2I_3 + 4a^2 I_4 \right) \right. \\
&\quad \left. + \frac{B}{5} \left(5I_6 + 80a^2 I_7 + 336a^4 I_8 + 512a^6 I_9 + 256a^8 I_{10} \right) \right], \\
I_n &= \frac{(2n - 3)!!}{(2n - 2)!![(b - \epsilon)^2 - a^2]^{n-1/2}} F\left(n - \frac{1}{2}, \frac{1}{2}; 1; -\frac{4\epsilon b}{(b - \epsilon)^2 - a^2} \right),
\end{aligned}
}
\tag{3.29}
$$

where η_1 and η_2 are the atomic surface densities for the cylinder and the sphere.

3.4.5 INTERACTION BETWEEN PARALLEL INFINITE CYLINDRICAL SURFACES

As for the cases of parallel lines and planes, we may calculate an interaction energy per unit length for infinite cylinders. However, in this case, it is a little more complicated, as the θ integral must still be evaluated for the secondary cylindrical surfaces. Therefore the interaction per unit length is equal to

$$
\begin{aligned}
E_{cc} &= \eta_2 b_2 \int_{-\pi}^{\pi} E_{ce} \, d\theta \\
&= 2\pi^2 b_1 b_2 \eta_1 \eta_2 \left(-A J_3 + B J_6 \right), \\
J_n &= \frac{(2n - 3)!!}{(2n - 2)!!} \int_{-\pi}^{\pi} \rho^{1-2n} F\left(n - 1/2, n - 1/2; 1; b_1^2/\rho^2 \right) d\theta,
\end{aligned}
$$

where b_1 and b_2 are the radii of the two cylinders, which are separated by a distance δ between the axes. Therefore

$$\rho^2 = (\delta - b_2)^2 + 4\delta b_2 \sin^2(\theta/2),$$

and there are two cases which must be considered. Firstly, when the two cylinders are external to each other (and in this case, $\delta > b_1 + b_2$) so that on taking the expression for J_n and expanding as a series, we may derive

$$J_n = \frac{(2n-3)!!}{(2n-2)!!} \sum_{\ell=0}^{\infty} \frac{(n-1/2)_\ell(n-1/2)_\ell}{(\ell!)^2} b_1^{2\ell} K_{n+\ell},$$

where the constants K_p are defined by

$$K_p = \int_{-\pi}^{\pi} [(\delta - b_2)^2 + 4\delta b_2 \sin^2(\theta/2)]^{1/2-p} \, d\theta,$$

and on making the substitution $t = \sin^2(\theta/2)$ gives

$$K_p = 2(\delta - b_2)^{1-2p} \int_0^t t^{-1/2}(1-t)^{-1/2}[1+\mu t]^{1/2-p} \, dt,$$

where $\mu = 4\delta b_2/(\delta - b_2)^2$. Using the Euler integral form for the hypergeometric function, we get

$$K_p = 2\pi(\delta - b_2)^{1-2p} F(p - 1/2, 1/2; 1; -\mu),$$

where again we may employ the quadratic transformation (Eq. 2.28), which yields

$$K_p = 2\pi\delta^{1-2p} F\left(p - 1/2, p - 1/2; 1; b_2^2/\delta^2\right).$$

Expanding $K_{n+\ell}$ as a series and substituting in J_n gives

$$J_n = \frac{2\pi(2n-3)!!}{\delta^{2n-1}(2n-2)!!} \sum_{\ell=0}^{\infty} \sum_{m=0}^{\infty} \left(\frac{(n-1/2)_{\ell+m}}{\ell!\, m!}\right)^2 \left(\frac{b_1}{\delta}\right)^{2\ell} \left(\frac{b_2}{\delta}\right)^{2m},$$

which is the series form of Appell's hypergeometric function of the fourth kind F_4. Therefore we may write

$$J_n = \frac{2\pi(2n-3)!!}{\delta^{2n-1}(2n-2)!!} F_4\left(n - 1/2, n - 1/2, 1, 1; \frac{b_1^2}{\delta^2}, \frac{b_2^2}{\delta^2}\right),$$

and therefore the interaction energy for two exterior cylinders is given by

$$\boxed{\begin{aligned} E_{cce} = 4\pi^3 b_1 b_2 \eta_1 \eta_2 &\left[-\frac{3A}{8\delta^5} F_4\left(5/2, 5/2, 1, 1; b_1^2/\delta^2, b_2^2/\delta^2\right) \right. \\ &\left. + \frac{63B}{256\delta^{11}} F_4\left(11/2, 11/2, 1, 1; b_1^2/\delta^2, b_2^2/\delta^2\right) \right]. \end{aligned}}$$

(3.30)

We now consider the case where one cylinder is inside the other. In this configuration, we assume that b_2 is the larger cylinder and that the two cylinders do not overlap; therefore $b_2 > \delta + b_1$. This calculation follows along similar lines to the previous case except that

$$\rho^2 = (b_2 - \delta)^2 + 4\delta b_2 \sin^2(\theta/2).$$

Therefore after the quadratic transformation, we have

$$K_p = 2\pi b_2^{1-2p} F\left(p - 1/2, p - 1/2; 1; \delta^2/b_2^2\right).$$

Thus

$$J_n = \frac{2\pi (2n-3)!!}{b_2^{2n-1}(2n-2)!!} F_4\left(n - 1/2, n - 1/2, 1, 1; \frac{b_1^2}{b_2^2}, \frac{\delta^2}{b_2^2}\right),$$

and therefore the interaction energy for the two nested cylinders is given by

$$
\boxed{
\begin{aligned}
E_{cci} = 4\pi^3 b_1 b_2 \eta_1 \eta_2 &\left[-\frac{3A}{8b_2^5} F_4\left(5/2, 5/2, 1, 1; b_1^2/b_2^2, \delta^2/b_2^2\right) \right. \\
&\left. + \frac{63B}{256 b_2^{11}} F_4\left(11/2, 11/2, 1, 1; b_1^2/b_2^2, \delta^2/b_2^2\right) \right].
\end{aligned}
}
\tag{3.31}
$$

When the two cylinders are coaxial, then Appell's F_4 function degenerates to the usual hypergeometric function; thus the energy for two coaxial cylinders is given by

$$
\boxed{
\begin{aligned}
E_{cci}^{\star} = 4\pi^3 b_1 b_2 \eta_1 \eta_2 &\left[-\frac{3A}{8b_2^5} F\left(5/2, 5/2; 1; b_1^2/b_2^2\right) \right. \\
&\left. + \frac{63B}{256 b_2^{11}} F\left(11/2, 11/2; 1; b_1^2/b_2^2\right) \right],
\end{aligned}
}
\tag{3.32}
$$

where it is understood that $b_1 < b_2$.

3.4.6 INTERACTION BETWEEN AN AXIAL POINT AND SEMIINFINITE CYLINDRICAL SURFACES

We now begin to consider cases involving the end of a cylindrical surface which may represent, for example, the uncapped end of a carbon nanotube. In these cases the short range nature of van der Waals forces are such that only one end of the finite structure needs to be considered; therefore a semiinfinite cylinder captures the essential behaviour of the system without having to take into account the length of the cylinder and the effects of the other end. We begin by considering a point $\mathbf{P} = (0, 0, Z)$ and a cylinder of radius b located with its axis on the z-axis and extending from the xy-plane into the positive z-direction; that is the cylinder is given parametrically by $(b\cos\theta, b\sin\theta, z)$, where $\theta \in (-\pi, \pi]$ and $z \in [0, \infty)$. In this configuration the distance between the point and a typical surface element on the cylinder is given by

$$\rho^2 = b^2 + (z - Z)^2,$$

and therefore the interaction energy is given by

$$E = \eta(-AI_3 + BI_6),$$

where

$$I_n = b \int_0^\infty \int_{-\pi}^{\pi} \frac{d\theta \, dz}{[b^2 + (z - Z)^2]^n}.$$

We comment that it is immediately apparent that the θ integration is trivial and therefore

$$I_n = 2\pi b \int_0^\infty \frac{dz}{[b^2 + (z - Z)^2]^n},$$

and we employ the substitution $u = z - Z$, which gives

$$I_n = 2\pi b \int_{-Z}^\infty \frac{du}{(b^2 + u^2)^n}$$

$$= 2\pi b \left\{ \int_0^\infty \frac{du}{(b^2 + u^2)^n} + \int_{-Z}^0 \frac{du}{(b^2 + u^2)^n} \right\}. \tag{3.33}$$

We now take each of these integrals in turn. The leftmost integral in Eq. (3.33), where we make the substitution $u = b \tan \psi$, giving

$$I_{n,1} = b^{1-2n} \int_0^{\pi/2} \cos^{2n-2} \psi \, d\psi$$

$$= \frac{B(n - 1/2, 1/2)}{2b^{2n-1}}.$$

Taking the second integral from Eq. (3.33), we see that it is an even function of u and therefore

$$I_{n,2} = \int_0^Z \frac{du}{(b^2 + u^2)^n}.$$

We now make the substitution of variable $u = Z\sqrt{1 - t}$ which leads to

$$I_{n,2} = \frac{Z}{2} \int_0^1 (1 - t)^{-1/2} (b^2 + Z^2 - tZ^2)^{-n} \, dt$$

$$= \frac{Z}{2(b^2 + Z^2)^n} \int_0^1 (1 - t)^{-1/2} \left[1 - \frac{tZ^2}{b^2 + Z^2} \right]^{-n} dt,$$

which is in integral form for the usual hypergeometric function with $\alpha = n$, $\beta = 1$, and $\gamma = 3/2$, so

$$I_{n,2} = \frac{Z}{(b^2 + Z^2)^n} F\left(n, 1; 3/2; \frac{Z^2}{b^2 + Z^2} \right).$$

Thus the total value of the integral I_n is given by

$$I_n = 2\pi b \left\{ \frac{B(n - 1/2, 1/2)}{2b^{2n-1}} + \frac{Z}{(b^2 + Z^2)^n} F\left(n, 1; 3/2; \frac{Z^2}{b^2 + Z^2} \right) \right\},$$

and hence the total interaction energy for an axial point and a semiinfinite cylinder is given by

$$
E_{psc} = 2\pi b\eta \left\{ -A \left[\frac{3\pi}{16b^5} + \frac{Z}{(b^2 + Z^2)^3} F\left(3, 1; 3/2; \frac{Z^2}{b^2 + Z^2}\right) \right] \right.
$$
$$
\left. + B \left[\frac{63\pi}{512b^{11}} + \frac{Z}{(b^2 + Z^2)^6} F\left(6, 1; 3/2; \frac{Z^2}{b^2 + Z^2}\right) \right] \right\},
$$

(3.34)

where η is the atomic surface density of the cylinder, and we have put in the values of $B(5/2, 1/2) = 3\pi/8$ and $B(11/2, 1/2) = 63\pi/256$.

NESTED CARBON NANOSTRUCTURES

4.1 INTRODUCTION

In the natural world, particles tend to organise themselves into their most stable structures, which are the minimum energy configurations. This is also true at the atomic level. When two atoms are far apart, an attractive force pulls them closer to each other, but if they are too close, a repulsive force pushes them away. There exists a certain separation distance where the two atoms are most likely to be, namely where the interaction energy between them is at a minimum. This distance is termed the equilibrium distance, and it is shown in Fig. 1.23. Ideally, atoms prefer to be apart by a distance equivalent to their equilibrium distance, as do molecules and other nanostructures.

This chapter focuses on the determination of stable equilibrium configurations for nested carbon nanostructures. A nested structure refers to a structure which comprises both inner and outer components. To begin with, we examine the case where the inner structure is simply a single atom. Then we extend the analysis to more complicated structures, including fullerene@fullerene (carbon onion), fullerene@nanotube, nanotube@nanotube, and carbon onion@nanotube. We note that we use the convention X@Y to denote a nested structure for which molecule X is nested inside molecule Y. In the final parts of this chapter, we look at the interaction of many body systems. For example, the case of a nested atom or molecule inside either a uniform concentric ring or a nanotube bundle. We ignore any complications arising from temperature variations, so tacitly we are assuming any changes in thermal energy are insignificant.

Throughout this chapter, we assume that a spherical fullerene C_N is of I_h type 1 symmetry, as discussed in Section 1.2.3. In Tables 1.5 and 1.6, we provide the radii and the mean surface densities for the first five fullerenes of this symmetry.

4.2 ATOM@FULLERENE—ENDOHEDRAL FULLERENE

Fullerenes are ideal traps for atoms or small clusters. The encapsulation of paramagnetic atoms like lanthanum, copper, nitrogen, or phosphorus in fullerenes leads to isolated paramagnetic systems. These systems may act like a shopping bag to transport single atoms. Various atoms have been shown to be contained within fullerenes, and Fig. 4.1A and B[1] illustrate the containment of single nitrogen and lanthanum atoms inside C_{60} and C_{82}, respectively.

---

[1] See http://homepage.mac.com/jschrier/endofullerenes-table.html.

Modelling and Mechanics of Carbon-based Nanostructured Materials. http://dx.doi.org/10.1016/B978-0-12-812463-5.00004-2

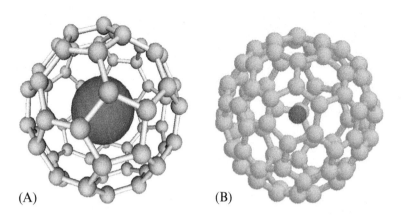

FIG. 4.1

Endohedral fullerenes.

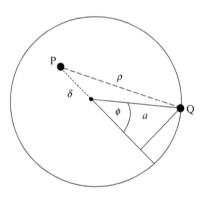

FIG. 4.2

Atom inside a spherical fullerene.

To model the interaction between an atom and a fullerene using the continuum approach, we assume that the carbon atoms on the fullerene are uniformly distributed over the surface of the molecule. Assuming that some other atom is located at the point **P**, as shown in Fig. 4.2, we find that the potential for a carbon atom interacting with a spherical fullerene is given by

$$E^* = -Q_6 + Q_{12},$$ (4.1)

where Q_n $(n = 6, 12)$ are defined by

$$Q_n = C_n \eta_{f1} \int_S \frac{1}{\rho^n} dS,$$ (4.2)

and ρ denotes the distance from the carbon atom to a typical surface element dS of the spherical molecules. The constants C_6 and C_{12} are the Lennard-Jones potential constants A and B respectively, and η_{f1} represents the surface density of carbon atoms for the fullerene. The distance ρ between the atom and a typical surface element of the fullerene is given by

$$\rho^2 = a^2 + \delta^2 + 2a\delta\cos\phi. \tag{4.3}$$

From Eq. (4.3), we also have $\rho\,d\rho = -a\delta\sin\phi\,d\phi$, thus Eq. (4.2) becomes

$$Q_n = C_n\eta_{f1}\int_0^\pi \frac{2\pi a\sin\phi}{\rho^n}a\,d\phi = \frac{2\pi aC_n\eta_{f1}}{\delta}\int_{a-\delta}^{a+\delta}\frac{1}{\rho^{n-1}}d\rho$$

$$= \frac{2\pi aC_n\eta_{f1}}{\delta(2-n)}\left\{\frac{1}{(a+\delta)^{n-2}} - \frac{1}{(a-\delta)^{n-2}}\right\}. \tag{4.4}$$

As a result, we find from Eq. (4.1) that the potential energy of a spherical fullerene interacting with a carbon atom is given by

$$E^* = \frac{2\pi\eta_{f1}a}{\delta}\left\{\frac{A}{4}\left(\frac{1}{(a+\delta)^4} - \frac{1}{(a-\delta)^4}\right) - \frac{B}{10}\left(\frac{1}{(a+\delta)^{10}} - \frac{1}{(a-\delta)^{10}}\right)\right\}, \tag{4.5}$$

which is equivalent to Eq. (3.15). In Fig. 4.3, we illustrate graphically $E^*(\delta)$ for a carbon atom inside fullerenes of different sizes. We note that the atom has a preferred position where its energy is a minimum, which is at the interatomic distance from the surface of the fullerene. We find that the atom is likely to be located approximately 3.4 Å away from the inner surface of the fullerene.

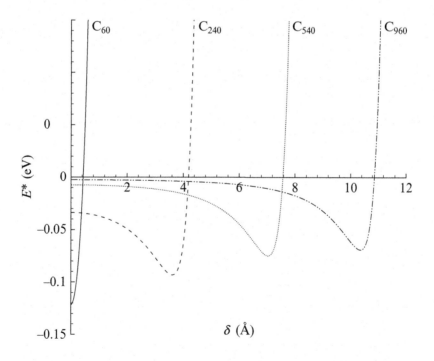

FIG. 4.3

Energy of an atom inside a spherical fullerene.

Another simple nested structure is that of the atom@carbon nanotube, and the determination of the interaction potential energy for this system is given as an exercise at the end of this chapter.

WORKED EXAMPLE 4.1

Solve the following three problems:

1. Find the attractive and the repulsive constants A and B for carbon and nitrogen given that $\epsilon_C = 0.002635\,\text{eV}$, $\sigma_C = 3.369\,\text{Å}$, $\epsilon_N = 0.006281\,\text{eV}$, and $\sigma_N = 3.365\,\text{Å}$.
2. Use the Lorentz–Berthelot mixing rules or empirical combining rules, $\epsilon_{ab} = \sqrt{\epsilon_a \epsilon_b}$ and $\sigma_{ab} = (\sigma_a + \sigma_b)/2$, to determine the attractive and repulsive Lennard-Jones constants A and B for each nonbonded C–N interaction.
3. For N@C_N, where C_N is a spherical fullerene with N carbon atoms, use Eq. (4.5) and the values of A_{CN} and B_{CN} from (2) to determine the appropriate size of a fullerene which gives rise to the nitrogen atom having its minimum energy position located at the centre of the fullerene.

Solution

1. From $\epsilon = A^2/(4B)$ and $2^{1/6}\sigma = (2B/A)^{1/6}$, we have the equations for A and B, namely

$$A = 4\epsilon\sigma^6, \quad B = 4\epsilon\sigma^{12}. \tag{4.6}$$

Thus by substituting ϵ_C and σ_C into Eq. (4.6), we obtain $A_C = 15.412\,\text{eV\,Å}^6$ and $B_C = 22{,}534.75\,\text{eV\,Å}^{12}$, and similarly for nitrogen, we deduce $A_N = 36.475\,\text{eV\,Å}^6$ and $B_N = 52{,}955.32\,\text{eV\,Å}^{12}$.
2. From the Lorentz–Berthelot mixing rules, we have $\epsilon_{CN} = \sqrt{\epsilon_C \epsilon_N}$ and $\sigma_{CN} = (\sigma_C + \sigma_N)/2$, which give rise to $\epsilon_{CN} = 0.004068\,\text{eV}$ and $\sigma_{CN} = 3.367\,\text{Å}$. As such from Eq. (4.6), we obtain the attractive and repulsive constants for C–N interaction, namely $A_{CN} = 23.709\,\text{eV\,Å}^6$ and $B_{CN} = 34{,}544.75\,\text{eV\,Å}^{12}$.
3. To determine the radius a of a spherical fullerene for each atom, we minimise the energy E^*/η_{f1}, which we refer to here as E_1:

$$E_1 = \frac{2\pi a}{\delta}\left\{\frac{A}{4}\left(\frac{1}{(a+\delta)^4} - \frac{1}{(a-\delta)^4}\right) - \frac{B}{10}\left(\frac{1}{(a+\delta)^{10}} - \frac{1}{(a-\delta)^{10}}\right)\right\},$$

when $\delta = 0$. If we define

$$f(\delta) = 2\pi a\left\{\frac{A}{4}\left(\frac{1}{(a+\delta)^4} - \frac{1}{(a-\delta)^4}\right) - \frac{B}{10}\left(\frac{1}{(a+\delta)^{10}} - \frac{1}{(a-\delta)^{10}}\right)\right\}, \quad g(\delta) = \delta,$$

then we find that at $\delta = 0$, we have the case of zero divided by zero. As such, by using l'Hôpital's Rule, the formula for E_1 at $\delta = 0$ can be given by

$$E_1 = \lim_{\delta \to 0} \frac{f'(\delta)}{g'(\delta)} = -\frac{4\pi A}{a^4} + \frac{4\pi B}{a^{10}}.$$

Next, to find the critical radius a which minimises E_1 we need to solve the following equation for a, thus

$$\frac{dE_1}{da} = \frac{16\pi A}{a^5} - \frac{40\pi B}{a^{11}} = 0,$$

to obtain that E_1 is a minimum when a satisfies

$$a = \left(\frac{5B}{2A}\right)^{1/6}. \tag{4.7}$$

By substituting the values of A_{CN} and B_{CN} for C–N interaction into Eq. (4.7), we obtain $a = 3.92\,\text{Å}$, which is the critical radius of a spherical fullerene for which the minimum energy configuration of the atom is at $\delta = 0$.

4.3 FULLERENE@FULLERENE—CARBON ONION

The synthesis by Ugarte (1992) of almost perfect spheres made up of concentric fullerenes, known as carbon onions (Fig. 4.4), under electron beam irradiation of carbon nanotubes and nanoparticles created an enormous impact on fullerene research. Since their discovery, many studies have been conducted to determine the mechanisms of carbon onion formation, their properties and structure. By simultaneously annealing and irradiating carbon nanotubes and polyhedral particles at temperatures of up to 700–800°C, Banhart and Ajayan (1996) and Banhart et al. (1997) discover the formation of perfect spherical onions and by further irradiating at this temperature, they find a reduction in the intershell spacing, which may then produce a diamond at the core of the carbon onion, as shown in Fig. 4.5. These experiments and others show that when fully annealed, the carbon onions are usually spherical with few obvious defects, and they normally possess a central shell of approximately 7–10 Å in diameter, which corresponds approximately to the diameter of a C_{60} fullerene. Other shapes of onion structures have also been found experimentally, such as ellipsoidal and tetrahedral onions. However, these shapes are not necessarily minimum-energy configurations, and it is argued that these structures are not energetically favourable when compared to spherical onions. The ab initio calculation of Lu and Yang (1994) supports the view that the more spherical the onion structure, the lower the energy and therefore the more stable the structure.

FIG. 4.4

Image by Banhart and Ajayan showing diamond core inside carbon onion particle.

Source: Harris, P.J.F., 1999. Carbon Nanotubes and Related Structures. Cambridge University Press, Cambridge. (Hardcover)

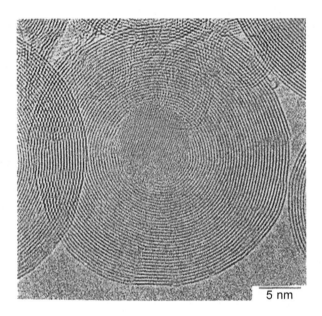

FIG. 4.5

Image by Banhart and Ajayan showing diamond core inside carbon onion particle.

Source: Harris, P.J.F., 1999. Carbon Nanotubes and Related Structures. Cambridge University Press, Cambridge. (Hardcover)

As suggested by Kroto and McKay (1988), a plausible model for carbon onions is that the structure is made up of concentric Goldberg type 1 fullerenes of I_h symmetry. From Eq. (1.4), the first five shells of the carbon onion of this type comprise C_{60}, C_{240}, C_{540}, C_{960}, and C_{1500}, and thus the intershell spacing is approximately 3.4 Å, which is very close to the interlayer spacing in graphite and to the spacing obtained experimentally and theoretically for carbon onions. The schematic of a carbon onion comprising the first five shells of Goldberg type 1 fullerenes of I_h symmetry is presented in Fig. 4.6.

By assuming that carbon onions are spherical molecules comprising concentric Goldberg type 1 fullerenes and by minimising the interaction energy between adjacent layers, we can determine the nested structure of carbon onions. Following Fig. 4.2 in the previous section, we assume that the atom at point **P** is now located on the surface of a second fullerene, which is nested inside the first, as illustrated in Fig. 4.7. As a result, the potential energy for the two concentric fullerenes can be obtained by integrating E^* in Eq. (4.1) over the surface of the new fullerene, which is assumed to have the parametric equation $(x, y, z) = (a_1 \sin \theta_1 \cos \phi_1, a_1 \sin \theta_1 \sin \phi_1, a_1 \cos \theta_1)$. This gives rise to

$$E_o = -P_6 + P_{12}, \tag{4.8}$$

where here P_n ($n = 6, 12$) are defined by

$$P_n = \eta_{f2} \int_0^{2\pi} \int_0^{\pi} Q_n a_1^2 \sin \theta_1 d\theta_1 d\phi_1, \tag{4.9}$$

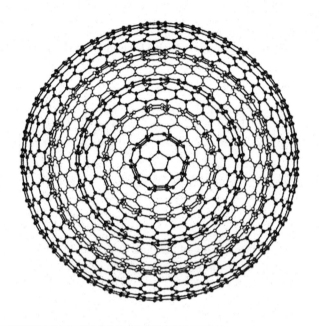

FIG. 4.6

Model of ve shelled carbon onion comprising $C_{60}@C_{240}C_{540}@C_{60}@C_{1500}$

Source: Lu, J.P., Yang, W., 1994. The shape of large single- and multiple-shell fullerenes. Phys. Rev. B 49, 11421–11424.

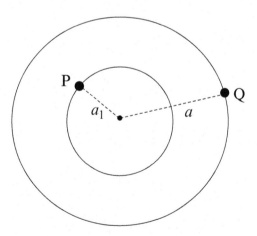

FIG. 4.7

Double-shell carbon onion.

Table 4.1 van der Waals interaction energy from continuum model given by (4.11) (Thamwattana and Hill, 2008) and compared with Lu and Yang (LY), (1994)

Energy of $C_{N1}@C_{N2}$	C_{240}		C_{540}		C_{960}		C_{1500}	
	LY	Eq. (4.11)	LY	Eq. (4.11)	LY	Eq. (4.11)	LY	Eq. (4.11)
C_{60}	−5.57	−5.571	−0.81	−0.807	−0.20	−0.216	−0.07	−0.071
C_{240}			−17.7	−17.827	−2.34	−2.682	−0.55	−0.553
C_{540}					−35.9	−36.028	−4.28	−4.157
C_{960}							−59.9	−54.968

Baowan, D., Thamwattana, N., Hill, J.M., 2008. Suction energy and offset configuration for double-walled carbon nanotubes. Commun. Nonlinear Sci. Numer. Simul. 13, 1431–1447.

where η_{f2} is the mean atomic surface density of the inner fullerene. In this case, the distance δ, which is the distance between the centre of the outer fullerene and an atom inside the fullerene, now becomes the radius of the inner fullerene, or in other words $\delta = a_1$. Thus Eq. (4.9) simplifies to give

$$P_n = \frac{4\pi^2 a a_1 C_n \eta_{f1} \eta_{f2}}{(2-n)} \left\{ \frac{1}{(a+a_1)^{n-2}} - \frac{1}{(a-a_1)^{n-2}} \right\} \int_0^\pi \sin\theta_1 \, d\theta_1$$
$$= \frac{8\pi^2 a a_1 C_n \eta_{f1} \eta_{f2}}{(2-n)} \left\{ \frac{1}{(a+a_1)^{n-2}} - \frac{1}{(a-a_1)^{n-2}} \right\}, \tag{4.10}$$

which upon substituting into Eq. (4.8), we obtain the potential interaction energy for the two-layer nested fullerene given by

$$E_o = 4\pi^2 a a_1 \eta_{f1} \eta_{f2} \left\{ \frac{A}{2} \left(\frac{1}{(a+a_1)^4} - \frac{1}{(a-a_1)^4} \right) \right.$$
$$\left. - \frac{B}{5} \left(\frac{1}{(a+a_1)^{10}} - \frac{1}{(a-a_1)^{10}} \right) \right\}. \tag{4.11}$$

In Table 4.1, we compare the interaction energy of a nested fullerene obtained from Eq. (4.11) with that obtained by the discrete atom–atom model of Lu and Yang (1994) (see also paper by Thamwattana and Hill, 2008).

WORKED EXAMPLE 4.2

For the nested fullerenes of I_h symmetries, the number of atoms in the nth shell is $N(n) = 60n^2$ and the average radius is approximated by $\bar{R}_n \approx 2.4 a_{C-C} n$, where a_{C-C} denotes the average bond length; here, we use $a_{C-C} = 1.421$ Å. For $n = 6 \ldots 10$, use these formulae to determine the mean radius and the mean density of the extended family of the spherical fullerenes, then find the potential energy of $C_{N(n)}@C_{N(n+1)}$ using Eq. (4.11).

Solution

```
>   restart;
>   N_array:= array(6..11):
>   for i from 6 to 11 do
>   N_array[i] := 60*i^2;
>   od;
```

$$N\_array_6 := 2160$$
$$N\_array_7 := 2940$$
$$N\_array_8 := 3840$$
$$N\_array_9 := 4860$$
$$N\_array_{10} := 6000$$
$$N\_array_{11} := 7260$$

```
>  R_array[i] := array(6..11):
>  a_CC:= 1.421:
>  for i from 6 to 11 do
>  R_array[i] := 2.4*a_CC*i;
>  od;
```

$$R\_array_6 := 20.4624$$
$$R\_array_7 := 23.8728$$
$$R\_array_8 := 27.2832$$
$$R\_array_9 := 30.6936$$
$$R\_array_{10} := 34.1040$$
$$R\_array_{11} := 37.5144$$

```
>  eta_array[i] := array(6..11):
>  for i from 6 to 11 do
>  eta_array[i] := evalf(N_array[i]/(4*Pi*R_array[i]^2));
>  od;
```

$$eta\_array_6 := 0.4105166223$$
$$eta\_array_7 := 0.4105166226$$
$$eta\_array_8 := 0.4105166226$$
$$eta\_array_9 := 0.4105166223$$
$$eta\_array_{10} := 0.4105166223$$
$$eta\_array_{11} := 0.4105166226$$

```
>  A:= 18.54: B:= 29*10^3:

>  Eo := (a1,a,eta1,eta2) -> 4*Pi^2*a*a1*eta1*eta2*((A/2)*(1/(a+a1)^4 -
>  1/(a-a1)^4) - (B/5)*( 1/(a+a1)^(10) - 1/(a-a1)^(10))):
>  Eo_array := array(6..10):
>  for i from 6 to 10 do
>  Eo_array[i] :=
>  evalf(Eo(R_array[i],R_array[i+1],eta_array[i],eta_array[i+1]));
>  od;
```

$$Eo\_array_6 := -134.1384082$$
$$Eo\_array_7 := -178.8557426$$
$$Eo\_array_8 := -229.9603531$$
$$Eo\_array_9 := -287.4524931$$
$$Eo\_array_{10} := -351.3323022$$

4.4 FULLERENE@CARBON NANOTUBE

In this section, we examine a single fullerene situated inside a carbon nanotube. The analysis in this chapter is extended in Chapter 7, where more than one fullerene may be located inside a carbon nanotube in a structure known as a nanopeapod. To calculate the interaction energy between a fullerene and a nanotube, we need the potential energy E between a fullerene and a single atom located at a distance δ away from the centre of the fullerene (see Fig. 3.5), which we have derived in Section 3.3.2. Now we assume that the atom at point **P** is located on the surface of a carbon nanotube, and therefore the potential energy between a fullerene and a carbon nanotube may be obtained by integrating over the surface of the cylindrical nanotube.

In axially symmetric cylindrical polar coordinates, we assume that a fullerene of radius a is located at $(\varepsilon, 0, 0)$, as shown in Fig. 4.8, and in a carbon nanotube of infinite extent with a parametric equation $(b\cos\theta, b\sin\theta, z)$. Noting that ε is the distance between the centre of the offset molecule and the central axis of the tube, b is the tube radius, $-\pi \leqslant \theta \leqslant \pi$ and $-\infty < z < \infty$. From Fig. 4.8, the distance from the centre of the fullerene to the wall of carbon nanotube is given by

$$\delta^2 = b^2 + \varepsilon^2 - 2b\varepsilon\cos\theta + z^2.$$

Thus the potential energy between the fullerene and the entire carbon nanotube is obtained by performing the surface integral (3.15) over the carbon nanotube and is given by

$$E^{tot} = b\eta_g \int_{-\pi}^{\pi} \int_{-\infty}^{\infty} E(\delta)\,dz\,d\theta. \tag{4.12}$$

Here, Eq. (3.15) can be rewritten in the form

$$E(\delta) = 4\pi a^2 \eta_f \left\{ \frac{B}{5} \left[\frac{5}{(\lambda^2 + z^2)^6} + \frac{80a^2}{(\lambda^2 + z^2)^7} + \frac{336a^4}{(\lambda^2 + z^2)^8} + \frac{512a^6}{(\lambda^2 + z^2)^9} \right. \right.$$
$$\left. \left. + \frac{256a^8}{(\lambda^2 + z^2)^{10}} \right] - A \left[\frac{1}{(\lambda^2 + z^2)^3} + \frac{2a^2}{(\lambda^2 + z^2)^4} \right] \right\},$$

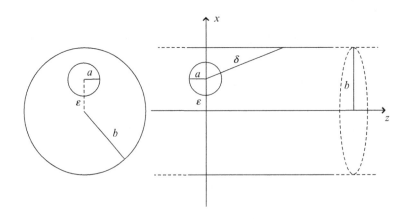

FIG. 4.8

An offset fullerene inside a single-walled carbon nanotube.

where $\lambda^2 = b^2 + \varepsilon^2 - 2b\varepsilon \cos \theta - a^2$, which gives $\delta^2 = \lambda^2 + a^2 + z^2$. By employing the substitution $z = \lambda \tan \psi$, and making use of

$$\int_{-\infty}^{\infty} \frac{dz}{(\lambda^2 + z^2)^{n+1}} = \frac{1}{\lambda^{2n+1}} \int_{-\pi/2}^{\pi/2} \cos^{2n} \psi \, d\psi = \frac{(2n-1)!!\pi}{(2n)!!\lambda^{2n+1}},$$

and

$$J_n = \int_{-\pi}^{\pi} \frac{d\theta}{(\alpha - \beta \cos \theta)^{n+1/2}}, \tag{4.13}$$

where $\alpha = b^2 + \varepsilon^2 - a^2$ and $\beta = 2b\varepsilon$, the potential energy Eq. (4.12) becomes

$$E^{tot} = 4\pi^2 a^2 b \eta_f \eta_g \left[\frac{B}{5} \left(\frac{315}{256} J_5 + \frac{1155a^2}{64} J_6 + \frac{9009a^4}{128} J_7 + \frac{6435a^6}{64} J_8 \right. \right.$$
$$\left. \left. + \frac{12155a^8}{256} J_9 \right) - \frac{A}{8} (3J_2 + 5a^2 J_3) \right], \tag{4.14}$$

where J_n are the integrals defined by Eq. (4.13), and they can be evaluated to obtain

$$J_n = \frac{2\pi}{\gamma^{n+1/2}} F\left(n + \frac{1}{2}, \frac{1}{2}; 1; -\omega/\gamma \right), \tag{4.15}$$

where $F(a, b; c, z)$ is the usual hypergeometric function, $\gamma = \alpha - \beta$ and $\omega = 2\beta$. We refer to Worked Example 4.3 for details of the evaluation of Eq. (4.13).

WORKED EXAMPLE 4.3

Evaluate the integral J_n shown in Eq. (4.13).

Solution

The integral (4.13) may be evaluated either in terms of elliptic integrals (see Cox et al., 2007a) or by using hypergeometric functions as follows.

Bisecting the interval of the integral J_n in Eq. (4.13) and reversing the sign of one of them and combining gives

$$J_n = 2 \int_0^{\pi} \frac{d\theta}{(\alpha - \beta \cos \theta)^{n+1/2}}. \tag{4.16}$$

Upon introducing $x = \theta/2$ and $m = n + 1/2$, Eq. (4.16) becomes

$$J_n = 4 \int_0^{\pi/2} \frac{dx}{(\gamma + \omega \sin^2 x)^m}, \tag{4.17}$$

where $\gamma = \alpha - \beta$ and $\omega = 2\beta$. By making the substitution $t = \cot x$ into Eq. (4.17), we obtain

$$J_n = \frac{4}{(\gamma + \omega)^m} \int_0^{\infty} \frac{(1 + t^2)^{m-1} dt}{(1 + \kappa t^2)^m}, \tag{4.18}$$

where $\kappa = \gamma/(\gamma + \omega)$. Thus Eq. (4.18) can be rewritten in the form

$$J_n = \frac{4}{(\gamma + \omega)^m} \int_0^{\infty} \frac{1}{[1 + (\kappa - 1)t^2/(1 + t^2)]^m} \frac{dt}{(1 + t^2)}, \tag{4.19}$$

which on making the substitutions $z = t/(1 + t^2)^{1/2}$ and $u = z^2$ obtains

$$J_n = \frac{2}{(\gamma + \omega)^m} \int_0^1 \frac{u^{-1/2}(1 - u)^{-1/2}}{[1 - (1 - \kappa)u]^m} du.$$

From Gradshteyn and Ryzhik (2000) (p. 995), we have

$$J_n = \frac{2\pi}{(\gamma + \omega)^m} F(m, 1/2; 1; 1 - \kappa), \tag{4.20}$$

where $F(p, q; s; z)$ denotes the usual hypergeometric function, and we have used $\Gamma(1/2) = \sqrt{\pi}$. Further, from Gradshteyn and Ryzhik (2000) (p. 998), we find that Eq. (4.20) can be written in the form

$$J_n = \frac{2\pi}{\gamma^m} F(m, 1/2; 1; -\omega/\gamma),$$

which is not a terminating hypergeometric series since $m = n + 1/2$, and therefore the parameters p and q are both noninteger.

As shown in Fig. 4.9, the preferred location of a C_{60} molecule inside a carbon nanotube $(10, 10)$ is where the centre of the buckyball lies on the tube axis. In the case of $(16, 16)$ ($b = 10.86\,\text{Å}$), $\varepsilon = 4.31\,\text{Å}$ is obtained (see the dashed line in Fig. 4.9). These results are equivalent to the distance between the centre of the buckyball and the wall of the nanotube of $6.784\,\text{Å}$ and $6.542\,\text{Å}$, respectively. Furthermore,

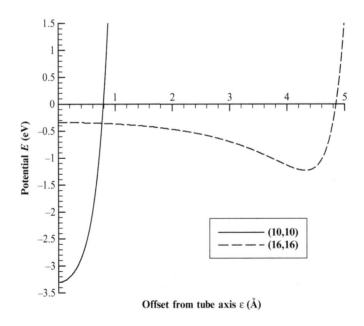

FIG. 4.9

The interaction potential of an offset C_{60} molecule inside a $(10, 10)$ and a $(16, 16)$ carbon nanotube, with respect to the radial distance ε from the tube axis.

it can be seen that as the radius of the tube gets larger, the location of minimum energy tends to be closer to the nanotube wall.

WORKED EXAMPLE 4.4

Use MAPLE to show that in a $(10, 10)$ nanotube, the preferred position of a C_{60} fullerene is at $\varepsilon = 0$ and in a $(16, 16)$, $\varepsilon = 4.314$ Å. Furthermore, find the preferred position of C_{60} inside a $(20, 20)$ carbon nanotube.

Solution

```
>  restart;
>  A := 17.4: B:= 29*10^3:
>  ng:= 0.3812: nf := 0.3789:
>  a := 3.55 :
>  alpha:= b^2 + epsilon^2 - a^2: beta := 2*b*epsilon:
>  J := n ->
>  2*Pi*hypergeom([n+1/2,1/2],[1],-2*beta/(alpha-beta))/(alpha-beta)^(n +
>  1/2):

>  E := 4*Pi^2*a^2*b*ng*nf*((B/5)*(315*J(5)/256 + 1155*a^2*J(6)/64 +
>  9009*a^4*J(7)/128 + 6435*a^6*J(8)/64 + 12155*a^8*J(9)/256) -
>  (A/8)*(3*J(2) + 5*a^2*J(3))):

>  plot(subs(b=6.784,E),epsilon=0..0.5);
```

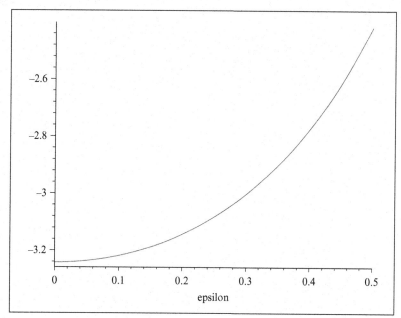

```
>  Eq1 := diff(subs(b=6.784,E),epsilon):
>  epsilon1 := fsolve(Eq1,epsilon);
```

$$\varepsilon 1 := 0.1564199462 \, 10^{-9}$$

```
>  plot(subs(b=10.86,E),epsilon=0..5);
```

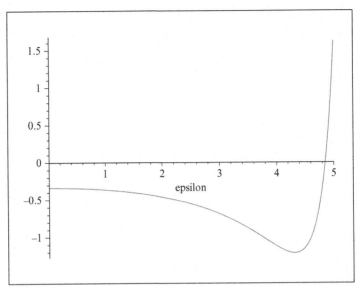

```
>  Eq2 := diff(subs(b=10.86,E),epsilon):
>  epsilon2:= fsolve(Eq2,epsilon=4);
```

$$\varepsilon 2 := 4.311820279$$

```
>  plot(subs(b=13.557,E),epsilon=0..8);
```

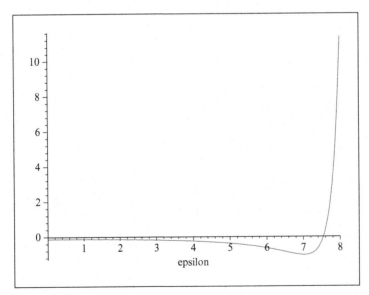

```
>  Eq3 := diff(subs(b=13.557,E),epsilon):
>  epsilon3:= fsolve(Eq3,epsilon=7);
```

$$\varepsilon 3 := 7.025219221$$

4.5 CARBON ONION@CARBON NANOTUBE

In a similar way to an atom or a fullerene, a carbon onion is likely to have a preferred location a certain distance from the surface of an encapsulating carbon nanotube. This position is that which minimises the potential energy of the nanostructure. We start by defining a geometry as shown in Fig. 4.10, where the carbon onion is offset with respect to the centre of the nanotube by ε in the x direction. Furthermore, the radii of the nested fullerenes making up the carbon onion are assumed to range from a_1 for the inner most fullerene to a_m for the outer most fullerene, where m is a positive number representing the number of shells in the carbon onion. To obtain the potential energy of an offset carbon onion inside a single-walled carbon nanotube, we use the Eqs (4.14) and (4.15), which give the potential energy between an offset spherical fullerene of radius a and a carbon nanotube of radius b and infinite length. Assuming that the carbon onion structure is itself at a minimum energy configuration, the total interaction energy of the carbon onion and the carbon nanotube can therefore be obtained from summing up only interactions between the carbon nanotube and each shell of the carbon onion. Thus the potential of an offset carbon onion inside a carbon nanotube is given by

$$E_c = \sum_{i=1}^{m} 4\pi^2 ba_i^2 \eta_{fi}\eta_g \left[\frac{B}{5}\left(\frac{315}{256}J_5 + \frac{1155a_i^2}{64}J_6 + \frac{9009a_i^4}{128}J_7 + \frac{6435a_i^6}{64}J_8 \right.\right.$$
$$\left.\left. + \frac{12155a_i^8}{256}J_9 \right) - \frac{A}{8}\left(3J_2 + 5a_i^2 J_3 \right) \right],$$

(4.21)

where m represents the number of shells in the carbon onion and J_n is given by

$$J_n = \frac{2\pi}{\gamma_i^{n+1/2}}F\left(n + \frac{1}{2}, \frac{1}{2}; 1; -\omega/\gamma_i \right),$$

(4.22)

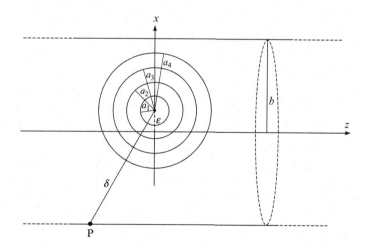

FIG. 4.10

Offset carbon onion inside a single-walled carbon nanotube.

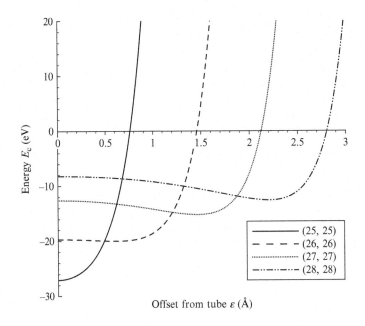

FIG. 4.11

Energy of an offset carbon onion $C_{60}@C_{240}@C_{540}@C_{960}$ inside a $(25, 25)$, $(26, 26)$, $(27, 27)$, and $(28, 28)$ carbon nanotube, with respect to the radial offset distance ε from the tube axis.

Table 4.2 Preferred Location of Carbon Onion $C_{60}@C_{240}@C_{540}@C_{960}$ Inside a Carbon Nanotube				
Carbon Nanotube	**(25, 25)**	**(26, 26)**	**(27, 27)**	**(28, 28)**
Radius, b (Å)	16.953	17.640	18.309	18.997
ε (Å)	0	0.609	1.462	2.204

where $\omega = 4b\varepsilon$ and $\gamma_i = (b - \varepsilon)^2 - a_i^2$.

In Fig. 4.11, we illustrate the potential energy E_c of an offset carbon onion with the structure $C_{60}@C_{240}@C_{540}@C_{960}$ with respect to the radial distance ε from the tube axis. Since the preferred position of the nanostructure is where its energy is a minimum, then it can be seen from Fig. 4.11 that as the tube gets larger, the preferred position tends to be further from the tube axis and closer to the surface of the tube. In particular, as shown in Table 4.2, the preferred position inside a $(25, 25)$ nanotube is such that the centre of the carbon onion lies on the tube axis.

WORKED EXAMPLE 4.5

Find the radius of a carbon nanotube which contains the carbon onion $C_{60}@C_{240}$ with its minimum energy position located at $\varepsilon = 0$.

Solution

In this case, we have $m = 2$, so from Eq. (4.21), we have that the total energy is given by

$$E_c = \sum_{i=1}^{2} 4\pi^2 b a_i^2 \eta_{fi} \eta_g \left[\frac{B}{5} \left(\frac{315}{256} J_5 + \frac{1155 a_i^2}{64} J_6 + \frac{9009 a_i^4}{128} J_7 + \frac{6435 a_i^6}{64} J_8 + \frac{12155 a_i^8}{256} J_9 \right) \right.$$
$$\left. - \frac{A}{8} \left(3J_2 + 5 a_i^2 J_3 \right) \right],$$

where $a_1 = 3.55$ Å and $\eta_{f1} = 0.3789$ Å^{-2} and $a_2 = 7.12$ Å and $\eta_{f2} = 0.3767$ Å^{-2} are radii and mean surface densities of C$_{60}$ and C$_{240}$, respectively. Given that $\varepsilon = 0$, J_n, which is given by Eq. (4.22), reduces to

$$J_n = \frac{2\pi}{(b^2 - a_i^2)^{n+1/2}}.$$

```
>   restart;
>   ng := 0.3812:
>   A:= 18.54: B:= 29*10^3:
>   J := (n) -> 2*Pi/(b^2 - a^2)^(n+1/2):
>   E :=
>   evalf(4*Pi^2*b*a^2*nf*ng*(B*(315*J(5)/256 + 1155*a^2*J(6)/64 +
>   9009*a^4*J(7)/128 + 6435*a^6*J(8)/64 + 12155*a^8*J(9)/265)/5 -
>   A*(3*J(2) + 5*a^2*J(3))/8)):

>   Ec := subs(a=3.55,nf=0.3789,E) + subs(a=7.12,nf=0.3767,E):

>   plot(Ec,b=9.5..12);
```

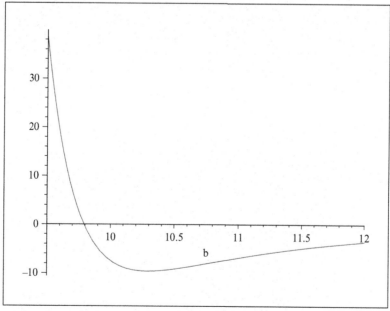

```
>   fsolve(diff(Ec,b),b=10);
```
$$10.29399591$$

Thus by using MAPLE, we find that the carbon onion $C_{60}@C_{240}$ will have a minimum energy at $\varepsilon = 0$ inside a carbon nanotube of radius $b = 10.29$ Å, which is approximately equal to a $(19, 19)$ tube.

Table 4.3 Radius of a Carbon Nanotube for Which the Preferred Location of Fullerene and Carbon Onions Is Along the z-Axis ($\varepsilon = 0$)

Nanostructure	C_{60}	$C_{60}@C_{240}$	$C_{60}@C_{240}@C_{540}$	$C_{60}@C_{240}@C_{540}C_{960}$
Radius b (Å)	6.74	10.29	13.66	16.96

We may extend the analysis in Worked Example 4.5 to different types of carbon onions. We find that the centre of a carbon onion is likely to lie on the nanotube axis ($\varepsilon = 0$) depending on the particular radius of the nanotube. In Table 4.3, we show the radii of carbon nanotubes that a fullerene C_{60} and carbon onions $C_{60}@C_{240}$, $C_{60}@C_{240}@C_{540}$, and $C_{60}@C_{240}@C_{540}@C_{960}$ have a preferred location of $\varepsilon = 0$ and a minimum interaction energy.

4.6 CARBON NANOTUBE@CARBON NANOTUBE—DOUBLE-WALLED CARBON NANOTUBE

As mentioned in Chapter 1, double-walled carbon nanotubes, as shown in Fig. 4.12, consist of two layers of graphite rolled up to form a tube. They possess very similar morphology and properties as single-walled carbon nanotubes, but have better resistance to chemicals. This is especially important when functionalisation is required to add new properties to the carbon nanotubes. Functionalisation refers to the grafting by covalent bonds of chemical functions to the nanotube surface. Covalent functionalisation of single-walled carbon nanotubes results in breaking some of the carbon–carbon

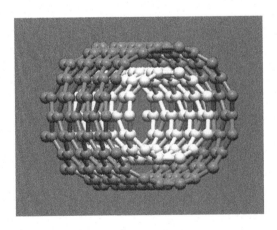

FIG. 4.12

Double-walled carbon nanotube.

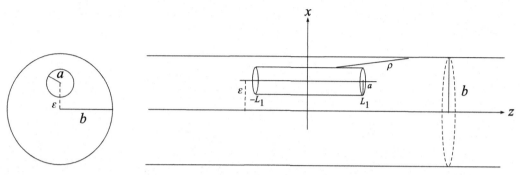

FIG. 4.13

An offset single-walled carbon nanotube inside a single-walled carbon nanotube.

bonds, leaving 'holes' in the nanotube structure, and thus, modifying both its mechanical and electrical properties. On the other hand, for covalent functionalisation of double-walled carbon nanotubes, only the outer wall is modified.

To examine the equilibrium configuration of a carbon nanotube inside another carbon nanotube, we assume that the parametric equation for an offset inner tube of radius a is $(a\cos\theta_1 + \varepsilon, a\sin\theta_1, z_1)$ and the parametric equation for an outer tube of radius b is $(b\cos\theta_2, b\sin\theta_2, z_2)$. The length of the inner tube is assumed to be $2L_1$ and the length of the outer tube is assumed to be $2L_2$, where L_2 tends to infinity and ε is the perpendicular distance between the axes of two tubes in the radial direction, as shown in Fig. 4.13. Thus, the distance ρ from any infinitesimal point on the surface of the inner tube to any infinitesimal point on the surface of the outer tube is given by

$$\rho^2 = [b\cos\theta_2 - (a\cos\theta_1 + \varepsilon)]^2 + (b\sin\theta_2 - a\sin\theta_1)^2 + (z_2 - z_1)^2,$$

and the total potential energy is given by

$$E^{tot} = ab\eta_g^2 \int_0^{2\pi} \int_0^{2\pi} (-AI_6 + BI_{12})d\theta_1 d\theta_2,$$

where I_6 is defined by

$$I_6 = \int_{-L_2}^{L_2} \int_{-L_1}^{L_1} \frac{1}{\rho^6} dz_1 dz_2 = \int_{-L_2}^{L_2} \int_{-L_1}^{L_1} \frac{1}{[\lambda^2 + (z_2 - z_1)^2]^3} dz_1 dz_2,$$

for which

$$\lambda^2 = [b\cos\theta_2 - (a\cos\theta_1 + \varepsilon)]^2 + (b\sin\theta_2 - a\sin\theta_1)^2$$
$$= a^2 + b^2 - 2ab\cos(\theta_1 - \theta_2) - 2\varepsilon(b\cos\theta_2 - a\cos\theta_1) + \varepsilon^2,$$

and on letting $x = (z_2 - z_1)$, we may deduce

$$I_6 = \int_{-L_1}^{L_1} \int_{-L_2 - z_1}^{L_2 - z_1} \frac{1}{(\lambda^2 + x^2)^3} dx dz_1 = \frac{1}{\lambda^5} \int_{-L_1}^{L_1} \int_{-\pi/2}^{\pi/2} \cos^4\phi\, d\phi dz_1,$$

which we obtain by substituting $x = \lambda \tan^2 \phi$ and letting L_2 tend to infinity. Finally, we may simplify I_6 to obtain

$$I_6 = \frac{1}{\lambda^5} \int_{-L_1}^{L_1} \frac{3\pi}{8} dz_1 = \frac{3\pi L_1}{4} \frac{1}{\lambda^5}.$$

By precisely the same method, I_{12} is given by

$$I_{12} = \int_{-L_2}^{L_2} \int_{-L_1}^{L_1} \frac{1}{\rho^{12}} dz_1 dz_2 = \frac{63\pi L_1}{128} \frac{1}{\lambda^{11}},$$

and therefore, the total potential energy for the double-walled carbon nanotube with an offset inner tube is given by

$$E^{tot} = abn_g^2 \pi L_1 \int_0^{2\pi} \int_0^{2\pi} \left(-\frac{3A}{4} \frac{1}{\lambda^5} + \frac{63B}{128} \frac{1}{\lambda^{11}} \right) d\theta_1 d\theta_2, \qquad (4.23)$$

where $\lambda^2 = a^2 + b^2 - 2ab \cos(\theta_1 - \theta_2) - 2\varepsilon(b \cos \theta_2 - a \cos \theta_1) + \varepsilon^2$.

The integrals (4.23) can be evaluated analytically in terms of elliptic functions, and we refer the reader to Baowan et al. (2008) for the details. Alternatively, we can use Eq. (3.31) to determine the interaction for the off-axis nested nanotubes. We note that E_{cci} in Eq. (3.31) is an interaction per unit length so for the configuration shown in Fig. 4.13, we need to multiply E_{cci}, by the length of the inner tube. Here, we use the algebraic computer package MAPLE to numerically evaluate Eq. (4.23), as

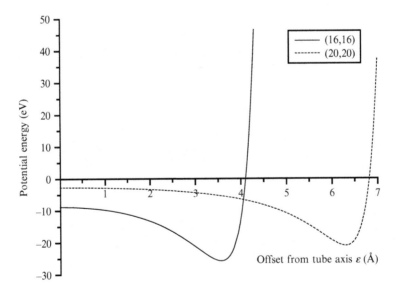

FIG. 4.14

Potential energy of an offset $(6, 6)$ carbon nanotube inside a $(16, 16)$ and a $(20, 20)$ carbon nanotube with respect to the radial distance ε.

shown in Fig. 4.14, for the relation of the potential energy and the offset position for a $(6, 6)$ carbon nanotube inside either a $(16, 16)$ or a $(20, 20)$ carbon nanotube. The values of ε for the $(6, 6)$ tube inside a $(16, 16)$ tube and a $(20, 20)$ tube are $3.567\,\text{Å}$ and $6.300\,\text{Å}$, respectively. This corresponds to the distances between the axis of the inner tube and the wall of the outer tube of $7.279\,\text{Å}$ and $7.257\,\text{Å}$, respectively. In general, we find that the larger the outer tube, the smaller the interspacing between the walls of the inner and the outer tubes.

WORKED EXAMPLE 4.6

Show that when $\varepsilon = 0$, Eq. (4.23) reduces to

$$E^{tot} = ab\eta_g^2 \pi L_1 \left(-\frac{3A}{4} J_5^* + \frac{63B}{128} J_{11}^* \right), \tag{4.24}$$

where

$$J_n^* = \frac{4\pi^2}{(a+b)^n} F\left(\frac{n}{2}, \frac{1}{2}; 1; \frac{4ab}{(a+b)^2} \right), \tag{4.25}$$

and $F(\alpha, \beta; \gamma; z)$ denotes the standard hypergeometric function.

Solution

We can rewrite Eq. (4.23) as

$$E^{tot} = ab\eta_g^2 \pi L_1 \left(-\frac{3A}{4} \int_0^{2\pi} \int_0^{2\pi} \frac{1}{\lambda^5} d\theta_1 d\theta_2 + \frac{63B}{128} \int_0^{2\pi} \int_0^{2\pi} \frac{1}{\lambda^{11}} d\theta_1 d\theta_2 \right), \tag{4.26}$$

where in the case of $\varepsilon = 0$ we have $\lambda^2 = a^2 + b^2 - 2ab\cos(\theta_1 - \theta_2)$.

Consider the integral

$$J_n^* = \int_0^{2\pi} \int_0^{2\pi} \frac{d\theta_1 d\theta_2}{\lambda^n}, \tag{4.27}$$

where $\lambda^2 = a^2 + b^2 - 2ab\cos(\theta_1 - \theta_2)$. Since the integrand is a symmetric function of $\theta_1 - \theta_2$, the intermediate integral J_n^{**} is defined by

$$J_n^{**} = \int_0^{2\pi} \frac{d\theta_1}{\{\alpha + \beta \sin^2[(\theta_1 - \theta_2)/2]\}^{n/2}},$$

where $\alpha = (a - b)^2$ and $\beta = 4ab$ can be shown by differentiation with respect to θ_2 to be independent of θ_2, namely

$$\frac{dJ_n^{**}}{d\theta_2} = \int_0^{2\pi} -\frac{\partial}{\partial \theta_1} \left(\frac{1}{\{\alpha + \beta \sin^2[(\theta_1 - \theta_2)/2]\}^{n/2}} \right) d\theta_1 = 0.$$

Thus θ_2 is set to zero, and one can trivially perform the θ_2 integration so that Eq. (4.27) becomes

$$J_n^* = 8\pi \int_0^{\pi/2} \frac{dx}{\lambda^n}, \tag{4.28}$$

where $\lambda^2 = (a - b)^2 + 4ab\sin^2 x$. Thus the integral J_n may be defined by

$$J_n = \int_0^{\pi/2} \frac{dx}{\lambda^n} = \int_0^{\pi/2} \frac{dx}{(\alpha + \beta \sin^2 x)^{n/2}}.$$

By making the substitution $t = \cot x$, we obtain

$$J_n = \int_0^\infty \frac{(1+t^2)^{n/2-1}}{(\beta + \alpha + \alpha t^2)^{n/2}} dt = \frac{1}{(\alpha + \beta)^{n/2}} \int_0^\infty \frac{(1+t^2)^{n/2-1}}{(1+\gamma t^2)^{n/2}} dt,$$

where $\gamma = \alpha/(\alpha + \beta)$. By writing this integral in the form

$$J_n = \frac{1}{(\alpha + \beta)^{n/2}} \int_0^\infty \frac{1}{[1 - (1-\gamma)t^2/(1+t^2)]^{n/2}} \frac{dt}{(1+t^2)},$$

leads to the substitution

$$z = \frac{t}{(1+t^2)^{1/2}}, \quad t = \frac{z}{(1-z^2)^{1/2}}, \quad dt = \frac{dz}{(1-z^2)^{3/2}}, \tag{4.29}$$

and the further substitution $u = z^2$ gives

$$J_n = \frac{1}{(\alpha + \beta)^{n/2}} \int_0^1 \frac{dz}{[1 - (1-\gamma)z^2]^{n/2}(1-z^2)^{1/2}}$$

$$= \frac{1}{2(\alpha + \beta)^{n/2}} \int_0^1 \frac{u^{-1/2}(1-u)^{-1/2}}{[1 - (1-\gamma)u]^{n/2}} du.$$

From Gradshteyn and Ryzhik (2000) (p. 995, eq. 9.111), we may deduce

$$J_n = \frac{\pi}{2(a+b)^n} F\left(\frac{n}{2}, \frac{1}{2}; 1; \frac{4ab}{(a+b)^2}\right), \tag{4.30}$$

where $F(a, b; c; z)$ is a hypergeometric function. Thus by substituting Eq. (4.30) back into Eq. (4.28), we obtain Eq. (4.25), and upon combining this with Eq. (4.26), we finally arrive at Eq. (4.24).

We note that it is possible to further simplify Eq. (4.30) to a Legendre function. From Erdélyi et al. (1953), since two of the numbers $\pm(1-c)$, $\pm(a-b)$, $\pm(a+b-c)$ are equal to each other, it can be shown that this result admits a quadratic transformation, which leads to a Legendre function. Using the transformation

$$F(a, b; 2b; 4z/(1+z)^2) = (1+z)^{2a} F(a, a+1/2 - b; b+1/2; z^2),$$

gives

$$J_n = \frac{\pi(1+\xi)^n}{2(a+b)^n} F(n/2, n/2; 1; \xi^2),$$

where $\xi = b/a$. Using the definitions from Gradshteyn and Ryzhik (2000) (pp. 960, 998)

$$P_\nu^\mu(z) = \frac{1}{\Gamma(1-\mu)} \left(\frac{z-1}{z+1}\right)^{-\mu/2} \left(\frac{z+1}{2}\right)^\nu F\left(-\nu, -\nu - \mu; 1-\mu; \frac{z-1}{z+1}\right),$$

and

$$F(a, b; c, z) = (1-z)^{c-a-b} F(c-a, c-b; c; z),$$

where $P_\nu^\mu(z)$ is a Legendre function of the first kind. In this case, μ is zero, and the integral in terms of the Legendre function is obtained and given by

$$J_n = \frac{\pi}{2(a^2 - b^2)^{n/2}} P_{n/2-1}\left(\frac{a^2 + b^2}{a^2 - b^2}\right).$$

4.7 NANOTUBE BUNDLES

A nanotube bundle refers to a closely packed array of aligned carbon tubes, as shown in Fig. 4.15. A molecular dynamics simulation of Kang et al. (2006) shows that a single nanotube can oscillate in the centre of the nested hexagonal nanotube bundles, generating a new type of sigahertz oscillator. By denoting one nanotube as the centre of the bundle, hexagonal rings of nanotubes can be defined in the network. In terms of modelling the interactions in a bundle, we adopt the following definition of a bundle. With reference to Fig. 4.16, we refer to a nanotube bundle as N carbon nanotubes, each of the same length $2L$ and radius r, and all lying parallel to each other. The axis of every tube is assumed to be equidistant from a common bundle axis, which is taken to be the z-axis, and this distance R is termed the bundle radius. The nanotube axes are also assumed to be symmetrically spaced around the cylinder defined by the bundle axis and the radius R. Therefore the ith tube in the nanotube bundle has a surface in Cartesian coordinates (x, y, z) given by

$$\left(R\cos\left(\frac{2\pi(i-1)}{N} \right) + r\cos\theta_i, R\sin\left(\frac{2\pi(i-1)}{N} \right) + r\sin\theta_i, z_i \right), \tag{4.31}$$

where $i \in \{1 \ldots N\}$, $0 \leqslant \theta_i \leqslant 2\pi$, and $-L \leqslant z_i \leqslant L$. However, since the van der Waals forces are short range, it is convenient for some of the analysis that follows to assume a semiinfinite tube, with $0 \leqslant z_i < \infty$, or a completely infinite tube, with $-\infty < z_i < \infty$. We note that the advantage of adopting

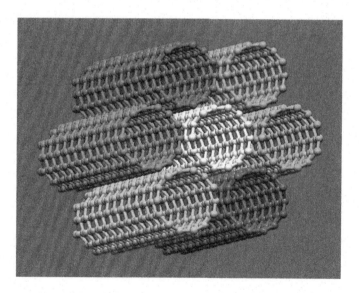

FIG. 4.15

Carbon nanotube bundle.

Source: Kang, J.W., Song, K.O., Hwang, H.J., Jiang, Q., 2006. Nanotube oscillator based on a short single-walled carbon nanotube bundle. Nanotechnology 17, 2250–2258.

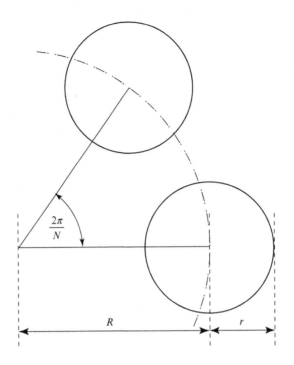

FIG. 4.16

Part of bundle of N tubes, showing an angle between two adjacent tubes, bundle radius R, and constituent tube radius r.

our definition of a bundle in terms of a ring of N nanotubes is that the results for the hexagonal $N = 6$ emerge from the analysis, which can then be used for nearest neighbour calculations for bundles in the usual triangular lattice. Moreover, future technology and fabrication techniques may well enable other nanotube bundle geometries to become available that are not currently realisable. Therefore it is useful to analyse the most general case so that interesting phenomena may be identified, even if it is not considered feasible within the context of present day technology.

4.7.1 INTERACTION POTENTIAL FOR TWO NANOTUBES

For the purpose of determining the equilibrium spacing in a nanotube bundle, we assume that the tubes are infinite in both directions. We begin by calculating the interaction between an infinite nanotube and an atom lying at a perpendicular distance ρ from the tube axis. The interaction energy is given by

$$E_{ta} = \frac{3r\pi^2 \eta_t}{4} \left[-\frac{AP_{3/2}(\gamma)}{\left(\rho^2 - r^2\right)^{5/2}} + \frac{21BP_{9/2}(\gamma)}{32\left(\rho^2 - r^2\right)^{11/2}} \right], \tag{4.32}$$

where $\gamma = (\rho^2 + r^2)/(\rho^2 - r^2)$ and $P_\nu(z)$ denotes the usual associated Legendre functions of degree ν.

WORKED EXAMPLE 4.7

Derive the interaction energy between an infinite nanotube and an atom as shown in Eq. (4.32).

Solution

The interaction energy for a nanotube and a single atom is given by the integral

$$E_{ta} = \eta \left(-AI_3 + BI_6 \right),$$ (4.33)

$$I_n = r \int_{-\infty}^{\infty} \int_0^{2\pi} \left[(\rho - r)^2 + 4r\rho \sin^2 \frac{\phi}{2} + z^2 \right]^{-n} d\phi \, dz.$$

On first evaluating the z integration, using the substitution $z = \lambda \tan \psi$, where $\lambda^2 = (\rho - r)^2 + 4r\rho \sin^2(\phi/2)$ we may show that

$$I_n = \frac{r\pi \, (2n - 3)!!}{2^{n-1} \, (n-1)!} \int_0^{2\pi} \left[(\rho - r)^2 + 4r\rho \sin^2 \frac{\phi}{2} \right]^{1/2-n} d\phi.$$

Then by substituting $t = \sin^2(\phi/2)$, we obtain

$$I_n = \frac{r\pi \, (2n-3)!!}{2^{n-2} \, (\rho - r)^{2n-1} \, (n-1)!} \int_0^1 t^{-1/2} (1-t)^{-1/2} \left\{ 1 - \left[1 - \left(\frac{\rho + r}{\rho - r} \right)^2 \right] t \right\}^{1/2-n} dt,$$

which is the fundamental integral form for the hypergeometric function, and therefore

$$I_n = \frac{r\pi^2 \, (2n-3)!!}{2^{n-2} \, (\rho - r)^{2n-1} \, (n-1)!} F \left(n - \frac{1}{2}, \frac{1}{2}; 1; 1 - \left(\frac{\rho + r}{\rho - r} \right)^2 \right),$$

where $F(a, b; c; z)$ is the usual hypergeometric function. Using the transformation from Erdélyi et al. (1953, Section 2.9, eq. (2)), we may show that

$$I_n = \frac{r\pi^2 \, (2n-3)!!}{2^{n-2} \, (\rho - r) \, (\rho + r)^{2n-2} \, (n-1)!} F \left(\frac{3}{2} - n, \frac{1}{2}; 1; 1 - \left(\frac{\rho + r}{\rho - r} \right)^2 \right).$$

Finally, by using Erdélyi et al. (1953, Section 3.2, eq. (28)), we may transform this expression from a hypergeometric function to a Legendre function and show that

$$I_n = \frac{r\pi^2 \, (2n-3)!!}{2^{n-2} \, (\rho^2 - r^2)^{n-1/2} \, (n-1)!} P_{n-3/2} \left(\frac{\rho^2 + r^2}{\rho^2 - r^2} \right),$$ (4.34)

where $P_\nu(z)$ is the Legendre function of the first kind of degree ν and by substituting Eq. (4.34) into Eq. (4.33) gives Eq. (4.32).

Using Eq. (4.32), we derive an analytical expression for the interaction potential per unit length for two parallel carbon nanotubes E_{tt}. The distance between the axis is denoted by δ and the radius of the tubes are r_1 and r_2. This can be calculated by substituting $\rho^2 = (\delta - r_2)^2 + 4\delta r_2 \sin^2(\theta/2)$ into Eq. (4.32) and then integrating over the circumference of the second cylinder. Two analytical expressions are derived for this potential, namely in the form of a series of associated Legendre functions or using Appell's hypergeometric functions of two variables, which can be expressed as

$$E_{tt} = \frac{3}{2}\eta_t^2 r_1 r_2 \pi^3 \alpha^{-5}\left[-AF_2\left(\frac{5}{2},-\frac{3}{2},\frac{1}{2},1,1;-\frac{r_1^2}{\alpha^2},-\frac{4r_2\delta}{\alpha^2}\right)\right.$$

$$\left. + \frac{21}{32}B\alpha^{-6}F_2\left(\frac{11}{2},-\frac{9}{2},\frac{1}{2},1,1;-\frac{r_1^2}{\alpha^2},-\frac{4r_2\delta}{\alpha^2}\right)\right],$$

(4.35)

where $F_2(\alpha,\beta,\beta',\gamma,\gamma';x,y)$ is Appell's hypergeometric function of two variables given in Section 2.7, and α is defined by $\alpha^2 = (\delta - r_2)^2 - r_1^2$.

4.7.2 EQUILIBRIUM POSITION FOR A BUNDLE OF NANOTUBES

As previously stated, we define a nanotube bundle as N carbon nanotubes of the same length and radius, evenly spaced and equidistant from a common axis, which we term the axis of the bundle. All the nanotubes making up the bundle are assumed to be aligned parallel to the bundle axis and of radius r. The distance from the bundle axis to each tube axis is termed the bundle radius and is denoted by R. We assume that the bundle adopts a lowest energy level configuration; that is we determine the value of the interspacing distance δ, for which the total potential energy E_B is minimised. This means that for a particular bundle number N and tube radius r, the interactions between tubes will prescribe the equilibrium distance δ between tubes and therefore the bundle radius R can be determined.

The total interaction potential for the bundle E_B is defined as the sum of all the constituent interactions, which is given by

$$E_B = \frac{N}{2}\sum_{k=1}^{N-1}E_{tt}\left(2R\sin\left(\frac{k\pi}{N}\right)\right),$$

(4.36)

where $E_{tt}(\delta)$ is given by Eq. (4.35) and which is the tube-tube energy for two nanotubes of radius r at a distance δ. The above sum is actually dominated by the nearest neighbour interactions, and so for most practical purposes, we can approximate Eq. (4.36) by

$$E_B \approx NE_{tt}\left(2R\sin\left(\frac{\pi}{N}\right)\right),$$

and a graph of the bundle energy E_B versus radius R is shown in Fig. 4.17 for three different sized carbon nanotubes.

For the purposes of actually constructing nanotube bundles, we are interested in the minimum energy configuration. Accordingly, in Fig. 4.18 we plot the bundle radius R against the nanotube radius r for various numbers of nanotubes N. It can be seen from this figure that the relationship between the bundle radius and nanotube radius is almost linear and scaled by a factor based on the number of nanotubes comprising the bundle N. If we term the equilibrium interstice λ as that distance between two tubes when the interaction energy is minimised, that distance varies slightly as a function of nanotube radius, but generally lies between $3.10\,\text{Å}$ and $3.16\,\text{Å}$. With this simplification and the previous comment

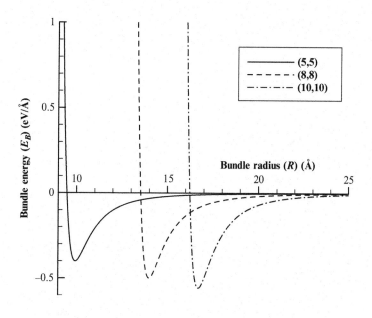

FIG. 4.17

Bundle energy versus radius for three different types of carbon nanotubes, with $N = 6$.

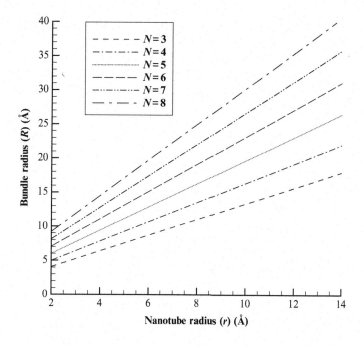

FIG. 4.18

Bundle radius versus nanotube radius for bundles with $N \in \{3, 4, \ldots, 8\}$.

that the nearest neighbour interactions dominate the bundle energy E_B, the bundle radius R can now be approximated by

$$R \approx \frac{2r + \lambda}{2\sin(\pi/N)}.$$

4.8 CARBON NANOTUBE@NANOTUBE BUNDLE

In this section, we consider a configuration comprising a single carbon nanotube located on the axis of a uniform concentric ring or bundle of carbon nanotubes. To find the equilibrium location of the nested nanotube, we first must calculate the interaction potential between the centred single-walled carbon nanotube of radius r_0, and a nanotube bundle formed from N nanotubes of radius r. The single nanotube is located with its axis parallel to the z-axis, but it is offset by a distance ϵ in the direction ϕ relative to the direction of the x-axis. That is, the axis of the nanotube may be given parametrically as $(\epsilon \cos \phi, \epsilon \sin \phi, z)$. Therefore the distance between the axis of the single nanotube and that of the kth nanotube in the nanotube bundle is given by

$$d_k^2 = \left(R\cos\left(\frac{2k\pi}{N}\right) - \epsilon \cos \phi\right)^2 + \left(R\sin\left(\frac{2k\pi}{N}\right) - \epsilon \sin \phi\right)^2$$
$$= (R - \epsilon)^2 + 4R\epsilon \sin^2\left(\frac{k\pi}{N} - \frac{\phi}{2}\right).$$

In this case, because the single nanotube radius r_0 is not necessarily the same as the radius r of the nanotubes forming the bundle, we need to calculate the interaction potential between two parallel tubes of differing radii as given by Eq. (4.35).

As in the previous section, we can approximate the bundle radius in terms of the radius of the centre nanotube r_0, the radius of the constituent tubes in the bundle r, and the bundle radius R. In this case, we may express the approximate relationship by

$$R \approx r_0 + r + \lambda,$$

where, as before, λ is the intertube equilibrium distance, which generally varies from 3.1 to 3.16 Å. Combining this with the expression for the minimised bundle energy allows us to derive the following approximate formula for the radius of constituent tubes in a bundle of number N to minimise the energy for a centred tube of radius r_0,

$$r \approx \frac{(r_0 + \lambda)\sin(\pi/N) - \lambda/2}{1 - \sin(\pi/N)}.$$

We note that for $N = 6$, it follows immediately that $r \approx r_0$, as expected.

4.9 FULLERENE@NANOTUBE BUNDLE

Here we are interested in the interaction potential between a fullerene of radius r_0 located at a position $(\varepsilon \cos \phi, \varepsilon \sin \phi, 0)$ relative to the centre of a nanotube bundle. To calculate the total energy W_f, we assume for the moment that the nanotubes are infinite in length. As before, the distance from the centre of the fullerene to the axis of the kth tube in the bundle d_k, is given by

$$d_k^2 = (R - \varepsilon)^2 + 4R\varepsilon \sin^2 \left(\frac{k\pi}{N} - \frac{\phi}{2} \right),$$

and we require that $d_k > r_0 + r$, for all $k \in \{1, 2, 3, \ldots, N\}$.

We denote the interaction energy between the fullerene and a single tube by $E_{ft}(d_k)$ so that the total suction energy of the fullerene W_f is given by

$$W_f = - \sum_{k=1}^{N} E_{ft} \left(\left[(R - \varepsilon)^2 + 4R\varepsilon \sin^2 \left(\frac{k\pi}{N} - \frac{\phi}{2} \right) \right]^{1/2} \right).$$

Note that the suction energy will be discussed in detail in Chapter 5 in Sections 5.1.2 and 5.2.2. For infinite nanotubes, an expression for the function E_{ft} is given by

$$E_{ft}(d_k) = \pi^2 r_0^2 r n_f n_t \left[\frac{B}{5} \left(\frac{315}{256} J_5 + \frac{1155}{64} r_0^2 J_6 + \frac{9009}{128} r_0^4 J_7 \right. \right.$$
$$\left. \left. + \frac{6435}{64} r_0^6 J_8 + \frac{12155}{256} r_0^8 J_9 \right) - \frac{A}{8} \left(3 J_2 + 5 r_0^2 J_3 \right) \right], \tag{4.37}$$

where the constants A and B are the attractive and repulsive constants, respectively, for graphene-C_{60} systems, and the integrals J_n are defined by

$$J_n = \int_{-\pi}^{\pi} \frac{d\theta}{(r^2 + d_k^2 - r_0^2 - 2rd_k \cos\theta)^{n+1/2}},$$

which can be expressed analytically as

$$J_n = \frac{2\pi}{[(r - d_k)^2 - r_0^2]^{n+1/2}} F \left(\frac{1}{2}, n + \frac{1}{2}; 1; -\frac{4rd_k}{(r - d_k)^2 - r_0^2} \right), \tag{4.38}$$

where $F(a, b; c; z)$ is the usual hypergeometric function. We note that the derivations of Eqs (4.37) and (4.38) are similar to those of Eqs (4.14) and (4.15) shown in Section 4.4, with the distance ρ in Eqs (4.14) and (4.15) replaced by d_k.

Assuming that the equilibrium position for the C_{60} fullerene is on the axis of the nanotube bundle, the total energy W_f can be simplified to $W_f = -NE_{ft}(R)$. Thus the minimum for any value of N is given by the minimum of the function E_{ft}, and this minimum is the equilibrium position for the fullerene on the outside of a single tube. By equating this distance with the bundle radius R, which optimises the suction energy for a particular tube radius r, we can thereby determine a relationship between these two radii, which would lead to an optimised configuration for any value of N. In Fig. 4.19, we graph the relationship between the nanotube radius r and the bundle radius R, which optimises the total energy for the C_{60} fullerene. On the same figure, we also show the tube versus bundle data from Fig. 4.18. The points where the lines intersect prescribe the specific values of r and R, which lead to optimised configurations for specific values of N. As can be seen from Fig. 4.19, limiting the tube radius r to the range 2–12 Å means that optimum configurations can only be constructed for $N \in \{4, 5, 6, 7\}$.

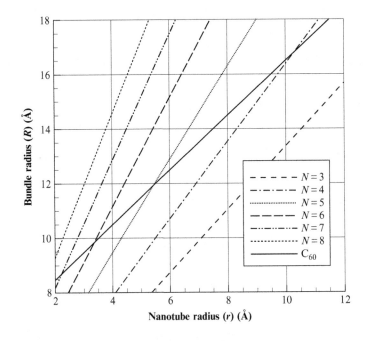

FIG. 4.19

Bundle radius versus nanotube radius for a C_{60} fullerene and nanotube bundles with $N \in \{3, 4, \ldots, 8\}$.

Table 4.4 Parameters for Optimised C_{60} Fullerene–Nanotube Bundle Energy

Number	Tube Radius	Bundle Radius	Suction Energy
N	r (Å)	R (Å)	W_f (eV)
4	10.294	16.792	2.314
5	5.439	11.930	2.551
6	3.355	9.838	2.731
7	2.219	8.692	2.850

More precise parameters for these ideal oscillators are given in Table 4.4, including the total energy for each configuration. We note that the total energy W_f increases with the number of tubes N forming the bundle; one way to optimise a bundle oscillator is to design the bundle to have as many tubes as possible. However, we also observe that the maximum value of total energy $W_f = 2.85$ eV is still less than the maximum energy that may be obtained for a fullerene inside a single-walled carbon nanotube, where we find the maximum value to be $W_{\max} = 3.242$ eV.

EXERCISES

4.1. (a) Find the attractive and repulsive constants A and B for atoms in Table 4.5.

Table 4.5 Well Depth ϵ and van der Waals Diameter σ for Various Elements.

X	C	Li$^+$	Na$^+$	K$^+$	He	Ne	Ar	Kr	Xe
σ (Å)	3.50	2.10	2.50	2.93	3.03	3.18	3.46	3.55	3.83
ϵ (meV)	2.1	74.4	43.6	28.0	1.4	2.2	7.7	5.3	6.3

Breton, J., Gonzalez-Platas, J., Girardet, C., 1994. Endohedral adsorption in graphitic nanotubules. J. Chem. Phys. 101, 3334–3340.

(b) Use the Lorentz–Berthelot mixing rules, $\epsilon_{ab} = \sqrt{\epsilon_a \epsilon_b}$ and $\sigma_{ab} = (\sigma_a + \sigma_b)/2$, to determine the attractive and repulsive Lennard-Jones constants A and B for each nonbonded C–X interaction, where X denotes an atom shown in Table 4.5.

(c) For each type of atom, use Eq. (4.5) and the values of A and B from (b) to determine the appropriate size of a spherical fullerene C_N, which gives rise to the atom having its minimum energy position located at the centre of the fullerene.

4.2. (a) Find the potential interaction energy for an offset single atom inside an infinite length carbon nanotube of radius b.

(b) For a $(15, 15)$ carbon nanotube, what is the equilibrium distance of a carbon atom with respect to the axial centre of the carbon nanotube?

4.3. Use the solution in Problem 1(a) to solve Problem 1(b) for other types of atoms shown in Table 4.5.

4.4. For fullerene C_{240} of I_h symmetry, find the radius of a nanotube which minimises the energy (Eq. 4.14) when the centre of the fullerene is located a distance $\varepsilon = 1$ Å from the nanotube's axial centre.

4.5. Given an innermost nanotube of $(5, 5)$ structure, find the next five layers which give rise to a minimum energy configuration of a concentric multiwalled carbon nanotube.

4.6. Determine the total interaction energy for an atom inside a nanotube bundle of N-fold symmetry, given that each tube comprising the bundle is of radius b and is infinite in length.

4.7. For a C_{240} fullerene of I_h symmetry inside a sixfold symmetry bundle comprising $(16, 16)$ nanotubes of infinite length, determine the bundle radius such that the fullerene has its minimum energy when its centre is on the bundle central axis.

ACCEPTANCE CONDITION AND SUCTION ENERGY

5.1 INTRODUCTION

This chapter outlines the concepts of an acceptance condition and the suction energy to predict whether a carbon nanostructure, such as a C_{60} fullerene, would be accepted into the interior of a carbon nanotube by van der Waals forces alone. The issue of suction of particles into containers is fundamental, particularly for applications where drugs or genes are encapsulated into nanocarriers for targeted delivery to tumour cells.

The suction energy (W) is defined as the total work performed by van der Waals interactions on a molecule entering a carbon nanotube. In certain cases the van der Waals force becomes repulsive as the molecule crosses the tube opening. In these cases the acceptance energy (W_a) is defined as the total work performed by van der Waals interactions on the molecule entering the nanotube, up until the point that the van der Waals force once again becomes attractive. In this book, we use elementary mechanical principles and classical applied mathematical modelling techniques to formulate explicit analytical criteria and ideal model behaviour in a scientific context, which was previously only elucidated through molecular dynamics simulation. While van der Waals interactions have been calculated previously using classical approaches, this chapter extends the analysis and provides explicit expressions for the acceptance condition and the suction force. We note again that temperature effects are assumed not to arise. In essence, we consider the case $T = 0\,\mathrm{K}$ for which the interaction potential is much larger than the thermal energy kT, where k is the Boltzmann constant.

The Lennard-Jones potential function and the continuum approximation are employed throughout this book to determine the interatomic potential for various carbon nanostructures. To remind the reader, the Lennard-Jones function (3.1) can be written as

$$\Phi(\rho) = -A\rho^{-6} + B\rho^{-12},$$

where A and B are the attractive and repulsive constants, respectively, and ρ denotes the distance between two typical surface elements. It is convenient for some of the analysis in this chapter to assume a semiinfinite tube because the van der Waals force operates at a short distance, and we are primarily concerned with the encapsulation of molecules at one end of the tube. In the following sections, we define the terms 'acceptance condition' and 'suction energy', which are used throughout this chapter and in subsequent chapters.

Modelling and Mechanics of Carbon-based Nanostructured Materials. http://dx.doi.org/10.1016/B978-0-12-812463-5.00005-4

5.1.1 ACCEPTANCE ENERGY

In Fig. 5.1, we show a typical interaction force F for a molecule located near an open end of a carbon nanotube, noting that $\pm Z_o$ is the roots of an equation $F(Z) = 0$. The integral of $F(Z)$ represents the work done by the van der Waals forces, which are imparted onto the molecule in the form of kinetic energy. In order for a particular molecule with a centre of mass located a distance Z from the nanotube end (negative Z in Fig. 5.1) to be accepted into the interior of a nanotube (positive Z), the sum of its initial kinetic energy and that received from moving from $-\infty$ to $-Z_o$ needs to be greater than that which is lost when the van der Waals force is negative (i.e. in the region $-Z_o < Z < Z_o$). We term this the acceptance energy (W_a), which allows us to write the acceptance condition as $E_K + W_a > 0$, where E_K is the initial kinetic energy of the molecule and W_a is the work done from $-\infty$ to Z_o.

Therefore for a molecule assumed initially at rest and located on the tube axis to be accepted into the interior of a nanotube, the condition becomes $W_a > 0$, and the following inequality must hold true:

$$\int_{-\infty}^{Z_o} F(Z)dZ > 0, \tag{5.1}$$

where Z_o is the positive root of the axial interaction force $F(Z)$, as shown in Fig. 5.1; this is the formal condition for a molecule to enter the nanotube. The left-hand side of Eq. (5.1) is termed the acceptance energy (W_a) and may be plotted against the nanotube radius to determine the minimum radius of

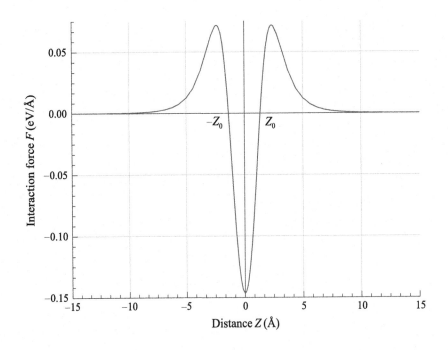

FIG. 5.1

Typical interaction force.

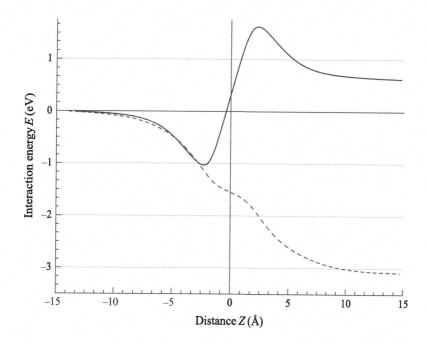

FIG. 5.2

Typical interaction energy, showing a molecule being accepted (- - -) and rejected (——) from two nanotubes.

nanotube a_0 that will accept a particular molecule. Alternatively, acceptance is further illustrated by the interaction energy, as shown in Fig. 5.2, in which a molecule will be accepted if the energy inside the tube (positive Z) is less than the energy outside the tube (negative Z).

In some cases, it is possible for the interaction force to exhibit two peaks, as shown in Fig. 5.3. Therefore the acceptance condition given by Eq. (5.1) is slightly modified. In particular, with reference to Fig. 5.3, the molecule will be accepted into the nanotube interior if and only if both the integral from $-\infty$ to the root Z_1 and the integral from $-\infty$ to the root Z_2 are positive. Thus the acceptance condition becomes

$$\int_{-\infty}^{Z_1} F(Z)dZ > 0 \quad \text{and} \quad \int_{-\infty}^{Z_2} F(Z)dZ > 0. \tag{5.2}$$

5.1.2 SUCTION ENERGY

The suction energy W is defined as the total energy or work done as generated by van der Waals interactions acquired by a particular molecule as a consequence of being sucked into the nanotube, or more formally with reference to Fig. 5.1, we have

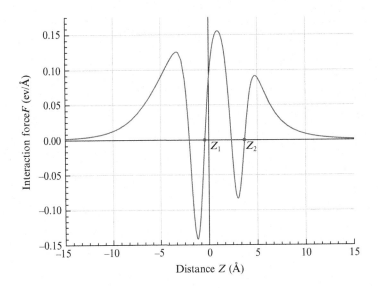

FIG. 5.3

Schematic of interaction force.

$$W = \int_{-\infty}^{\infty} F(Z)dZ, \tag{5.3}$$

which is the work done that is transformed into kinetic energy. Alternatively, the suction energy is written as

$$W = -\int_{-\infty}^{\infty} \frac{dE}{dz} dz = E(-\infty) - E(\infty).$$

Similarly, it is possible to plot the suction energy (W) against the nanotube radius to determine the radius which provides maximum suction b_{max} and the radius at which it becomes energetically favourable for the entering molecule to be inside the nanotube b_{min}. At the point where the curve of the suction energy crosses the horizontal axis (i.e. $W = 0$), it becomes energetically favourable for the molecule to be inside the nanotube. However, in the region of this energetically favourable radius b_{min} to the minimum radius for acceptance b^*, the suction energy is not sufficient to overcome the barrier energy at the tube end. Therefore at $T = 0\,K$, for the molecule to be encapsulated inside the nanotube when the radius falls within this range ($b_{min} < b < b^*$), some additional energy is required. We note, however, that in practice it may be difficult to fire one nanostructure along the axis of another, especially while maintaining the system at constant temperature.

In this chapter, we provide examples of the acceptance condition and the suction energy for three different systems of carbon nanostructures, namely a C_{60} fullerene inside a carbon nanotube, double-walled carbon nanotubes, and a nanobundle.

5.2 C_{60} FULLERENE INSIDE A CARBON NANOTUBE

C_{60} fullerenes and carbon nanotubes are the most commonly studied nanostructures due to their simple geometric configurations, namely a sphere for the C_{60} fullerenes and a cylinder for the nanotubes. The encapsulation of a C_{60} fullerene inside a carbon nanotube has been studied using computer simulation by a number of researchers, including Hodak and Girifalco (2001), Qian et al. (2001), and Okada et al. (2001). These studies determined the nanotube radii which will accept a C_{60} fullerene into its interior, primarily to examine the oscillatory behaviour for a C_{60} fullerene inside a carbon nanotube and to assemble the nanopeapod (the details are given in Chapter 7). In this chapter, we are interested in using elementary applied mathematical modelling to examine whether a carbon nanotube will encapsulate a C_{60} fullerene into its interior by van der Waals force alone as detailed below.

5.2.1 ACCEPTANCE CONDITION

The interaction between an approximately spherical fullerene and a cylindrical carbon nanotube is modelled using the continuum approximation obtained by averaging atoms over the surface of each entity, as discussed in Chapter 1.3. To start, the calculation for the interaction energy between a carbon atom and a C_{60} fullerene is reviewed, utilising the Lennard-Jones potential function and the continuum approximation, as given in Section 3.3.2. The potential energy for an atom on the tube interacting with all atoms of the sphere with radius a is given by

$$E(\delta) = -Q_6(\delta) + Q_{12}(\delta),$$

where Q_n is defined by

$$Q_n = C_n \eta_f \int_S \frac{1}{\delta^n} dS,$$

and δ denotes the distance between a typical tube surface element and the centre of the fullerene, as shown in Fig. 5.4 and given by $\delta^2 = b^2 + (Z - z)^2$. The constants C_6 and C_{12} are the Lennard-Jones potential constants A and B, respectively, and η_f represents the atomic surface density of a C_{60} fullerene. Upon integrating, we obtain

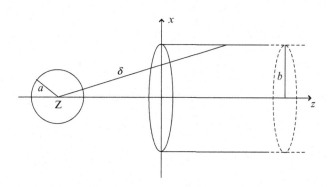

FIG. 5.4

Geometry of a fullerene molecule entering a carbon nanotube.

$$Q_n(\delta) = \frac{2C_n \eta_f \pi a}{\delta(2-n)} \left(\frac{1}{(\delta+a)^{n-2}} - \frac{1}{(\delta-a)^{n-2}} \right). \tag{5.4}$$

Substituting Eq. (5.4) into $E(\delta)$ and simplifying, the interaction energy between an atom and a C_{60} fullerene is obtained and can be written as

$$E(\delta) = \frac{\eta_f \pi a}{\delta} \left[\frac{A}{2} \left(\frac{1}{(\delta+a)^4} - \frac{1}{(\delta-a)^4} \right) - \frac{B}{5} \left(\frac{1}{(\delta+a)^{10}} - \frac{1}{(\delta-a)^{10}} \right) \right], \tag{5.5}$$

which is equivalent to Eq. (3.15). The total potential energy between a C_{60} fullerene and a carbon nanotube is obtained by performing another surface integral on Eq. (5.5) over a cylindrical nanotube.

From Fig. 5.4, the van der Waals interaction force between the fullerene molecule and an atom on the tube is of the form $F_{vdW} = -\nabla E$. Therefore the axial force is obtained by differentiating the energy with respect to the axial direction z and is given by

$$F_z = -\frac{(Z-z)}{\delta} \frac{dE}{d\delta}.$$

The total axial force between the entire carbon nanotube and the fullerene is obtained by integrating the van der Waals force over the cylindrical nanotube surface. In this case, such a force is independent of the cylindrical angle, and therefore we may deduce

$$F_z^{tot}(Z) = -2\pi b\eta_g \int_0^\infty \frac{dE}{d\delta} \frac{(Z-z)}{\delta} dz, \tag{5.6}$$

where η_g is the mean atomic density of carbon atoms in a graphene structure, such as a carbon nanotube, and because $\delta^2 = b^2 + (Z-z)^2$, we have $d\delta = -((Z-z)/\delta)dz$. Thus Eq. (5.6) can be simplified to give

$$F_z^{tot}(Z) = 2\pi b\eta_g \int_{\sqrt{b^2+Z^2}}^\infty \frac{dE}{d\delta} d\delta,$$

$$= -2\pi^2 \eta_f \eta_g a^2 b \left[\frac{A}{2\delta a} \left(\frac{1}{(\delta+a)^4} - \frac{1}{(\delta-a)^4} \right) \right.$$

$$\left. - \frac{B}{5\delta a} \left(\frac{1}{(\delta+a)^{10}} - \frac{1}{(\delta-a)^{10}} \right) \right]_{\delta=\sqrt{b^2+Z^2}}. \tag{5.7}$$

By placing the fractions over common denominators, expanding and reducing to fractions in terms of powers of $(\delta^2 - a^2)$, it can be shown that

$$\frac{A}{2\delta a} \left(\frac{1}{(\delta+a)^4} - \frac{1}{(\delta-a)^4} \right) = -4A \left(\frac{1}{(\delta^2-a^2)^3} + \frac{2a^2}{(\delta^2-a^2)^4} \right), \tag{5.8}$$

$$\frac{B}{5\delta a} \left(\frac{1}{(\delta+a)^{10}} - \frac{1}{(\delta-a)^{10}} \right) = -\frac{4B}{5} \left(\frac{5}{(\delta^2-a^2)^6} + \frac{80a^2}{(\delta^2-a^2)^7} + \frac{336a^4}{(\delta^2-a^2)^8} \right.$$

$$\left. + \frac{512a^6}{(\delta^2-a^2)^9} + \frac{256a^8}{(\delta^2-a^2)^{10}} \right). \tag{5.9}$$

Substituting these identities into Eq. (5.7) gives a precise expression for the z component of the van der Waals force experienced by a fullerene located at a position Z on the z-axis as

$$F_z^{tot}(Z) = \frac{8\pi^2 \eta_f \eta_g b}{a^4 \lambda^3} \left[A \left(1 + \frac{2}{\lambda} \right) - \frac{B}{5a^6\lambda^3} \left(5 + \frac{80}{\lambda} + \frac{336}{\lambda^2} + \frac{512}{\lambda^3} + \frac{256}{\lambda^4} \right) \right], \quad (5.10)$$

where $\lambda = (b^2 - a^2 + Z^2)/a^2$. Analytically determining the roots of $F_z^{tot}(Z)$ is not a simple task due to the complexity of the expression and the order of the polynomial involved. The function for the spherical C_{60} fullerene inside various radii (b) of carbon nanotubes is shown in Fig. 5.5, and we may see that there will be at most two real roots of the form $Z = \pm Z_o$. These roots will only exist when the value of b is less than some critical value b_o for some particular value of the parameter a. In the case of a C_{60} fullerene ($a = 3.55$ Å), the value of $b_o \approx 6.509$ Å.

The integral of $F_z^{tot}(Z)$, as defined by Eq. (5.10), represents the work imparted onto the fullerene and equates directly to the kinetic energy. Therefore the integral of Eq. (5.10) from $-\infty$ to Z_o represents the acceptance energy (W_a) for the system and would need to be positive for a nanotube to accept a fullerene by the suction force alone. If the acceptance energy is negative, then this represents the magnitude of initial kinetic energy needed by the fullerene in the form of the inbound initial velocity for it to be accepted into the nanotube.

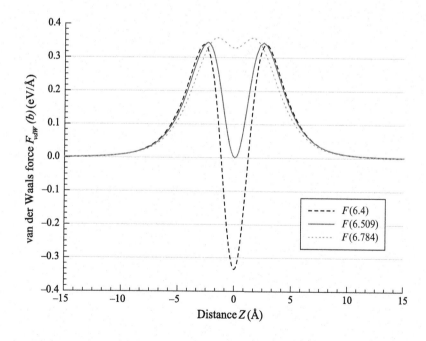

FIG. 5.5

Force in positive z direction experienced by a C_{60} fullerene due to van der Waals interaction with a semiinfinite carbon nanotube.

To calculate this acceptance energy, a change of variable of $Z = \sqrt{b^2 - a^2} \tan \psi$ is made so that $\lambda = (b^2 - a^2) \sec^2 \psi / a^2$ and $dZ = \sqrt{b^2 - a^2} \sec^2 \psi \, d\psi$ and the limits of the integration change to $-\pi/2$ and $\psi_o = \tan^{-1}(Z_o/\sqrt{b^2 - a^2})$, which yields

$$W_a = \frac{8\pi^2 \eta_f \eta_g b}{a^2 \sqrt{b^2 - a^2}} \left[A (J_2 + 2J_3) \right.$$

$$\left. - \frac{B}{5a^6} (5J_5 + 80J_6 + 336J_7 + 512J_8 + 256J_9) \right], \qquad (5.11)$$

where $J_n = a^{2n}(b^2 - a^2)^{-n} \int_{-\pi/2}^{\psi_o} \cos^{2n} \psi \, d\psi$. In this case a value of Z_o cannot be specified explicitly and must be determined numerically. Once determined, it can be substituted in the expression for W_a for any value of parameters where $b < b_o$. In Fig. 5.6, the acceptance energy is graphed for a C_{60} fullerene and a nanotube of radii in the range $6.1 < b < 6.5 \text{ Å}$, using the values of Z_o, as graphed in Fig. 5.7. The graphs show that $W_a = 0$ when $b \approx 6.338 \text{ Å}$ and nanotubes which are smaller than this will not accept C_{60} fullerenes by suction force alone. Therefore this model predicts that a $(10, 10)$ nanotube $(b = 6.784 \text{ Å})$ will accept a C_{60} fullerene from rest, but a $(9, 9)$ nanotube $(b = 6.106 \text{ Å})$ will not. As an example, for the $(8, 8)$ nanotube with a radius $b = 5.428 \text{ Å}$, the acceptance energy predicted by this model is $W_a = -252 \text{ eV}$. This equates to firing the C_{60} fullerene at the unlikely speed of greater than 8200 ms^{-1} for the molecule to be accepted into the nanotube interior, which would be very difficult to achieve experimentally. Also, when $b > b_o$, the force graph does not cross the axis, and therefore Z_o is not real. In this case the fullerene will always be accepted by the nanotube.

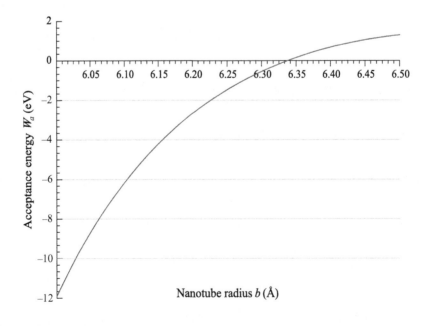

FIG. 5.6

Acceptance energy threshold for a C_{60} fullerene to be sucked into a carbon nanotube.

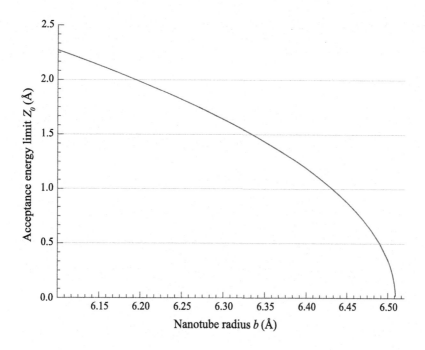

FIG. 5.7

Upper limit of integration Z_o used to determine the acceptance energy for a C$_{60}$ fullerene and carbon nanotube.

WORKED EXAMPLE 5.1

Use the algebraic package MAPLE to plot the relation between the acceptance condition and the distance Z for $b = 6.27$, 6.338, and 6.4 Å.

Solution

From Eq. (5.5), we have the energy for a C$_{60}$ molecule interacting with an atom. We need to perform the surface integral for a cylindrical nanotube to obtain the total potential of the system. In MAPLE, we can use `Int()` command to do the numerical integration. We start by calling the `restart` command to clear the internal memory, then we define all parameters which are used in the model and are given in Tables 1.5–1.7. In this case, we are interested in the interaction energy between a C$_{60}$ fullerene and a carbon nanotube, so we have used the attractive constant $A = 17.4$ eVÅ^6 and the repulsive constant $B = 29 \times 10^3$ eV Å^{12}.

```
> restart;
> A:=17.4:
> B:=29000:
> n_f:=0.3789:
> n_g:=0.3812:
> a:=3.55:
```

The distance between the centre of the C$_{60}$ fullerene and an atom is defined by δ, and it is assumed to be a function of the tube radius b and the distance Z.

```
> d := (b,Z) -> sqrt(b^2 +(Z-z)^2);
```

The interaction energy between the carbon atom and the C$_{60}$ fullerene is given by Eq. (5.5).

```
> E := (b,Z) -> n_f*Pi*a*(A*(1/(d(b,Z)+a)^4 - 1/(d(b,Z)-a)^4)/2
- B*(1/(d(b,Z)+a)^10 - 1/(d(b,Z)-a)^10)/5)/d(b,Z):
```

Using the continuum approximation, the surface integral for the tube is obtained by multiplying Eq. (5.5) by the radius b and the mean atomic surface density n_g. Furthermore, due to the symmetry of the system, we are interested in the total potential energy in the z-direction, and this energy is calculated by integrating once with respect to z from zero to infinity.

```
> Etot:= (b,Z) -> b*2*Pi*n_g*(Int(E(b,Z),z=0..infinity)):
```

Finally, we use the `plot` command to show the graphs.

```
> plot([Etot(6.27,Z),Etot(6.338,Z),Etot(6.4,Z)],Z=-15..15);
```

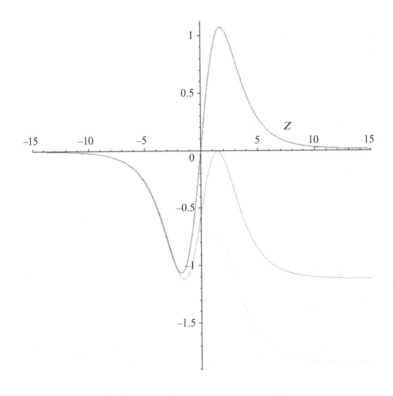

5.2.2 SUCTION ENERGY

The suction energy (W) for a fullerene can be determined by the same integral used to obtain the acceptance energy (W_a), Eq. (5.11), except with the upper limit changed from ψ_o to $\pi/2$. In this case the value for J_n becomes

$$J_n = \frac{a^{2n}}{(b^2 - a^2)^n} \frac{(2n-1)!!}{(2n)!!} \pi,$$

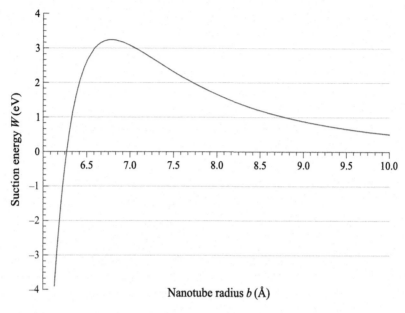

FIG. 5.8

Suction energy for a C$_{60}$ fullerene entering a carbon nanotube.

where !! represents the double factorial notation such that $(2n - 1)!! = (2n - 1)(2n - 3)\ldots 3 \cdot 1$ and $(2n)!! = (2n)(2n - 2)\ldots 4 \cdot 2$. Substitution and simplification gives

$$W = \frac{\pi^3 \eta_f \eta_g a^2 b}{(b^2 - a^2)^{5/2}} \left[A\,(3 + 5\mu) \right.$$
$$\left. - \frac{B\left(315 + 4620\mu + 18018\mu^2 + 25740\mu^3 + 12155\mu^4\right)}{160(b^2 - a^2)^3} \right], \tag{5.12}$$

where $\mu = a^2/(b^2 - a^2)$. In Fig. 5.8, the suction energy W is plotted for a C$_{60}$ fullerene entering a nanotube with radii in the range $6 < b < 10$ Å. The graph shows that W is positive whenever $b > 6.27$ Å and has a maximum value of $W = 3.243$ eV when $b = b_{max} = 6.783$ Å. It is worth noting that a $(10, 10)$ carbon nanotube with $b \approx 6.784$ Å is almost exactly the optimal size to maximise W and therefore have a C$_{60}$ fullerene accelerate to a maximum velocity upon entering the nanotube.

Since the suction energy can be converted directly to the kinetic energy, assuming no energy loss, we have $W = (Mv^2)/2$, where M is the mass of fullerene and v is its velocity. Thus, we find $v = \sqrt{2W/M}$. In Table 5.1, we provide the values of the suction energy and the velocity for various C$_{60}$−nanotube oscillators. From the table, we can see that the C$_{60}$−$(10, 10)$ nanotube oscillator has maximum velocity, which is due to the maximum suction energy.

Table 5.1 Suction Energy and Velocity for Various Oscillator Configurations

Oscillator Configuration	Tube Radius b (Å)	Energy W (eV)	Velocity v (m/s)
C_{60}–(10, 10)	6.784	3.243	932
C_{60}–(11, 11)	7.463	2.379	798
C_{60}–(12, 12)	8.141	1.512	636
C_{60}–(13, 13)	8.820	0.982	513

5.3 DOUBLE-WALLED CARBON NANOTUBES

As mentioned in Chapters 1 and 4, double-walled carbon nanotubes consist of two layers of graphite rolled up to form a cylindrical tube. Suction forces for double-walled carbon nanotubes have been examined by Rivera et al. (2005) using molecular dynamics simulations. Their studies investigate a $(7, 0)$ and $(9, 9)$ double-walled carbon nanotube system for specific nanotube lengths. In this section, the acceptance condition and the suction energy are determined for the double-walled carbon nanotube system and unlike molecular dynamics simulations, employs a mathematical model applying to any combination of length and diameter.

5.3.1 ACCEPTANCE CONDITIONS

Here, the energy of a carbon nanotube being sucked symmetrically into another carbon nanotube is examined. For convenience, Tube 1 refers to the carbon nanotube entering the open end of the semiinfinite tube, Tube 2, as shown in Fig. 5.9, and the origin of the coordinate system is assumed to be located at the left opening of Tube 2. We assume that Tube 1 is of radius a and length $2L_1$ and its centre is located at $(0, Z)$, which might be inside or outside Tube 2. Furthermore, Tube 2 of radius b is assumed to be semiinfinite in length. In an axially symmetric cylindrical polar coordinate system (r, θ, z), the parametric equations for Tube 1 and Tube 2 are given by $(a\cos\theta_1, a\sin\theta_1, z_1)$ and

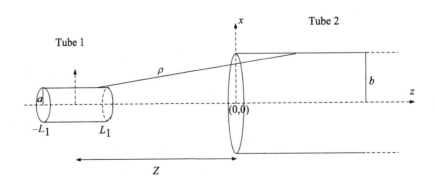

FIG. 5.9

Geometry of a single-walled carbon nanotube symmetrically entering another single-walled carbon nanotube.

$(b \cos \theta_2, b \sin \theta_2, z_2)$, respectively. In this case, the distance ρ between two typical surface elements on Tube 1 and Tube 2 is given by

$$\rho^2 = (b \cos \theta_2 - a \cos \theta_1)^2 + (b \sin \theta_2 - a \sin \theta_1)^2 + (z_2 - z_1)^2,$$

which may be simplified to yield

$$\rho^2 = a^2 + b^2 - 2ab \cos(\theta_1 - \theta_2) + (z_2 - z_1)^2. \tag{5.13}$$

Using the Lennard-Jones potential together with the continuum approximation, the total potential energy can be written as

$$E^{tot} = ab\eta_g^2 \int_0^{2\pi} \int_0^{2\pi} (-AI_6 + BI_{12})d\theta_1 d\theta_2, \tag{5.14}$$

where η_g represents the mean surface density of the carbon nanotube assumed to be the same for both tubes. In this case, the integrals I_n $(n = 6, 12)$ are defined as follows

$$I_n = \int_0^\infty \int_{Z-L_1}^{Z+L_1} \frac{dz_1 dz_2}{\rho^n} = \int_0^\infty \int_{Z-L_1}^{Z+L_1} \frac{dz_1 dz_2}{[\lambda^2 + (z_2 - z_1)^2]^{n/2}},$$

where λ^2 denotes $a^2 + b^2 - 2ab \cos(\theta_1 - \theta_2)$. On letting $x = z_2 - z_1$, the integral I_6 becomes

$$I_6 = \int_{Z-L_1}^{Z+L_1} \int_{-z_1}^\infty \frac{dx\, dz_1}{(\lambda^2 + x^2)^3} = \frac{1}{\lambda^5} \int_{Z-L_1}^{Z+L_1} \int_{-\tan^{-1}(z_1/\lambda)}^{\pi/2} \cos^4 \phi\, d\phi\, dz_1,$$

where on substituting $x = \lambda \tan \phi$, we have

$$I_6 = \frac{1}{\lambda^5} \int_{Z-L_1}^{Z+L_1} \left(\frac{3\pi}{16} + \frac{3}{8} \tan^{-1}(z_1/\lambda) + \frac{3}{8} \frac{\lambda z_1}{(\lambda^2 + z_1^2)} + \frac{1}{4} \frac{\lambda^3 z_1}{(\lambda^2 + z_1^2)^2} \right) dz_1.$$

Using the substitution $x = z_1/\lambda$, the above equation for I_6 becomes

$$I_6 = \frac{1}{\lambda^4} \left[\frac{3\pi L_1}{8} \frac{1}{\lambda} + \frac{3(Z + L_1)}{8\lambda} \tan^{-1} \left(\frac{Z + L_1}{\lambda} \right) - \frac{\lambda^2}{8[\lambda^2 + (Z + L_1)^2]} \right.$$

$$\left. - \frac{3(Z - L_1)}{8\lambda} \tan^{-1} \left(\frac{Z - L_1}{\lambda} \right) + \frac{\lambda^2}{8[\lambda^2 + (Z - L_1)^2]} \right]. \tag{5.15}$$

Similarly, I_{12} is given by

$$I_{12} = \frac{1}{\lambda^{10}} \left[\frac{63\pi L_1}{256} \frac{1}{\lambda} + \frac{63(Z + L_1)}{256\lambda} \tan^{-1} \left(\frac{Z + L_1}{\lambda} \right) - \frac{21\lambda^2}{256[\lambda^2 + (Z + L_1)^2]} \right.$$

$$- \frac{21\lambda^4}{640[\lambda^2 + (Z + L_1)^2]^2} - \frac{3\lambda^6}{160[\lambda^2 + (Z + L_1)^2]^3} - \frac{\lambda^8}{80[\lambda^2 + (Z + L_1)^2]^4}$$

$$- \frac{63(Z - L_1)}{256\lambda} \tan^{-1} \left(\frac{Z - L_1}{\lambda} \right) + \frac{21\lambda^2}{256[\lambda^2 + (Z - L_1)^2]}$$

$$\left. + \frac{21\lambda^4}{640[\lambda^2 + (Z - L_1)^2]^2} + \frac{3\lambda^6}{160[\lambda^2 + (Z - L_1)^2]^3} + \frac{\lambda^8}{80[\lambda^2 + (Z - L_1)^2]^4} \right]. \tag{5.16}$$

Thus from Eqs (5.15) and (5.16), there are three types of integrals which need to be determined, and they are given by

$$J_n^* = \int_0^{2\pi} \int_0^{2\pi} \frac{d\theta_1 d\theta_2}{\lambda^n},$$

$$K_n^* = \int_0^{2\pi} \int_0^{2\pi} \frac{d\theta_1 d\theta_2}{\lambda^m (\lambda^2 + P_j^2)^n},$$

$$L_n^* = \int_0^{2\pi} \int_0^{2\pi} \frac{1}{\lambda^n} \tan^{-1}\left(\frac{P_j}{\lambda}\right) d\theta_1 d\theta_2, \tag{5.17}$$

where m and n are certain positive integers and P_j ($j = 1, 2$) is the abbreviation used for $P_1 = Z + L_1$ and $P_2 = Z - L_1$. As shown in Worked Examples 5.2, 5.3, and 5.4, Eq. (5.17) can be integrated to yield

$$J_n^* = \frac{4\pi^2}{(a+b)^{2n}} F\left(\frac{n}{2}, \frac{1}{2}; 1; \frac{4ab}{(a+b)^2}\right),$$

$$K_n^* = \frac{4\pi^2}{(a+b)^m[(a+b)^2 + P_j^2]^n} \sum_{i=0}^{\infty} \frac{(1/2)_i (m/2)_i}{(i!)^2} \left(\frac{4ab}{(a+b)^2}\right)^i$$

$$\times F\left(\frac{1}{2} + i, n; 1 + i; \frac{4ab}{(a+b)^2 + P_j^2}\right),$$

$$L_n^* = 4\pi^2 \sum_{k=0}^{\infty} \sum_{i=0}^{\infty} \frac{P_j^{2k+1} (2k)!}{2^{2k}(k!)^2 (2k+1)(a+b)^n[(a+b)^2 + P_j^2]^{k+1/2}}$$

$$\times \frac{(1/2)_i (n/2)_i}{(i!)^2} \left(\frac{4ab}{(a+b)^2}\right)^i F\left(\frac{1}{2} + i, k + \frac{1}{2}; 1 + i; \frac{4ab}{(a+b)^2 + P_j^2}\right),$$

where $F(a, b; c; z)$ denotes the usual hypergeometric function. Although complicated, numerical values for these integrals may be readily evaluated using the algebraic computer package MAPLE, or a similar mathematical software package.

WORKED EXAMPLE 5.2

Derive the integral J_n^*, as defined in Eq. (5.17)$_1$.

Solution

Consider the integral

$$J_n^* = \int_0^{2\pi} \int_0^{2\pi} \frac{d\theta_1 d\theta_2}{\lambda^n}, \tag{5.18}$$

where $\lambda^2 = a^2 + b^2 - 2ab \cos(\theta_1 - \theta_2)$. As the integrand is a symmetric function of $\theta_1 - \theta_2$, the intermediate integral J_n^{**} is defined by

$$J_n^{**} = \int_0^{2\pi} \frac{d\theta_1}{\{\alpha + \beta \sin^2[(\theta_1 - \theta_2)/2]\}^{n/2}},$$

where $\alpha = (a - b)^2$ and $\beta = 4ab$ can be shown by differentiation with respect to θ_2 to be independent of θ_2, namely

$$\frac{dJ_n^{**}}{d\theta_2} = \int_0^{2\pi} -\frac{\partial}{\partial\theta_1}\left(\frac{1}{\{\alpha + \beta \sin^2[(\theta_1 - \theta_2)/2]\}^{n/2}}\right)d\theta_1 = 0.$$

Thus θ_2 is set to zero and one can trivially perform the θ_2 integration so that Eq. (5.18) becomes

$$J_n^* = 8\pi \int_0^{\pi/2} \frac{dx}{\lambda^n},$$

where $\lambda^2 = (a - b)^2 + 4ab \sin^2 x$, and the integral J_n may be defined by

$$J_n = \int_0^{\pi/2} \frac{dx}{\lambda^n} = \int_0^{\pi/2} \frac{dx}{(\alpha + \beta \sin^2 x)^{n/2}}.$$

Making the substitution $t = \cot x$ to obtain

$$J_n = \int_0^\infty \frac{(1 + t^2)^{n/2-1}}{(\beta + \alpha + \alpha t^2)^{n/2}}dt = \frac{1}{(\alpha + \beta)^{n/2}}\int_0^\infty \frac{(1 + t^2)^{n/2-1}}{(1 + \gamma t^2)^{n/2}}dt,$$

where $\gamma = \alpha/(\alpha + \beta)$, and writing this integral in the form

$$J_n = \frac{1}{(\alpha + \beta)^{n/2}}\int_0^\infty \frac{1}{[1 - (1 - \gamma)t^2/(1 + t^2)]^{n/2}}\frac{dt}{(1 + t^2)},$$

leads to the substitution given by Eq. (4.29), and to remind the reader

$$z = \frac{t}{(1 + t^2)^{1/2}}, \quad t = \frac{z}{(1 - z^2)^{1/2}}, \quad dt = \frac{dz}{(1 - z^2)^{3/2}},$$

and the substitution $u = z^2$ gives

$$J_n = \frac{1}{(\alpha + \beta)^{n/2}}\int_0^1 \frac{dz}{[1 - (1 - \gamma)z^2]^{n/2}(1 - z^2)^{1/2}}$$

$$= \frac{1}{2(\alpha + \beta)^{n/2}}\int_0^1 \frac{u^{-1/2}(1 - u)^{-1/2}}{[1 - (1 - \gamma)u]^{n/2}}du.$$

From Gradshteyn and Ryzhik (2000) (p. 995, eq. 9.111) it may be deduced

$$J_n = \frac{\pi}{2(a + b)^n}F\left(\frac{n}{2}, \frac{1}{2}; 1; \frac{4ab}{(a + b)^2}\right),$$

where $F(a, b; c; z)$ is a hypergeometric function. From Erdélyi et al. (1953), because two of the numbers $\pm(1 - c)$, $\pm(a - b)$, $\pm(a + b - c)$ are equal to each other, it can be shown that this result admits a quadratic transformation, which leads to a Legendre function. Using the transformation

$$F(a, b; 2b; 4z/(1 + z)^2) = (1 + z)^{2a}F(a, a + 1/2 - b; b + 1/2; z^2),$$

gives

$$J_n = \frac{\pi(1 + \xi)^n}{2(a + b)^n}F(n/2, n/2; 1; \xi^2),$$

where $\xi = b/a$. Finally, we obtain

$$J_n^* = \frac{4\pi^2}{(a + b)^2}F\left(\frac{n}{2}, \frac{1}{2}; 1; \frac{4ab}{(a + b)^2}\right).$$

Alternatively, using the definitions from Gradshteyn and Ryzhik (2000) (pp. 960, 998)

$$P_\nu^\mu(z) = \frac{1}{\Gamma(1-\mu)} \left(\frac{z-1}{z+1}\right)^{-\mu/2} \left(\frac{z+1}{2}\right)^\nu F\left(-\nu, -\nu - \mu; 1 - \mu; \frac{z-1}{z+1}\right),$$

and

$$F(a, b; c, z) = (1 - z)^{c-a-b} F(c - a, c - b; c; z),$$

where $P_\nu^\mu(z)$ is a Legendre function of the first kind. In this case, μ is zero, and the integral in terms of the Legendre function is obtained and given by

$$J_n = \frac{\pi}{2(a^2 - b^2)^{n/2}} P_{n/2-1}\left(\frac{a^2 + b^2}{a^2 - b^2}\right).$$

WORKED EXAMPLE 5.3

Derive the integral K_n^*, as defined in Eq. (5.17)$_2$.

As λ^2 is an even function of $\theta_1 - \theta_2$, as mentioned in Worked Example 5.2, either intermediate integral is independent of the other variables. Therefore this second variable is assigned a zero value. In this event, one integration may be trivially performed, and it may be deduced

$$K_n^* = 8\pi \int_0^{\pi/2} \frac{dx}{\lambda^m(\lambda^2 + P_j^2)^n},$$

where $\lambda^2 = (a - b)^2 + 4ab \sin^2 x$. Instead of considering the above equation, only the integral K_n is considered, which is given by

$$K_n = \int_0^{\pi/2} \frac{dx}{\lambda^m(\lambda^2 + P_j^2)^n}. \tag{5.19}$$

Letting $\mu = (a - b)^2$, $\sigma = (a - b)^2 + P_j^2$, and $\nu = 4ab$, Eq. (5.19) becomes

$$K_n = \int_0^{\pi/2} \frac{dx}{(\mu + \nu \sin^2 x)^{m/2}(\sigma + \nu \sin^2 x)^n}.$$

Making the substitution $t = \cot x$ to obtain

$$K_n = \int_0^\infty \frac{(1 + t^2)^{n + \frac{m}{2} - 1}}{(\nu + \mu + \mu t^2)^{m/2}(\nu + \sigma + \sigma t^2)^n} dt$$

$$= \frac{1}{(\mu + \nu)^{m/2}(\sigma + \nu)^n} \int_0^\infty \frac{(1 + t^2)^{n + \frac{m}{2} - 1}}{(1 + \beta t^2)^{m/2}(1 + \gamma t^2)^n} dt,$$

where $\beta = \mu/(\mu + \nu)$ and $\gamma = \sigma/(\sigma + \nu)$. By writing this integral in the form

$$K_n = \frac{1}{(\mu + \nu)^{m/2}(\sigma + \nu)^n}$$

$$\times \int_0^\infty \frac{1}{[1 - (1 - \beta)t^2/(1 + t^2)]^{m/2}[1 - (1 - \gamma)t^2/(1 + t^2)]^n} \frac{dt}{(1 + t^2)},$$

and making the substitution

$$z = \frac{t}{(1+t^2)^{1/2}}, \quad t = \frac{z}{(1-z^2)^{1/2}}, \quad dt = \frac{dz}{(1-z^2)^{3/2}},$$

and the substitution $u = z^2$, we obtain

$$K_n = \frac{1}{2(\mu + v)^{m/2}(\sigma + v)^n} \int_0^1 \frac{u^{-1/2}(1-u)^{-1/2}}{[1-(1-\beta)u]^{m/2}[1-(1-\gamma)u]^n} du.$$

Note that $\mu + v = (b+a)^2$, $\sigma + v = (b+a)^2 + P_j^2$, $\beta = (b-a)^2/(b+a)^2$, and $\gamma = [(b-a)^2 + P_j^2]/[(b+a)^2 + P_j^2]$. According to Bailey (1972) (p. 73), the definition of an Appell hypergeometric function of two variables and of the first kind is defined by

$$F_1\left(\alpha : \beta, \beta'; \gamma; x, y\right) = \sum_{n=0}^{\infty} \sum_{m=0}^{\infty} \frac{(\alpha)_{m+n}(\beta)_m(\beta')_n}{m!n!(\gamma)_{m+n}} x^m y^n.$$

Also from Bailey (1972), the expression for the function F_1 in terms of a definite integral and a series involving the ordinary hypergeometric function (pp. 77 and 79), are

$$\frac{\Gamma(\alpha)\Gamma(\gamma - \alpha)}{\Gamma(\gamma)} F_1\left(\alpha; \beta, \beta'; \gamma; x, y\right) = \int_0^1 \frac{u^{\alpha-1}(1-u)^{\gamma-\alpha-1}}{(1-ux)^\beta(1-uy)^{\beta'}} du,$$

(5.20)

$$F_1\left(\alpha; \beta, \beta'; \gamma; x, y\right) = \sum_{i=0}^{\infty} \frac{(\alpha)_i(\beta)_i}{i!(\gamma)_i} F\left(\alpha + i, \beta'; \gamma + i; y\right) x^i,$$

so that K_n becomes

$$K_n = \frac{\pi}{2(a+b)^m[(a+b)^2 + P_j^2]^n} \sum_{i=0}^{\infty} \frac{(1/2)_i(m/2)_i}{(i!)^2}$$

$$\times F\left(\frac{1}{2} + i, n; 1 + i; \frac{4ab}{(a+b)^2 + P_j^2}\right) \left[\frac{4ab}{(a+b)^2}\right]^i,$$

where $F(a, b; c; z)$ denotes the usual hypergeometric function.

WORKED EXAMPLE 5.4

Derive the integral L_n^*, as defined in Eq. (5.17)$_3$.

Similar to Worked Example 5.3, the integral L_n^* can be written as

$$L_n^* = 8\pi \int_0^{\pi/2} \frac{1}{\lambda^n} \tan^{-1}\left(\frac{P_j}{\lambda}\right) dx,$$

where $\lambda^2 = (a-b)^2 + 4ab \sin^2 x$. For convenience, L_n is defined by

$$L_n = \int_0^{\pi/2} \frac{1}{\lambda^n} \tan^{-1}\left(\frac{P_j}{\lambda}\right) dx.$$

As $(P_j/\lambda)^2 < \infty$, from Gradshteyn and Ryzhik (2000) (p. 59, eq. 1.644.1), it is obtained

$$\tan^{-1}\left(\frac{P_j}{\lambda}\right) = \frac{P_j}{\sqrt{\lambda^2 + P_j^2}} \sum_{k=0}^{\infty} \frac{(2k)!}{2^{2k}(k!)^2(2k+1)} \left(\frac{P_j^2}{\lambda^2 + P_j^2}\right)^k,$$

and thus

$$L_n = \sum_{k=0}^{\infty} \frac{P_j^{2k+1}(2k)!}{2^{2k}(k!)^2(2k+1)} \int_0^{\pi/2} \frac{1}{\lambda^n(\lambda^2 + P_j^2)^{k+1/2}} dx.$$

From the result for K_n in Worked Example 5.3, it may be deduced

$$L_n = \frac{\pi}{2} \sum_{k=0}^{\infty} \frac{P_j^{2k+1}(2k)!}{2^{2k}(k!)^2(2k+1)} \cdot \frac{1}{(\mu + \nu)^{n/2}(\sigma + \nu)^{k+1/2}}$$

$$\times F_1\left(\frac{1}{2}; \frac{n}{2}, k + \frac{1}{2}; 1; 1 - \beta, 1 - \gamma\right),$$

and using the reduction of Appell's hypergeometric functions (5.20), the formula for L_n is given by

$$L_n = \frac{\pi}{2} \sum_{k=0}^{\infty} \sum_{i=0}^{\infty} \frac{P_j^{2k+1}(2k)!}{2^{2k}(k!)^2(2k+1)(a+b)^n[(a+b)^2 + P_j^2]^{k+1/2}} \frac{(1/2)_i(n/2)_i}{(i!)^2}$$

$$\times F\left(\frac{1}{2} + i, k + \frac{1}{2}; 1 + i; \frac{4ab}{(a+b)^2 + P_j^2}\right)\left[\frac{4ab}{(a+b)^2}\right]^i.$$

We note that Colavecchia et al. (2001) examine in some detail the numerical evaluation of the usual hypergeometric and Appell hypergeometric functions.

From the symmetry of the problem, only the force in the axial direction (z-direction) is considered, and the resultant axial force is obtained by differentiating the total energy with respect to the axial direction Z. Noting that due to the complexity of the expression, obtaining an analytical expression for F_Z^{tot} is not a simple task. Using the algebraic computer package MAPLE, together with the constants given in Tables 1.5–1.7, the numerical solutions for the total potential energy E^{tot} are illustrated in Fig. 5.10, where the inner tube is assumed to be a $(5, 5)$ carbon nanotube ($a = 3.392$ Å) and its length is $2L_1 = 100$ Å. It can be seen in Fig. 5.10 that the acceptance of an inner carbon nanotube into an outer carbon nanotube strongly depends on the outer radius b. From Fig. 5.10 for $b = 6.310$ Å, the inner tube will not be accepted into the outer tube due to the high energy barrier both outside and inside the tube. By increasing b to 6.313 Å, we see that there is an energy well which will suck an inner tube inside. However, as the other end of the nanotube approaches the outer tube edge, the inner tube encounters an energy barrier which prevents the inner tube from becoming completely taken up by the outer tube. For $b = 6.317$ Å, the inner tube is accepted into the outer tube and experiences a small repulsion as the far end of the inner tube crosses over into the outer tube.

Worked Example 5.5 illustrates the axial van der Waals force experienced by a $(5, 5)$ nanotube upon entering nanotubes with radii $b = 6.310$, 6.313 and 6.317 Å. The area under the graph represents the work done by the van der Waals force. For the inner tube to be sucked into the outer tube, the sum of the work which is obtained by moving the inner tube from $Z = -\infty$ to Z_o, which is the acceptance condition as defined by Eq. (5.1), needs to be greater than zero.

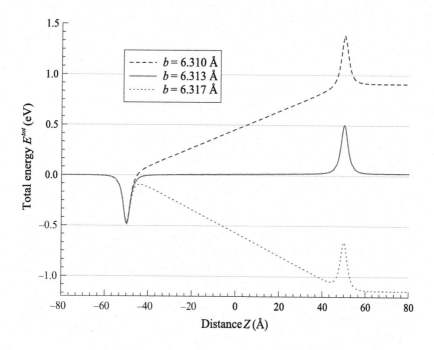

FIG. 5.10

Total potential energy for an inner $(5, 5)$ carbon nanotube interaction with an outer semiinfinite carbon nanotube $b = 6.310, 6.313,$ and $6.317 Å$.

WORKED EXAMPLE 5.5

Use the algebraic package MAPLE to plot the relation between the axial van der Waals force and the distance Z for a $(5, 5)$ carbon nanotube of the length $2L_1 = 100 Å$ entering into an outer nanotube with a radius $b = 6.310, 6.313,$ and $6.317 Å$.

Solution

We start by calling the `restart` command to clear internal memory, then we define all parameters which are used in the model and are given in Tables 1.5–1.7. In this case, we use η_g for the mean atomic surface density for carbon nanotubes, which correspond to the attractive constant $A = 15.2 \, eVÅ^6$, and the repulsive constant $B = 24.1 \times 10^3 \, eVÅ^{12}$.

```
>  restart;
>  A:=15.2:
>  B:=24100:
>  eta:= 0.3812:
```

We define λ as a function of the inner tube radius a and an outer tube radius b.

```
>  lambda:= (a,b)-> sqrt((a-b)^2 + 4*a*b*sin(x)^2):
```

We input Eq. (5.15) for I_6, which is a function of an inner tube radius a, an outer tube radius b, a length of an inner tube L_1 and a distance Z.

```
>  I6:= (a,b,L1,Z) -> ( 3*Pi*L1/(8*lambda(a,b))+
3*(L1+Z)*arctan((L1+Z)/lambda(a,b))/(8*lambda(a,b))-
lambda(a,b)^2/(8*(lambda(a,b)^2 + (L1+Z)^2)) -
3*(-L1+Z)*arctan((-L1+Z)/lambda(a,b))/(8*lambda(a,b)) +
lambda(a,b)^2/(8*(lambda(a,b)^2 + (-L1+Z)^2)))/lambda(a,b)^4:
```

Now, we input Eq. (5.16) for I_{12}, which is also a function of an inner tube radius a, an outer tube radius b, a length of an inner tube L_1 and a distance Z.

```
> I12:= (a,b,L1,Z) -> ( 63*Pi*L1/(256*lambda(a,b)) +
63*(L1+Z)*arctan((L1+Z)/lambda(a,b))/(256*lambda(a,b))   -
21*lambda(a,b)^2/(256*(lambda(a,b)^2 + (L1+Z)^2)) -
21*lambda(a,b)^4/(640*(lambda(a,b)^2+(L1+Z)^2)^2) -
3*lambda(a,b)^6/(160*(lambda(a,b)^2+(L1+Z)^2)^3) -
lambda(a,b)^8/(80*(lambda(a,b)^2+(L1+Z)^2)^4) -
63*(-L1+Z)*arctan((-L1+Z)/lambda(a,b))/(256*lambda(a,b)) +
21*lambda(a,b)^2/(256*(lambda(a,b)^2+(-L1+Z)^2)) +
21*lambda(a,b)^4/(640*(lambda(a,b)^2+(-L1+Z)^2)^2) +
3*lambda(a,b)^6/(160*(lambda(a,b)^2+(-L1+Z)^2)^3) +
lambda(a,b)^8/(80*(lambda(a,b)^2+(-L1+Z)^2)^4))/lambda(a,b)^10:
```

We need to do double integrals of I_6 and I_{12} from 0 to 2π and 0 to 2π. However, due to the symmetry of the system, we may multiply I_6 and I_{12} by 8 and integrate both equations from 0 to $\pi/2$.

```
> rho6:= (a,b,L1,Z) -> Int(I6(a,b,L1,Z),x=0..Pi/2):
> rho12:= (a,b,L1,Z) -> Int(I12(a,b,L1,Z),x=0..Pi/2):
> Energy:= (a,b,L1,Z) -> 8*Pi*a*b*eta^2*(-A*rho6(a,b,L1,Z)+
B*rho12(a,b,L1,Z)):
```

The axial van der Waals force is obtained by differentiating the total potential energy with respect to Z.

```
> Force:= (a,b,L1,Z)-> -diff(Energy(a,b,L1,Z),Z):
```

We use plot() command to plot all three graphs.

```
> plot([Force(3.392,6.31,50,Z),Force(3.392,6.313,50,Z),
Force(3.392,6.317,50,Z)],Z=-80..80);
```

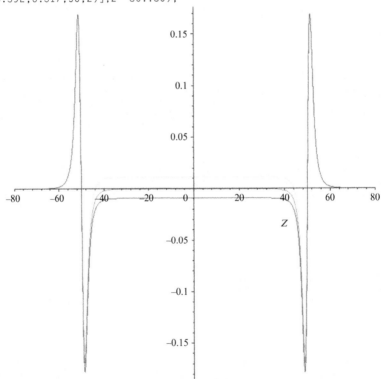

5.3.2 **SUCTION ENERGY**

In this section, we consider the suction energy, which is the total work performed by van der Waals interactions on the inner tube upon entering the outer tube, and it can be represented mathematically by Eq. (5.3). If we define W as the suction energy per unit length for the interaction between two carbon nanotubes, we may write

$$W = -E_{tt}, \tag{5.21}$$

where E_{tt} is the interaction energy per unit length for two carbon nanotubes, as defined by Eq. (4.35).

In Fig. 5.11, the suction energy for a $(5, 5)$ carbon nanotube entering an outer nanotube with radii in the range $6.2 < b < 8.2\,\text{Å}$ is depicted. From the figure, W is positive when $b > 6.313\,\text{Å}$, and it has a maximum value when $b = 6.822\,\text{Å}$, for which the difference in radii of the inner and outer tubes is $3.34\,\text{Å}$. This value is the interspacing distance between two graphene sheets.

5.4 **NANOTUBE BUNDLE**

As mentioned previously, a nanotube bundle is defined as a closely packed array of aligned tubes in a triangular lattice. For the purposes of this study, we take the bundle axis to be collinear with the z-axis, and the bundle radius to be denoted by R, which is the perpendicular distance from the central axis to the axis of each constituent tube, as shown in Figs 4.16 and 5.12. Each constituent tube is assumed to

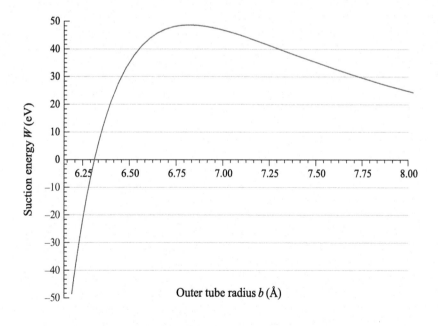

FIG. 5.11

Suction energy for an inner tube $(5, 5)$ entering an outer tube.

be of the same radius, which we denote by r, equivalent to the tube radius b in previous sections and of half-length L. The number of constituent tubes is assumed to be N. Therefore the ith tube in the nanotube bundle has a surface in Cartesian coordinates (x, y, z) given by Eq. (4.31), and to remind the reader

$$\left(R\cos\left(\frac{2\pi(i-1)}{N}\right) + r\cos\theta_i, R\sin\left(\frac{2\pi(i-1)}{N}\right) + r\sin\theta_i, z_i\right),$$

where $i \in \{1 \ldots N\}$, $0 \leqslant \theta_i \leqslant 2\pi$, and $-L \leqslant z_i \leqslant L$. However, it is convenient for some of the analysis to assume a semiinfinite tube, with $0 \leqslant z_i < \infty$. The numerical values of the various constants used in this model are shown in Tables 1.5–1.7.

5.4.1 ACCEPTANCE CONDITION

Now we will determine a condition in which a carbon nanotube, which is initially outside a nanotube bundle, will be accepted into the middle of the bundle. For bundles of sixfold symmetry, $N = 6$, the configuration of the problem is as shown in Fig. 5.12. We assume that the carbon nanotubes forming the bundle are of semiinfinite length and that the central nanotube is of length $2L_0$ and is centred on the z-axis at a position Z, which can be inside or outside the bundle. The Cartesian coordinates of a typical point on the middle tube is $(r_0\cos\theta_0, r_0\sin\theta_0, z_0 + Z)$, where r_0 is the tube radius, $0 \leqslant \theta_0 < 2\pi$, and $-L_0 \leqslant z_0 \leqslant L_0$. The Cartesian coordinates (x, y, z) of the nanotubes in the bundle are given in Eq. (4.31), where in this case $0 \leqslant z_i < \infty$, as shown in Fig. 5.12.

Due to the assumed symmetry of the problem, to obtain the potential energy we need only consider the interaction between the middle tube and one of the carbon nanotubes in the bundle. As shown in Fig. 5.13, we look at the tube with coordinates $(r_0\cos\theta_0, r_0\sin\theta_0, z_0 + Z)$, which is in the middle of the bundle, and the tube $i = 1$, which has coordinates $(r\cos\theta_1 + R, r\sin\theta_1, z_1)$. If E denotes the energy of the interaction between the centre tube and the first tube, then the total interaction energy of the system can be obtained by $E^{tot} = NE$, noting here that N is the number of tubes in the bundle, which is symmetrically located around the centre tube.

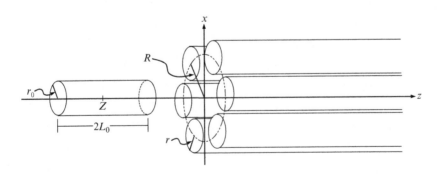

FIG. 5.12

Geometry of a single nanotube entering a bundle.

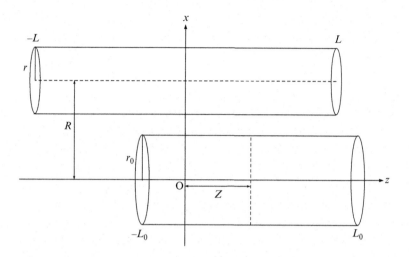

FIG. 5.13

Geometry of two interacting tubes.

Using the Lennard-Jones potential and the continuum approach, we obtain the interaction energy E as

$$E = rr_0 \eta_g^2 \int_0^{2\pi} \int_0^{2\pi} \left[\int_0^{\infty} \int_{-L_0}^{L_0} \left(-\frac{A}{\rho^6} + \frac{B}{\rho^{12}} \right) dz_0 \, dz_1 \right] d\theta_0 \, d\theta_1, \tag{5.22}$$

where η_g is the mean atomic density on a nanotube and ρ denotes the distance between two typical elements on each nanotube, which is given by

$$\rho^2 = (r\cos\theta_1 + R - r_0\cos\theta_0)^2 + (r\sin\theta_1 - r_0\sin\theta_0)^2 + (z_1 - z_0 - Z)^2. \tag{5.23}$$

Here, we write Eq. (5.22) as

$$E = rr_0 \eta_g^2 \int_0^{2\pi} \int_0^{2\pi} (-AI_3 + BI_6) \, d\theta_0 \, d\theta_1, \tag{5.24}$$

where the integrals I_n $(n = 3, 6)$ are defined by

$$I_n = \int_0^{\infty} \int_{Z-L_0}^{Z+L_0} \frac{dz_0^* dz_1}{[\lambda^2 + (z_1 - z_0^*)^2]^{n/2}}, \tag{5.25}$$

which is identical to Eq. (5.15), where in this case, $z_0^* = z_0 + Z$ and $\lambda^2 = (r\cos\theta_1 + R - r_0\cos\theta_0)^2 + (r\sin\theta_1 - r_0\sin\theta_0)^2$. We note that formally the case $R = 0$ corresponds to the suction of one nanotube into another concentric nanotube, and that in this case the above integrals can be evaluated analytically in terms of hypergeometric functions. Here, where $R > 0$, these integrals are determined numerically. We also note that on looking at an offset nanotube inside an infinite carbon nanotube, an integral of the form $\int_0^{2\pi} \int_0^{2\pi} \lambda^{-m} d\theta_0 d\theta_1$ arises.

In Figs 5.14 and 5.15 we plot the total potential energy E^{tot} and the van der Waals force F_{vdW} for a $(5,5)$ carbon nanotube of length $2L_0 = 100\,\text{Å}$ interacting with a sixfold symmetry $(5,5)$ carbon nanotube bundle, for which the tubes in the bundle are assumed to be semiinfinite in length. It can be seen from Figs 5.14 and 5.15 that the acceptance of a single carbon nanotube into a bundle strongly depends on the bundle radius R. From Fig. 5.14 for $R = 8.465\,\text{Å}$, the single nanotube will not be accepted into the bundle due to the high energy barrier both outside and inside the bundle. By increasing R to $8.485\,\text{Å}$, we see that there is an energy well which will suck a single nanotube inside. However, as the other end of the nanotube approaches the bundle edge, the nanotube encounters an energy barrier which prevents the nanotube from becoming completely taken up by the nanotube bundle. For $R = 8.500\,\text{Å}$, the single nanotube is accepted into the bundle and experiences a small repulsion as the far end of the nanotube crosses over into the bundle. As obtained from Fig. 5.16, the optimal bundle radius R for a single $(5,5)$ carbon nanotube interacting with a bundle of sixfold symmetry is such that $R = 9.9283\,\text{Å}$. From Worked Example 5.6, we observe that the tube is accepted into the bundle without any repulsion effects at the tube extremities. We note that the area under the graphs in Fig. 5.15 represents the work (energy) done by the van der Waals forces. For the single nanotube to be completely accepted into the bundle, this area must be greater than zero.

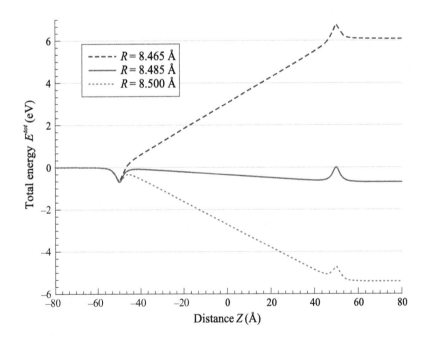

FIG. 5.14

Total energy of $(5,5)$ nanotube of length $2L_0 = 100\,\text{Å}$ entering into middle of sixfold symmetry bundle of $(5,5)$ nanotubes semiinfinite in length and of varying bundle radius R.

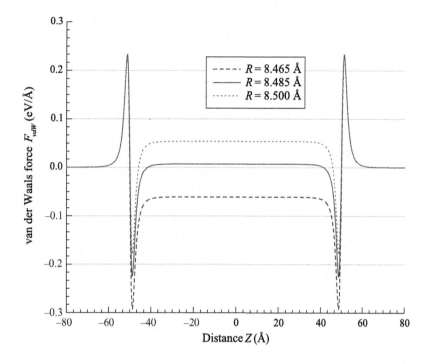

FIG. 5.15

van der Waals force of (5, 5) nanotube of length $2L_0 = 100$ Å entering into middle of sixfold symmetry bundle of (5, 5) nanotubes semiinfinite in length and of varying bundle radius R.

WORKED EXAMPLE 5.6

On assuming that $R = 9.9283$ Å, plot the relationship between the total energy and the distance Z and the van der Waals force and the distance Z of a (5, 5) nanotube of length $2L_0 = 100$ Å entering into the middle of a sixfold symmetry bundle of (5, 5) nanotubes which are semiinfinite in length.

Solution

Using the algebraic package MAPLE, we start by defining the constants for the bundle system.

```
> restart:
> A :=15.2:
> B :=24100:
> eta :=0.3812:
```

We define λ as a function of a single nanotube radius r_0, a constituent tube radius r and a bundle radius R, which corresponds to the distance between two typical elements on each nanotube.

```
> lambda := (r0,r,R) -> sqrt((r*cos(theta2) + R -
r0*cos(theta1))^2 + (r*sin(theta2) - r0*cos(theta1))^2):
```

Functions I_6 and I_{12} are assumed to be functions of r_0, r, R, a nanotube length L_0 and a distance Z, and they are precisely the same as given in Worked Example 5.5. The total potential energy may be deduced by integrating I_6 and I_{12}, with respect to angles θ_1 and θ_2, and it is assumed to be a function of a number of constituent tubes N.

```
> Energy := (r0,r,R,L0,N,Z) -> N*eta^2*r0*r*Int(Int(-A*I6(r0,r,R,L0,Z) +
B*I12(r0,r,R,L0,Z),theta1=0..2*Pi),theta2=0..2*Pi):
```

Since the radius of a $(5, 5)$ carbon nanotube is 3.392 Å with the half-length $L_0 = 50$ Å, R is assumed to be $R = 9.9283$ Å, and the bundle is assumed to consist of six nanotubes, the relation between the energy and the distance Z is obtained by using `plot()` and is given by

```
> plot(Energy(3.392,3.392,9.9283,50,6,Z),Z=-80..80);
```

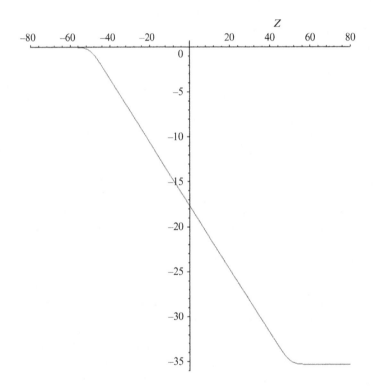

The van der Waals axial force is obtained by differentiating the negative total potential energy with respect to the distance Z, and by using the same constant values, the relation between the van der Waals axial force and the distance Z is illustrated.

```
Force:= (r0,r,R,L0,N,Z) -> -diff(Energy(r0,r,R,L0,N,Z),Z):
plot(Force(3.392,3.392,9.9283,50,6,Z),Z=-80..80);
```

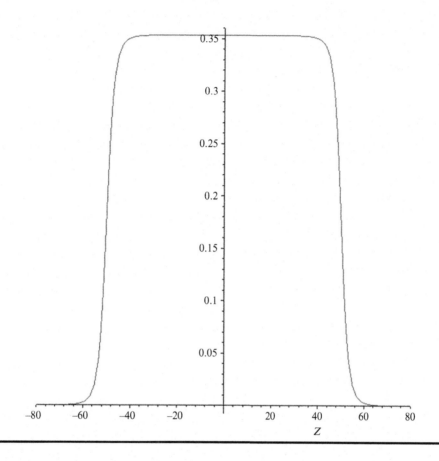

5.4.2 **SUCTION ENERGY**

If we define W as the suction energy per unit length for the single carbon nanotube interacting with the bundle, then we may write

$$W = -\sum_{k=1}^{N} E_{tt}\left(\left[(R-\epsilon)^2 + 4R\epsilon \sin^2\left(\frac{k\pi}{N} - \frac{\phi}{2}\right)\right]^{1/2}\right),$$ (5.26)

where E_{tt} is given by Eq. (4.35). Assuming that the equilibrium position for the single nanotube is located along the bundle axis, then W becomes simply

$$W = -NE_{tt}(R),$$

which means that N, the number of nanotubes in the bundle, determines the magnitude of the suction energy, but not the location of any extrema. Therefore we may determine a relationship between the radii of the nanotubes forming the bundle r and the bundle radius R, for which the suction energy W is maximised for a particular single nanotube radius r_0. By plotting this against the data derived in

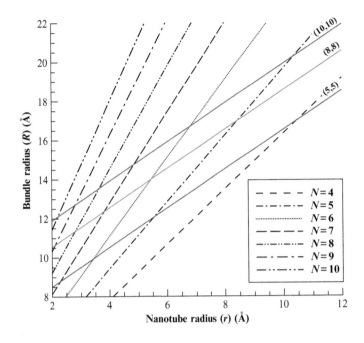

FIG. 5.16

Bundle radius versus nanotube radius for $(5, 5)$, $(8, 8)$, and $(10, 10)$ carbon nanotubes and nanotube bundles with $N \in \{4, 5, \ldots, 10\}$.

Section 4.8 for optimal nanotube bundles, we can determine optimum configurations for nanotube–nanobundle oscillators, in the sense that the suction energy per unit length is a maximum.

In Fig. 5.16, we show a graph of the relationship between nanotube radius r and bundle radius R for three nanotubes of type $(5, 5)$, $(8, 8)$, and $(10, 10)$. The point of interaction with the optimal bundle configurations from Section 5.4.1 prescribes the optimum oscillators in terms of a maximised suction energy per unit length of the single nanotube.

EXERCISES

5.1. Explain the terms 'acceptance condition' and 'suction energy'. What are the main differences between these two terms?

5.2. From Fig. 5.8, find where this suction energy plot crosses the axis and its peak.

5.3. Determine the acceptance condition for an atom entering a carbon nanotube and plot the relation between the acceptance condition and nanotube radii.

5.4. Determine the suction energy for an atom entering a carbon nanotube and plot the relation between the suction energy and nanotube radii.

5.5. Show that the suction energy for a C_{60} fullerene entering a carbon nanotube is given by Eq. (5.12), and to remind the reader is

$$W = \frac{\pi^3 \eta_f \eta_g a^2 b}{(b^2 - a^2)^{5/2}} \left[A(3 + 5\mu) - \frac{B(315 + 4620\mu + 18018\mu^2 + 25740\mu^3 + 12155\mu^4)}{160(b^2 - a^2)^3} \right],$$

where $\mu = a^2/(b^2 - a^2)$.

5.6. What is the optimal nanotube radius of a nanotube which is sucked into a nanotube bundle in terms of maximising its suction energy, for a bundle consisting of: (a) $6 \times (10, 10)$, (b) $5 \times (5, 5)$, and (c) $7 \times (8, 8)$ tubes?

5.7. Determine the suction energy for a C_{60} fullerene entering a nanotube bundle.

NANO-OSCILLATORS

6.1 INTRODUCTION

The discovery of carbon nanotubes by Iijima (1991) has given rise to the possible creation of many new nanodevices. Due to their unique mechanical properties such as high strength, low weight and flexibility, both multiwalled and single-walled carbon nanotubes promise many new applications of nanomechanical systems. One device that has attracted much attention is the creation of nanoscale oscillators, or the so-called gigahertz oscillators. While there are difficulties for micromechanical oscillators, or resonators, to reach frequencies in the gigahertz range, it is possible for nanomechanical systems to achieve this. Cumings and Zettl (2000) experimented on multiwalled carbon nanotubes by removing the cap from one end of the outer shell to attach a moveable nanomanipulator to the core in a high-resolution transmission electron microscope. By pulling the core out and allowing it to retract back into the outer shell, Cumings and Zettl found an ultra-low frictional force against the intershell sliding. They also observed that the extruded core, after release, quickly and fully retracts inside the outer shell due to the restoring force resulting from the van der Waals interaction acting on the extruded core. These results led Zheng et al. (2002) to study molecular gigahertz oscillators, for which the sliding of the inner-shell inside the outer-shell of a multiwalled carbon nanotube can generate oscillatory frequencies up to several gigahertz.

Based on the results of Zheng et al. (2002), the shorter the inner core nanotube, the higher the frequency. As a result, instead of using multiwalled carbon nanotubes, Liu et al. (2005) investigated the high frequencies generated by using a C_{60} fullerene, which is oscillating inside a single-walled carbon nanotube. Furthermore, in contrast to the multiwalled carbon nanotube oscillator, the C_{60} nanotube oscillator does not suffer the rocking motion, as a consequence of a reduced frictional effect. While Qian et al. (2001) and Liu et al. (2005) use molecular dynamics simulations to study this problem, this book employs elementary mechanical principles to provide models for the C_{60} single-walled carbon nanotube, double-walled carbon nanotubes, and single-walled carbon nanotube–nanobundle oscillators. The oscillatory behaviour of these systems was investigated by utilising the continuum approximation arising from the assumption that the discrete atoms can be smeared across each surface. However, we comment that both the molecular dynamics simulation and the analysis presented here assume both constant temperature $T = 0$ K and Newton's second law.

6.1.1 NEWTON'S SECOND LAW

Newton's second law, first proposed in 1687 by Sir Isaac Newton, is adopted here to describe the oscillatory motion of a particle (a C_{60} fullerene or a carbon nanotube) in a carbon nanotube or in a nanobundle. Newton's second law states that the acceleration of an object depends directly upon the total force acting upon the object and is inversely proportional to the mass of the object. As the force acting upon the object is increased, the acceleration is increased, but as the mass of the object is increased, the acceleration is decreased.

In this present study, we are primarily interested in the force in the z-direction. Generally, there are two main force components driving the system, which are the force that we apply to the system and the frictional force. In the nano-oscillation system, we assume that the only force which contributes to the system is the intermolecular force, and Newton's second law in the z-direction can be written as

$$M\frac{d^2Z}{dt^2} = F_{vdW}(Z) - F_r(Z), \tag{6.1}$$

where Z is the distance between the centres of the particle and the carbon nanotube, M is the mass of the particle and $F_{vdW}(Z)$ and $F_r(Z)$ are the restoring force and the frictional force, respectively. Following Cumings and Zettl (2000), we assume that the frictional force may be neglected and that the oscillation occurs due to the intermolecular force alone. However, as an example in Section 6.2.2, we illustrate the frictional effect for the oscillation of a C_{60} molecule inside a carbon nanotube. We note that Chapter 5 contains calculations for the suction force required to attract the molecule into the tube and begin oscillating, and the magnitude of the energy which is imparted to the molecule by this force is utilised in this chapter.

6.1.2 OSCILLATORY BEHAVIOUR

Imagine an oscillating system, such as the pendulum of a clock. If we displace the pendulum in one direction from its equilibrium position, the gravitational force will push it back towards equilibrium. If we displace it in the other direction, the gravitational force still acts towards the equilibrium position. No matter the direction of the displacement, the force always acts in a direction to restore the system to its equilibrium position. This is also true for nano-oscillator systems, where the oscillating molecule will return to its equilibrium location from where it was perturbed.

The frequency f is a measure of the number of occurrences of a repeating event per unit time, namely the period T. The period is the duration of one cycle in a repeating event and also the reciprocal of the frequency, $f = 1/T$. The period is measured in seconds (s), while frequency is measured in the SI unit of Hertz (Hz). In a system of constant velocity V, we have $V = X/T$, where X is the total displacement, so that we may derive $f = V/X$.

In the following section the analysis for the oscillatory behaviour of a C_{60} fullerene inside a carbon nanotube is illustrated. Subsequently, similar techniques are employed to study the oscillatory behaviour of double-walled carbon nanotubes and a carbon nanotube in a nanobundle.

6.2 OSCILLATION OF A FULLERENE C$_{60}$ INSIDE A SINGLE-WALLED CARBON NANOTUBE

6.2.1 OSCILLATORY BEHAVIOUR

In an axially symmetric cylindrical polar coordinate system (r, θ, z), it is assumed that a fullerene C$_{60}$ is located inside a carbon nanotube of length $2L$, centred around the z-axis and of radius b. As shown in Fig. 6.1, it is also assumed that the centre of the C$_{60}$ molecule is on the z-axis. Again, this is justified for the carbon nanotube $(10, 10)$, as previously described in Section 4.6.

From the symmetry of the problem, only the force in the axial direction needs to be considered. As a result, Newton's second law, neglecting the frictional force, gives

$$M\frac{d^2Z}{dt^2} = F_z^{tot}(Z),\tag{6.2}$$

where M is the total mass of a C$_{60}$ molecule and $F_z^{tot}(Z)$ is the total axial van der Waals interaction force between the C$_{60}$ molecule and the carbon nanotube, given by

$$F_z^{tot}(Z) = 2\pi b\eta_g[E(\delta_2) - E(\delta_1)],\tag{6.3}$$

where $E(\delta)$ is the potential function given by Eq. (5.5), which is restated here

$$E(\delta) = \frac{\eta_f \pi a}{\rho}\left[\frac{A}{2}\left(\frac{1}{(\delta+a)^4} - \frac{1}{(\delta-a)^4}\right) - \frac{B}{5}\left(\frac{1}{(\delta+a)^{10}} - \frac{1}{(\delta-a)^{10}}\right)\right],$$

and where $\delta_1 = \sqrt{b^2 + (Z+L)^2}$ and $\delta_2 = \sqrt{b^2 + (Z-L)^2}$. We refer the reader to Section 5.2.1 and Eq. (5.7) for the derivation of $F_z^{tot}(Z)$. Here, it is assumed that the effect of the frictional force may be neglected, which is reasonable for certain chiralities and diameters of the tube. For example, the preferred position of the C$_{60}$ molecule inside the carbon nanotube $(10, 10)$ is when the centre is located on the z-axis. Thus the molecule tends to move along the axial direction and tends not to suffer a rocking motion.

In Fig. 6.2, $F_z^{tot}(Z)$, as given by Eq. (6.3), is plotted for the case of the carbon nanotube $(10, 10)$ $(b = 6.784\,\text{Å})$ with the length of $2L = 100\,\text{Å}$. It can be observed that the force is very close to zero

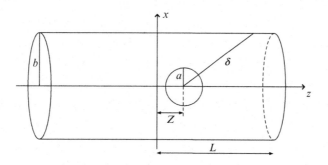

FIG. 6.1

Geometry for the fullerene C$_{60}$ oscillation.

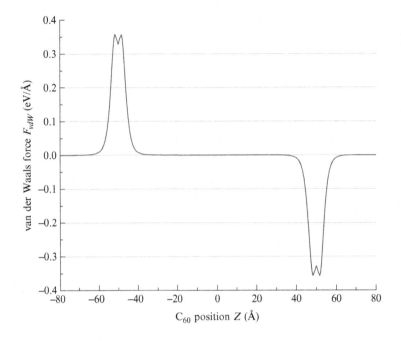

FIG. 6.2

Plot of $F_z^{tot}(Z)$ as given by Eq. (6.3) for the buckyball oscillating inside the carbon nanotube (10, 10).

everywhere except at both ends of the tube, where there is a pulse-like force which attracts the buckyball back towards the centre of the tube. For $a < b \ll L$, we assume that $F_z^{tot}(Z)$ can be estimated by using the Dirac delta function $\delta(x)$ and thus, Eq. (6.2) reduces to give

$$M\frac{d^2Z}{dt^2} = W\left[\delta(Z+L) - \delta(Z-L)\right], \tag{6.4}$$

where W is the suction energy for the C_{60} molecule which is given by Eq. (5.12) from Section 5.2.2. Now, multiplying Eq. (6.4) by dZ/dt on both sides produces

$$M\frac{d^2Z}{dt^2}\frac{dZ}{dt} = W[\delta(Z+L) - \delta(Z-L)]\frac{dZ}{dt}. \tag{6.5}$$

From $dH(x)/dx = \delta(x)$, where $H(x)$ is the usual Heaviside unit step function, Eq. (6.5) can be written as

$$\frac{M}{2}\frac{d}{dt}\left(\frac{dZ}{dt}\right)^2 = W\frac{d}{dZ}\left[H(Z+L) - H(Z-L)\right]\frac{dZ}{dt}. \tag{6.6}$$

By integrating both sides of Eq. (6.6) with respect to t and upon using that $H(Z+L) - H(Z-L) = 1$ for $-L \leqslant Z \leqslant L$ and zero elsewhere, yields

$$\frac{M}{2}\left(\frac{dZ}{dt}\right)^2 = W[H(Z+L) - H(Z-L)] + \frac{M}{2}v_0^2, \tag{6.7}$$

where v_0 is the assumed initial velocity with which the fullerene is fired on the z-axis towards the open end of the carbon nanotube in the positive z-direction. An initial velocity v_0 may be necessary for the case where the C_{60} molecule is not sucked into the carbon nanotube due to the strong repulsion force. From Eq. (6.7) for $-L \leqslant Z \leqslant L$, it can be shown that

$$\frac{M}{2} \left(\frac{dZ}{dt} \right)^2 = W + \frac{M}{2} v_0^2, \tag{6.8}$$

which implies that the buckyball travels inside the carbon nanotube at a constant speed $dZ/dt = v = (2W/M + v_0^2)^{1/2}$. Alternatively, Eq. (6.8) can also be formally obtained using the Lorentzian limit, as shown in the following Worked Example 6.1.

WORKED EXAMPLE 6.1

Derive Eq. (6.8) using the Lorentzian limit

$$\delta(x) = \lim_{\epsilon \to 0} \frac{1}{\pi} \frac{\epsilon}{\epsilon^2 + x^2}. \tag{6.9}$$

From Eqs (6.4) and (6.9) in the limit ϵ tending to zero, we have

$$M \frac{d^2 Z}{dt^2} = \frac{W\epsilon}{\pi} \left\{ \frac{1}{\epsilon^2 + (Z+L)^2} - \frac{1}{\epsilon^2 + (Z-L)^2} \right\}. \tag{6.10}$$

Multiplying both sides of Eq. (6.10) by dZ/dt produces

$$\frac{M}{2} \frac{d}{dt} \left(\frac{dZ}{dt} \right)^2 = \frac{W\epsilon}{\pi} \left\{ \frac{1}{\epsilon^2 + (Z+L)^2} - \frac{1}{\epsilon^2 + (Z-L)^2} \right\} \frac{dZ}{dt}, \tag{6.11}$$

and upon integrating with respect to t and applying the initial condition $t = 0$ (assuming the buckyball is at infinity) yields

$$\frac{M}{2} \left(\frac{dZ}{dt} \right)^2 = \frac{W\epsilon}{\pi} \int_{\infty}^{Z} \left\{ \frac{1}{\epsilon^2 + (\xi+L)^2} - \frac{1}{\epsilon^2 + (\xi-L)^2} \right\} d\xi + C, \tag{6.12}$$

where $C = M v_0^2 / 2$. On substitution of $\xi = -L + \epsilon \tan \psi$ and $\xi = L + \epsilon \tan \psi$, respectively, into the first and the second terms of Eq. (6.12) gives

$$\frac{M}{2} \left(\frac{dZ}{dt} \right)^2 = \frac{W}{\pi} \left(\int_{\pi/2}^{\tan^{-1}((Z+L)/\epsilon)} d\psi - \int_{\pi/2}^{\tan^{-1}((Z-L)/\epsilon)} d\psi \right) + C, \tag{6.13}$$

which simply gives rise to

$$\frac{M}{2} \left(\frac{dZ}{dt} \right)^2 = \frac{W}{\pi} \left\{ \tan^{-1} \left(\frac{Z+L}{\epsilon} \right) - \tan^{-1} \left(\frac{Z-L}{\epsilon} \right) \right\} + C. \tag{6.14}$$

As

$$\tan \left\{ \tan^{-1} \left(\frac{Z+L}{\epsilon} \right) - \tan^{-1} \left(\frac{Z-L}{\epsilon} \right) \right\} = \frac{2L\epsilon}{\epsilon^2 + Z^2 - L^2},$$

so that as ϵ tends to zero, then $\tan^{-1}((Z+L)/\epsilon) - \tan^{-1}((Z-L)/\epsilon) = \pi$, and as such Eq. (6.14) becomes Eq. (6.8).

WORKED EXAMPLE 6.2

Use the algebraic package MAPLE to plot the relationship between the energy and the distance between the centres of the C_{60} molecule and the nanotube Z for a $(10, 10)$ nanotube with a radius $b = 6.784$ Å and length $2L = 100$ Å.

Solution

Note that this Worked Example is similar to Worked Example 5.1, but with a finite length of carbon nanotube. Here, we integrate z where $z \in (Z - L, Z + L)$

```
> restart:
> A:=17.4:
> B:=29000:
> n_f:=0.3789:
> n_g:=0.3812:
> a:=3.55:
```

The distance between the centre of the C_{60} fullerene and an atom is defined by δ and is assumed to be a function of the tube radius b and z.

```
> d:= (b,z) -> sqrt(b^2 +z^2):
```

The interaction energy between the carbon atom and the C_{60} fullerene is given by Eq. (5.5).

```
>Energy := (b,z) -> n_f*Pi*a*( A*(1/(d(b,z)+a)^4
- 1/(d(b,z)-a)^4)/2 -B*(1/(d(b,z)+a)^10
- 1/(d(b,z)-a)^10)/5)/d(b,z);
```

We do the surface integral where $z \in (Z - L, Z + L)$ for the total potential energy of a finite tube length.

```
Etot:= (b,L,Z) -> b*2*Pi*n_g*(Int(Energy(b,z),z=0..Z+L)
-Int(Energy(b,z),z=0..Z-L)):
plot([Etot(6.784,50,Z)],Z=-80..80);
```

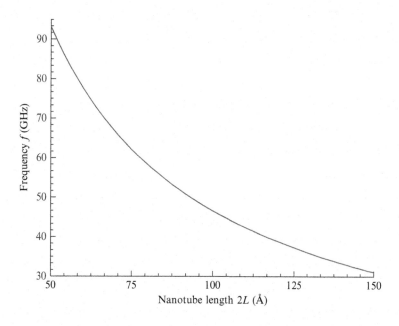

FIG. 6.3

The variation of the oscillatory frequency of the buckyball with respect to the length of the carbon nanotube $(10, 10)$.

On using Eq. (6.8), one obtains the velocity $v = 932$ ms^{-1} for the case when the C_{60} molecule is initially at rest outside the carbon nanotube $(10, 10)$ and the molecule gets sucked into the tube due to the attractive force. This gives rise to the frequency $f = v/(4L) = 46.6$ GHz, where the length of the nanotube is assumed to be $2L = 100$ Å. In Fig. 6.3, the oscillatory frequency is plotted with respect to the nanotube length. The result obtained agrees with the molecular dynamics study of Liu et al. (2005). Also considered is the case where the C_{60} molecule is fired on the tube axis towards the open end of the carbon nanotube of radius $a < 6.338$ Å, which does not accept a C_{60} molecule by suction forces alone due to the strong repulsive force of the carbon nanotube. For the carbon nanotube $(9, 9)$ ($b = 6.106$ Å), the initial velocity v_0 needs to be approximately 1152 ms^{-1} for the C_{60} molecule to penetrate into the tube. The C_{60} molecule cannot penetrate into either $(8, 8)$, $(7, 7)$, $(6, 6)$, or $(5, 5)$ nanotubes, even though it is fired into the tube with an initial velocity as high as 1600 ms^{-1}. In addition, it is found in the present model that for an $(8, 8)$ nanotube with $a = 5.428$ Å, the minimum initial velocity must be approximately 8210 ms^{-1} for the C_{60} molecule to penetrate into the tube.

WORKED EXAMPLE 6.3

Show that the velocity for the C_{60} fullerene initially at rest outside the $(10, 10)$ carbon nanotube is $v = 932$ ms^{-1}, where the molecule is sucked into the tube due to the attractive force, and the tube length is assumed to be $2L = 100$ Å.

Solution

We consider the equation

$$v = \sqrt{2W/M + v_0^2},$$

where M is the mass of the C_{60} fullerene, W is the suction energy given by Eq. (5.12) and, in this case, we assume that $v_0 = 0$. We first calculate the suction energy Eq. (5.12) by using the parameter values given in Tables 1.5–1.7. The suction energy is obtained as $W = 6.27\,eV$, where $1\,eV = 1.602 \times 10^{-19}\,kg\,m^2\,s^{-2}$. Therefore the velocity $v = 932\,ms^{-1}$, is obtained. The frequency is given by $f = v/(4L) = 932/(2 \times 100 \times 10^{-10}) = 46.6\,GHz$.

6.2.2 FRICTIONAL FORCE

In this section, the frictional force $F_r(Z)$ is incorporated into Newton's second law Eq. (6.1). Following Zheng et al. (2002), it is assumed to be a periodic interatomic-locking conservative force

$$F_r(Z) = \kappa_0 \sin(2\pi Z/\ell), \tag{6.15}$$

acting in the direction opposite to that of the motion, where ℓ is the spatial period of the interatomic locking. Typically for an armchair or (n, n) carbon nanotube, where n is a positive integer, ℓ is the distance between opposite bonds of the carbon ring, and $\kappa_0 = \tau_s\alpha$, where τ_s denotes the resistance strength and α denotes the area of a ring of contact of a certain prescribed length of the sphere. Here, we assume α is given by $\alpha = 4\pi b^2 \sin(\theta_0/2)$, for a certain angle θ_0. For multiwalled carbon nanotube oscillators, Zheng et al. (2002) state that the spatial period is dependent on the helicities of both tubes in the intershell sliding, and that the resistance force increases with the degree of commensurability of the two shells. Here, it is assumed that there is high commensurability between a C_{60} molecule and a single-walled carbon nanotube, which then leads to assuming $\ell = \sqrt{3}\sigma$, where σ is the bond length (see Worked Example 6.4 below). With reference to two tubes, commensurability refers to the two nanotubes having the same chirality (i.e. armchair/armchair or zigzag/zigzag), and incommensurability refers to the chirality's being different. By introducing Eq. (6.15) into Eq. (6.4), one obtains

$$M\frac{d^2Z}{dt^2} = W\left[\delta(Z + L) - \delta(Z - L)\right] - \kappa_0 \sin(2\pi Z/\ell). \tag{6.16}$$

Again, by multiplying both sides of Eq. (6.16) by dZ/dt, it can be deduced that

$$\frac{M}{2}\frac{d}{dt}\left(\frac{dZ}{dt}\right)^2 = W\frac{d}{dZ}\left[H(Z + L) - H(Z - L)\right]\frac{dZ}{dt} + \frac{\kappa_0\ell}{2\pi}\frac{d}{dZ}\cos(2\pi Z/\ell)\frac{dZ}{dt},$$

which upon integrating with respect to t, applying the zero velocity initial condition, and assuming that there is no frictional force at $t = 0$ as the C_{60} molecule is assumed to be outside the nanotube, the above equation becomes

$$\frac{M}{2}\left(\frac{dZ}{dt}\right)^2 = W\left[H(Z + L) - H(Z - L)\right] + \frac{\kappa_0\ell}{2\pi}\cos(2\pi Z/\ell),$$

which for the buckyball inside the carbon nanotube $(-L \leqslant Z \leqslant L)$, gives rise to

$$\frac{M}{2}\left(\frac{dZ}{dt}\right)^2 = W + \frac{\kappa_0\ell}{2\pi}\cos(2\pi Z/\ell),$$

or

$$dt = \frac{(M\pi)^{1/2}dZ}{\{2W\pi + \kappa_0\ell\cos(2\pi Z/\ell)\}^{1/2}}.$$

(6.17)

Integrating Eq. (6.17), it produces the time T taken by the buckyball to travel from $-L$ to L, as given by

$$T = (M\pi)^{1/2}\int_{-L}^{L}\frac{dZ}{\{2W\pi + \kappa_0\ell - 2\kappa_0\ell\sin^2(\pi Z/\ell)\}^{1/2}}.$$

(6.18)

From Eq. (6.18), making the substitution $\xi = \pi Z/\ell$, yields

$$T = \left(\frac{2M\ell k^2}{\pi\kappa_0}\right)^{1/2}\int_0^{\pi L/\ell}\frac{d\xi}{(1 - k^2\sin^2\xi)^{1/2}},$$

(6.19)

where the modulus k^2 is defined by

$$k^2 = \left(\frac{1}{2} + \frac{W\pi}{\kappa_0\ell}\right)^{-1}.$$

The integral appearing in Eq. (6.19) is the normal elliptic integral of the first kind, usually denoted by $F(\phi, k)$ and defined in Section 2.10. In terms of $F(\phi, k)$ Eq. (6.19) becomes

$$T = \left(\frac{2M\ell k^2}{\pi\kappa_0}\right)^{1/2}F(\pi L/\ell, k),$$

where we refer the reader to Section 2.10 for the general definition of elliptic functions, and to Byrd and Friedman (1971) for further details. We note that because of the frictional force we might expect the frequency of the motion ($f = 1/T$) to decrease with time. However, from Eq. (6.19) this is clearly not the case, and this is because there is no energy dissipation in the frictional term adopted in this model.

Generally, the position of the buckyball is determined from

$$t = \left(\frac{M\ell k^2}{2\kappa_0\pi}\right)^{1/2}\int_{-\pi Z/\ell}^{\pi L/\ell}\frac{d\xi}{(1 - k^2\sin^2\xi)^{1/2}},$$

which on nondimensionalising by T produces

$$\frac{t}{T} = \frac{1}{2}\left\{1 - \frac{F(-\pi Z/\ell, k)}{F(\pi L/\ell, k)}\right\},$$

(6.20)

which is shown graphically in Fig. 6.4, in which it is shown to be almost a straight line due to the small value of k^2.

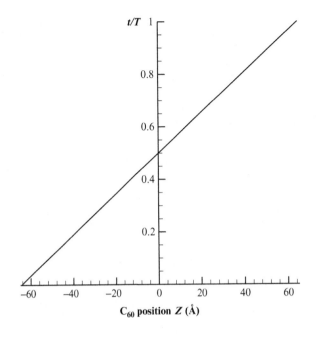

FIG. 6.4

Plot of Eq. (6.20) for the carbon nanotube (10, 10) from $-L$ to L.

WORKED EXAMPLE 6.4

Establish the relation $\ell = \sqrt{3}\sigma$ for the spatial period of the interatomic locking of a C_{60} fullerene in motion inside a carbon nanotube, where σ is the ideal carbon–carbon bond length.

Solution

We consider the case of an armchair carbon nanotube, as shown in Fig. 6.5. The period for the C_{60} fullerene traveling inside the armchair carbon nanotube is $\ell = \sqrt{3}\sigma$.

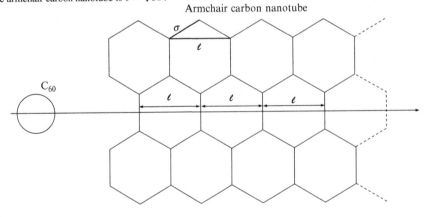

FIG. 6.5

The relation $\ell = \sqrt{3}\sigma$ for a C_{60} fullerene entering an armchair carbon nanotube.

6.3 OSCILLATION OF DOUBLE-WALLED CARBON NANOTUBES
6.3.1 INTERACTION POTENTIAL FOR FINITE LENGTH DOUBLE-WALLED CARBON NANOTUBES

With reference to a rectangular Cartesian coordinate system (x_1, y_1, z_1), with its origin located at the centre of the outer tube, a typical point on the surface of the inner tube has the coordinates $(a \cos \theta_1, a \sin \theta_1, z_1)$, where a is the assumed radius of the inner tube. Similarly, with reference to a rectangular Cartesian coordinate system (x_2, y_2, z_2) with the origin located at the centre of the outer tube, a typical point on the surface of the outer tube has the coordinates $(b \cos \theta_2, b \sin \theta_2, z_2)$, where b is the assumed radius of the outer tube, as shown in Fig. 6.6. Now assuming that the two tubes are concentric and that the distance between their centres is Z, the distance ρ between two typical points is given by Eq. (5.13), which we restate here

$$\rho^2 = a^2 + b^2 - 2ab \cos(\theta_1 - \theta_2) + (z_2 - z_1)^2.$$

The total potential energy for all atoms of the inner tube interacting with all atoms of the outer tube is given by

$$E^{tot} = \eta_g^2 ab \int_0^{2\pi} \int_0^{2\pi} (-AI_6 + BI_{12}) d\theta_1 d\theta_2, \tag{6.21}$$

where η_g represents the mean surface density of carbon atoms, and a and b are the radii of the inner and outer tubes, respectively. Note that Eq. (6.21) is equivalent to Eq. (5.14). Similarly, we define the integrals I_n $(n = 6, 12)$ as follows

$$I_n = \int_{-L_2}^{L_2} \int_{Z-L_1}^{Z+L_1} \frac{dz_1 dz_2}{\rho^n} = \int_{-L_2}^{L_2} \int_{Z-L_1}^{Z+L_1} \frac{dz_1 dz_2}{[\lambda^2 + (z_2 - z_1)^2]^{n/2}}, \tag{6.22}$$

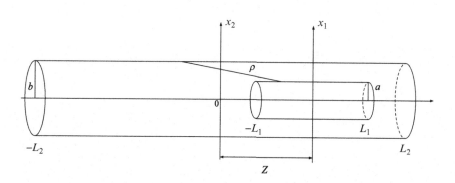

FIG. 6.6

Double-walled carbon nanotubes of lengths $2L_1$ and $2L_2$.

where $\lambda^2 = a^2 + b^2 - 2ab\cos(\theta_1 - \theta_2)$. We comment that Eq. (6.22) is similar to Eq. (5.15), but here both two tubes are assumed to be finite in length. On letting $x = z_2 - z_1$, the integral I_6 becomes

$$I_6 = \int_{Z-L_1}^{Z+L_1} \int_{-L_2-z_1}^{L_2-z_1} \frac{dxdz_1}{(\lambda^2 + x^2)^3} = \frac{1}{\lambda^5} \int_{Z-L_1}^{Z+L_1} \int_{-\tan^{-1}[(L_2+z_1)/\lambda]}^{\tan^{-1}[(L_2-z_1)/\lambda]} \cos^4\phi \, d\phi dz_1,$$

where by making the substitution $x = \lambda \tan\phi$ we obtain

$$I_6 = \frac{1}{\lambda^5} \int_{Z-L_1}^{Z+L_1} \left[\frac{3}{8}\tan^{-1}\left(\frac{L_2-z_1}{\lambda}\right) + \frac{3}{8}\frac{\lambda(L_2-z_1)}{[\lambda^2+(L_2-z_1)^2]} \right.$$
$$+ \frac{1}{4}\frac{\lambda^3(L_2-z_1)}{[\lambda^2+(L_2-z_1)^2]^2} + \frac{3}{8}\tan^{-1}\left(\frac{L_2+z_1}{\lambda}\right) + \frac{3}{8}\frac{\lambda(L_2+z_1)}{[\lambda^2+(L_2+z_1)^2]}$$
$$\left. + \frac{1}{4}\frac{\lambda^3(L_2+z_1)}{[\lambda^2+(L_2+z_1)^2]^2} \right] dz_1.$$

Finally, using the two substitutions $x = (L_2 - z_1)/\lambda$ and $y = (L_2 + z_1)/\lambda$ gives

$$I_6 = \sum_{i=1}^{4}(-1)^{i+1}\left[\frac{3}{8}\frac{(Z-\ell_i)}{\lambda^5}\tan^{-1}\left(\frac{Z-\ell_i}{\lambda}\right) - \frac{1}{8\lambda^2[\lambda^2+(Z-\ell_i)^2]}\right],$$

and by precisely the same method, I_{12} becomes

$$I_{12} = \sum_{i=1}^{4}(-1)^{i+1}\left[\frac{63}{256}\frac{(Z-\ell_i)}{\lambda^{11}}\tan^{-1}\left(\frac{Z-\ell_i}{\lambda}\right) - \frac{21}{256\lambda^8[\lambda^2+(Z-\ell_i)^2]}\right.$$
$$- \frac{21}{640\lambda^6[\lambda^2+(Z-\ell_i)^2]^2} - \frac{3}{160\lambda^4[\lambda^2+(Z-\ell_i)^2]^3}$$
$$\left. - \frac{1}{80\lambda^2[\lambda^2+(Z-\ell_i)^2]^4}\right],$$

where the four lengths ℓ_i ($i = 1, 2, 3, 4$) are defined by $\ell_1 = -(L_1+L_2)$, $\ell_2 = -(L_2-L_1)$, $\ell_3 = L_1+L_2$, and $\ell_4 = L_2 - L_1$. These are the locations for the four critical positions for the oscillation, as shown in Fig. 6.7.

The equations for I_6 and I_{12} given above exhibit a similar form to those of Eqs (5.15) and (5.16), respectively. Therefore by the same observation, there are two types of integrals which need to be

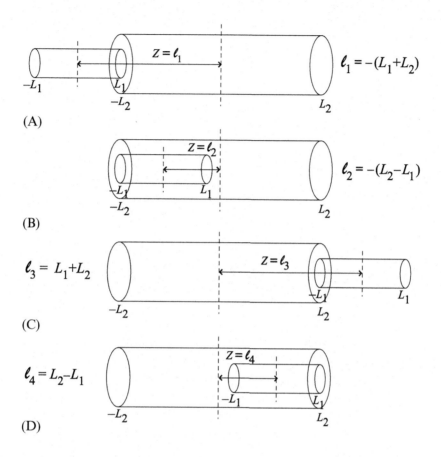

FIG. 6.7

Four critical positions for two concentric nanocylinders.

evaluated, namely the double integrals K_n^* and L_n^*, as defined by Eqs (5.17)$_2$ and (5.17)$_3$, respectively, and to remind the reader, these are restated here:

$$K_n^* = \int_0^{2\pi} \int_0^{2\pi} \frac{d\theta_1 d\theta_2}{\lambda^m(\lambda^2 + P_i^2)^n}, \qquad L_n^* = \int_0^{2\pi} \int_0^{2\pi} \frac{1}{\lambda^n} \tan^{-1}\left(\frac{P_i}{\lambda}\right) d\theta_1 d\theta_2,$$

where m and n denote certain positive integers and in this case $P_i = Z - \ell_i$ ($i = 1, 2, 3, 4$), and their evaluation is detailed in Worked Examples 5.3 and 5.4.

The parameter values for double-walled carbon nanotubes shown in Table 1.5 are employed. Using the algebraic computer package MAPLE, the potential function and intermolecular force versus the difference between the centres of the tubes Z for varying inner tube lengths are shown in Figs 6.8 and 6.9, respectively.

As a result of the four critical positions for the distance between the centres of the tubes, there are three regions, namely (ℓ_1, ℓ_2), (ℓ_2, ℓ_4), and (ℓ_4, ℓ_3), which reflect the behaviours of inner nanotube.

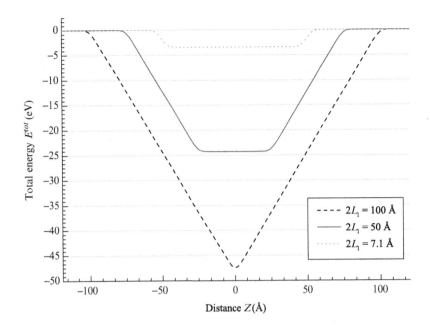

FIG. 6.8

Total potential energy of a (5, 5) nanotube with various $2L_1$ entering into a (10, 10) nanotube with length $2L_2 = 100\,\text{Å}$.

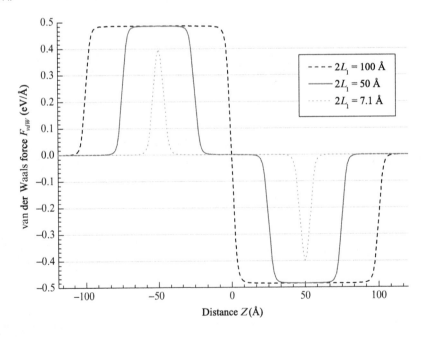

FIG. 6.9

Force distribution of a (5, 5) nanotube with various $2L_1$ entering into a (10, 10) nanotube with length $2L_2 = 100\,\text{Å}$.

We can describe these behaviours in terms of the potential function such that the inner tube will travel with decreasing potential energy in the first region to reach a constant minimum energy in the second region, after which the energy will increase until it becomes zero at the position that the inner tube leaves the outer tube. In terms of the force distribution, the force is almost zero in the second region, and there are two strong attractive forces in the first and the third regions, which tend to keep the inner tube inside. This means that once inside the outer cylinder, the inner tube will tend to oscillate rather than escape from the outer tube because of the forces at the ends tending to reverse the direction of the motion. However, not every inner tube will necessarily be sucked in by the interatomic van der Waals force alone as studied in Section 5.3, and it may be necessary to either initiate the oscillatory motion by initially extruding the inner cylinder or by giving the inner tube an initial velocity. We also observe that when $L_1 \ll L_2$, we obtain peak-like forces which are similar to those obtained in Section 6.2 for a C_{60} oscillating inside a single-walled nanotube. The force distribution for double-walled carbon nanotubes, as shown in Fig. 6.9, may be approximated by the Heaviside function $H(Z)$ to obtain

$$F_Z^{tot} = W\,[H(Z + L_2 + L_1) - H(Z + L_2 - L_1) - H(Z - L_2 + L_1)$$
$$+ H(Z - L_2 - L_1)],$$

where W is the suction energy which is given by Eq. (5.21). Detail for the Heaviside function is given in Section 2.3.

6.3.2 OSCILLATORY BEHAVIOUR

Newton's second law Eq. (6.1) is also adopted here to describe the oscillation behaviour of double-walled carbon nanotubes with the inner tube oscillating. The frequency of the oscillation for the case where the inner tube is pulled out a distance d, as shown in Fig. 6.10, and released is investigated. The frictional force is assumed to be negligible for the movement of the inner tube.

We approximate the van der Waals force experienced by the oscillating nanotube as the Heaviside function, and we assume that the inner nanotube travels through three regions as illustrated in Fig. 6.11.

Newton's second law for the first region can be written as

$$M\frac{d^2Z}{dt^2} = -W,$$

where M denotes the mass of the inner tube and W is the suction energy for double-walled carbon nanotubes which is given by Eq. (5.21). We assume that the inner tube is initiated at rest at $Z = Z_0 =$

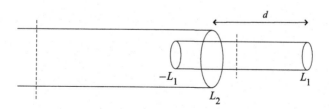

FIG. 6.10

The extrusion distance d for the inner tube oscillating inside the outer tube.

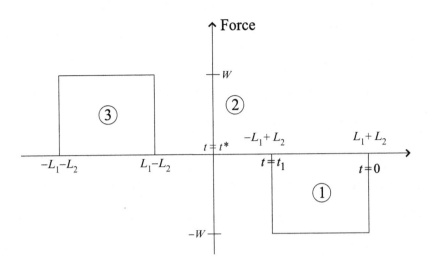

FIG. 6.11

Three regions for the idealised van der Waals force for a double-walled carbon nanotube oscillator.

$L_2 - L_1 + d$, where d is an extrusion distance as illustrated in Fig. 6.10. We may derive the velocity equation in this region as given by

$$\frac{dZ}{dt} = -\frac{W}{M}t, \tag{6.23}$$

so that the displacement equation can be obtained as

$$Z(t) = \frac{-W}{2M}t^2 + (L_2 - L_1 + d), \quad -L_1 + L_2 < Z < Z_0,$$

and the time for the inner tube to travel the length of this region is

$$t_1 = \sqrt{\frac{2Md}{W}}.$$

The travelling velocity for the inner tube in the second region is a constant which can be determined from the velocity of the first region Eq. (6.23) at $t = t_1$. By a straightforward method, the displacement equation for the second region becomes

$$Z(t) = -\sqrt{\frac{2Wd}{M}}t + (L_2 - L_1 + d), \quad L_1 - L_2 < Z < -L_1 + L_2.$$

The period T for the oscillating inner tube can be obtained as $T = 4t^*$, thus $f = 1/(4t^*)$ and t^* can be determined by the above equation at $Z = 0$. Finally, the oscillation frequency in a particular case $v_0 = 0$ becomes

$$f = \frac{1}{4}\sqrt{\frac{2W}{M}}\left(\frac{\sqrt{d}}{(L_2 - L_1 + 2d)}\right), \tag{6.24}$$

which has a maximum value at $d = (L_2 - L_1)/2$. In addition, the extrusion distance d must be less than the length of the inner tube $2L_1$. Physically, the frequency only applies when the length lies within the limits $L_1 + 2d_{min} < L_2 < 5L_1$, where d_{min} denotes the practical limitation on the minimum extrusion distance.

The frequency always increases when the initial velocity is increased. Furthermore, the shorter the inner tube, the higher frequency. This is because the force is a constant in each case, so that the lighter the weight of the shorter tube, the higher the velocity and therefore the higher the frequency. Moreover, the longer extrusion distance tends to increase the oscillatory frequency because it leads to a higher potential energy level. This in turn gives a higher van der Waals force and a higher velocity, which as a result gives rise to a higher frequency. In the case of equal lengths, when the extrusion distance is increased, it also increases the distance for the inner tube to move from one end to the other, which leads to a lower oscillation frequency. The longer length also leads to a larger mass which tends to slow down the movement, as shown in Fig. 6.12. Noting that when $L_1 = L_2$, small oscillations occur near a stable equilibrium point. Moreover, when the extrusion and the initial velocity are equal to zero, the system becomes static.

By using this model, we consider for example the oscillating $(5, 5)$ nanotube in the $(10, 10)$ nanotube, where both tubes have the same half-length $L_1 = L_2 = 50\,\text{Å}$ and the inner tube is initiated at rest. We obtain $W = 0.4851\,\text{eV}\,\text{Å}^{-1}$, which for an extrusion distance of $d = 50\,\text{Å}$ produces a frequency of $f = 17.12\,\text{GHz}$. Details of similar calculations are given in Worked Example 6.5.

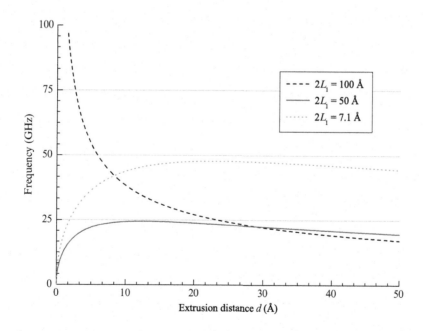

FIG. 6.12

The frequency profile for $(5, 5)$ nanotube of length $2L_1$ oscillating inside $(10, 10)$ nanotube versus extrusion distance d when the initial velocity is assumed to be zero.

WORKED EXAMPLE 6.5

Plot the relation between the oscillation frequency f and the extrusion distance d for double-walled carbon nanotubes with the outer $(10, 10)$ carbon nanotube having a length of $2L_2 = 100$ Å. Investigate the oscillation of an inner $(5, 5)$ carbon nanotube initially at rest having a length of $2L_1 = 80, 60, 40$, and 20 Å.

Solution

We utilise the algebraic package MAPLE to plot the graph, and we start by defining the constants.

```
> restart;
> A:=15.2: B:=24100:
> a:=3.392: b:=6.779: L2:=50:
> eta:= 0.3812:
> M:=9.35495E-26:
```

For convenience, we define A_i $(i = 1, 2, 3, 4)$, which corresponds to the four lengths ℓ_i $(i = 1, 2, 3, 4)$, to be a function of the inner tube half-length L_1.

```
> A1:= L1 -> L1+L2+Z:
> A2:= L1 -> -L1+L2+Z:
> A3:= L1 -> -L1+L2-Z:
> A4:= L1 -> L1+L2-Z:
> 14:= L1 -> L2-L1:
> 13:= L1 -> L1+L2:
```

Now, we input the equations for I_6 and I_{12}, which are also functions of the inner tube half-length L_1.

```
> I6:= L1 -> (3*A1(L1)*arctan(A1(L1)/lambda)/(8*lambda) -
lambda^2/(8*(lambda^2+A1(L1)^2)) -
3*A2(L1)*arctan(A2(L1)/lambda)/(8*lambda) +
lambda^2/(8*(lambda^2+A2(L1)^2))
-3*A3(L1)*arctan(A3(L1)/lambda)/(8*lambda) +
lambda^2/(8*(lambda^2+A3(L1)^2)) +
3*A4(L1)*arctan(A4(L1)/lambda)/(8*lambda) -
lambda^2/(8*(lambda^2+A4(L1)^2)) )/lambda^4:

>I12:= L1 -> ( 63*A1(L1)*arctan(A1(L1)/lambda)/(256*lambda) -
21*lambda^2/(256*(lambda^2+A1(L1)^2)) -
21*lambda^4/(640*(lambda^2+A1(L1)^2)^2) -
3*lambda^6/(160*(lambda^2+A1(L1)^2)^3) -
lambda^8/(80*(lambda^2+A1(L1)^2)^4) -
63*A2(L1)*arctan(A2(L1)/lambda)/(256*lambda) +
21*lambda^2/(256*(lambda^2+A2(L1)^2)) +
21*lambda^4/(640*(lambda^2+A2(L1)^2)^2) +
3*lambda^6/(160*(lambda^2+A2(L1)^2)^3) +
lambda^8/(80*(lambda^2+A2(L1)^2)^4) -
63*A3(L1)*arctan(A3(L1)/lambda)/(256*lambda) +
21*lambda^2/(256*(lambda^2+A3(L1)^2)) +
21*lambda^4/(640*(lambda^2+A3(L1)^2)^2) +
3*lambda^6/(160*(lambda^2+A3(L1)^2)^3) +
lambda^8/(80*(lambda^2+A3(L1)^2)^4) +
63*A4(L1)*arctan(A4(L1)/lambda)/(256*lambda) -
21*lambda^2/(256*(lambda^2+A4(L1)^2)) -
21*lambda^4/(640*(lambda^2+A4(L1)^2)^2) -
3*lambda^6/(160*(lambda^2+A4(L1)^2)^3) -
lambda^8/(80*(lambda^2+A4(L1)^2)^4) )/lambda^10:
```

where

```
> lambda:= sqrt((a-b)^2 + 4*a*b*sin(x)^2):
```

Then we use `Int()` to numerically integrate the total potential energy:

```
> rho6:= L1 -> Int(I6(L1),x=0..Pi/2):
> rho12:=L1 -> Int(I12(L1),x=0..Pi/2):
> total:= L1 -> 8*Pi*a*b*eta^2*(-A*rho6(L1) + B*rho12(L1)):
```

Now, we determine the parameters for the oscillation frequency. The frequency is assumed to be a function of the inner tube half-length L_1, the extrusion distance d and the initial velocity v_0; it also has the units of gigahertz. We also note that we need to convert all the units for every parameter to be SI units.

```
> E0:= L1 -> eval(total(L1),Z=0):
> alpha:= L1 -> sqrt(1.602*10^(-9)*(2*E0(L1))/
(M(L1)*(14(L1)-13(L1))));
> fre:= (L1,d,v0) -> 10^(-4)*((alpha(L1)^2*sqrt(v0^2+d*alpha(L1)^2))/
(4*(2*v0^2-2*v0*sqrt(v0^2+alpha(L1)^2*d)+alpha(L1)^2*(2*d+14(L1)))));
```

Finally, we plot the relationship between the oscillation frequency and the extrusion distance d by varying the inner tube half-length.

```
plot([fre(40,d,0),fre(30,d,0),fre(20,d,0),fre(10,d,0)],d=0..40);
```

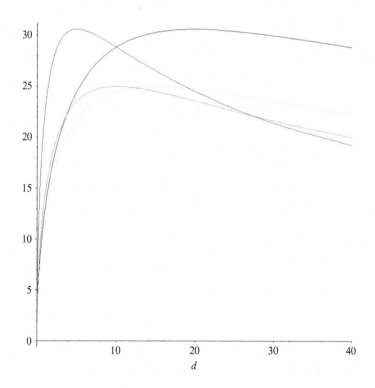

We comment that the maximum frequency occurs at $d = (L_2 - L_1)/2$.

AN ALTERNATIVE APPROACH

Alternatively, here we outline a method to determine the frequency which does not use the Heaviside function. On neglecting the frictional force and using the relation for the potential energy and force, Newton's second law Eq. (6.1) becomes

$$M\frac{d^2Z}{dt^2} = -\frac{\partial E^{tot}}{\partial Z},$$

(6.25)

where Z is the distance between the centres of the tubes and M is the mass of the inner tube. By multiplying both sides of Eq. (6.25) by dZ/dt and integrating, we obtain

$$\frac{M}{2}\left(\frac{dZ}{dt}\right)^2 + E(Z) = \frac{M}{2}v_0^2 + E_0,$$

(6.26)

where v_0 is the prescribed initial velocity and E_0 is the initial potential energy, which is a function of the extrusion length d. Due to the symmetry of the oscillating inner tube, two regions for its motion shown in Fig. 6.13 are considered. In the first region, the inner tube is pulled out a distance d and released, then the distance between centres becomes $Z_0 = L_2 - L_1 + d = \ell_4 + d$, which leads to the geometric constraint $d \leq 2L_1$. On assuming that the prescribed initial velocity at this point is v_0, it can be deduced

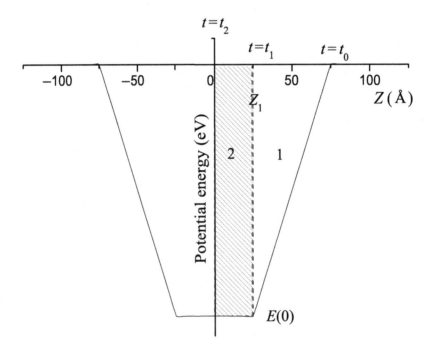

FIG. 6.13

Two regions for the idealised potential energy for double-walled carbon nanotubes oscillator.

$$\frac{dZ}{dt} = -\sqrt{v_0^2 + \alpha^2(Z_0 - Z)}, \quad \ell_4 \le Z \le Z_0,$$

where $\alpha^2 = 2E(0)/[M(\ell_4 - \ell_3)]$, and $E(0)$ is the potential energy at $Z = 0$. We note that $\ell_1 = -(L_1 + L_2)$, $\ell_2 = -(L_2 - L_1)$, $\ell_3 = L_1 + L_2$, and $\ell_4 = L_2 - L_1$, as illustrated in Fig. 6.7. Thus the displacement equation in this region can be written as

$$Z(t) = Z_0 - \frac{\alpha^2}{4}t^2 - v_0 t, \quad \ell_4 \le Z \le Z_0,$$

and the time for the inner tube to travel the length of this region is

$$t_1 = \frac{2}{\alpha^2}\left(\sqrt{v_0^2 + \alpha^2 d} - v_0\right).$$

In the second region, from Eq. (6.26) we may deduce

$$\frac{dZ}{dt} = -\sqrt{v_0^2 + \alpha^2 d}, \quad 0 \le Z \le \ell_4.$$

At $t = t_1$, $Z(t_1) = \ell_4$, the displacement equation becomes

$$Z(t) = \ell_4 + \frac{2}{\alpha^2}\left(v_0^2 + \alpha^2 d\right) - \left(t + \frac{2v_0}{\alpha^2}\right)\sqrt{v_0^2 + \alpha^2 d}, \quad 0 \le Z \le \ell_4.$$

Thus the overall time for the inner tube to travel in these two regions is given by

$$t_2 = \frac{2v_0^2 - 2v_0\sqrt{v_0^2 + \alpha^2 d} + \alpha^2(\ell_4 + 2d)}{\alpha^2\sqrt{v_0^2 + \alpha^2 d}}.$$

The period for the inner tube to move is $4t_2$, so that the oscillation frequency can be written as

$$f = \frac{\alpha^2\sqrt{v_0^2 + \alpha^2 d}}{4[2v_0^2 - 2v_0\sqrt{v_0^2 + \alpha^2 d} + \alpha^2(2d + \ell_4)]}, \tag{6.27}$$

which for prescribed v_0 has a maximum frequency at $d = (\alpha^2\ell_4 - 2v_0^2)/(2\alpha^2)$. Of particular practical interest is the case where $v_0 = 0$, for which Eq. (6.27) simplifies to give $f = \alpha\sqrt{d}/[4(2d + L_2 - L_1)]$, which has a maximum value at $d = (L_2 - L_1)/2$. This method gives rise to the same result as the use of the Heaviside function.

6.4 OSCILLATION OF NANOTUBES IN BUNDLES

6.4.1 INTERACTION POTENTIAL FOR FINITE LENGTHS OF NANOTUBES AND BUNDLES

In this section, we consider a single carbon nanotube oscillating in the middle of a bundle of finite length carbon nanotubes of N-fold symmetry. We assume that the centre of the oscillating tube remains on the

z-axis during its motion. In a cylindrical polar coordinate system, a typical point on the oscillating tube has the coordinates $(r_0 \cos \theta_0, r_0 \sin \theta_0, z_0 + Z)$, where r_0 is the tube radius, $-L_0 \leqslant z_0 \leqslant L_0$, and Z is the distance between the centre of the oscillating tube and the origin. The coordinates of the nanotubes in the bundle are given by Eq. (4.31), where in this case $-L \leqslant z_i \leqslant L$. In this analysis, we assume that the friction is negligible throughout. Energy dissipation through radial breathing modes and other secondary modes of vibration will have a dampening effect, but these are all ignored in the present model.

Due to the assumed symmetry of the problem, we need only consider the interaction between the oscillating tube, which is located at the centre of the bundle, and one of the carbon nanotubes in the bundle surrounding the oscillating tube. As shown in Fig. 5.13, we examine the tube with coordinates $(r_0 \cos \theta_0, r_0 \sin \theta_0, z_0 + Z)$, which is in the middle of the bundle and the tube $i = 1$ with coordinates $(r \cos \theta_1 + R, r \sin \theta_1, z_1)$. Again, the total interaction energy of the oscillating nanotube inside the bundle is given by $E^{tot} = NE$, where E is the energy of the interaction between the two nanotubes and N is the number of tubes in the bundle, which are located symmetrically around the inner oscillating nanotube.

Using the Lennard-Jones potential and the continuum approach, we obtain the interaction energy E as

$$E = r r_0 \eta^2 \int_0^{2\pi} \int_0^{2\pi} \left[\int_{-L}^{L} \int_{-L_0}^{L_0} \left(-\frac{A}{\rho^6} + \frac{B}{\rho^{12}} \right) dz_0 dz_1 \right] d\theta_0 d\theta_1, \tag{6.28}$$

where η is the mean atomic density of a nanotube, ρ denotes the distance between two typical surface elements on each nanotube, which is given by Eq. (5.23) and restated here:

$$\rho^2 = (r \cos \theta_1 + R - r_0 \cos \theta_0)^2 + (r \sin \theta_1 - r_0 \sin \theta_0)^2 + (z_1 - z_0 - Z)^2.$$

We can rewrite Eq. (6.28) as

$$E = r r_0 \eta^2 \int_0^{2\pi} \int_0^{2\pi} (-A I_6 + B I_{12}) d\theta_0 d\theta_1, \tag{6.29}$$

where the integrals I_n ($n = 6, 12$) are defined by

$$I_n = \int_{-L}^{L} \int_{Z-L_0}^{Z+L_0} \frac{dz_0^* dz_1}{[\lambda^2 + (z_1 - z_0^*)^2]^{n/2}}, \tag{6.30}$$

where $z_0^* = z_0 + Z$ and $\lambda^2 = (r \cos \theta_1 + R - r_0 \cos \theta_0)^2 + (r \sin \theta_1 - r_0 \sin \theta_0)^2$. We note that Eq. (6.30) is identical to Eq. (6.22).

We note again that formally for the case $R = 0$, we have the scenario of an oscillating nanotube inside another nanotube, and the integrals I_n can be evaluated analytically in terms of hypergeometric functions as shown in Section 6.3.1 for the case of double-walled carbon nanotubes. In other words, this system describes the oscillation of double-walled carbon nanotubes. Since $R > 0$, we need to evaluate these integrals numerically.

In Figs. 6.14 and 6.15 we plot the total energy E^{tot} and the van der Waals force F_{vdW} for a $(5, 5)$ carbon nanotube oscillating in a sixfold symmetry $(5, 5)$ carbon nanotube bundle. Noting that we use $R = 9.9283$ Å, which is the optimal bundle radius for the interaction of a $(5, 5)$ nanotube and a bundle of sixfold symmetry (see Fig. 5.16). A similar behaviour to that of the double-walled carbon nanotube oscillators, as studied in Section 6.3, is obtained here. A single carbon nanotube has a minimum energy

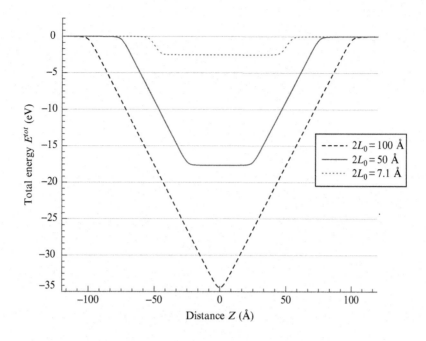

FIG. 6.14

Potential energy of $(5, 5)$ nanotube of length $2L_0$ oscillating inside sixfold symmetry bundle comprising $(5, 5)$ nanotubes of length $2L = 100\,\text{Å}$.

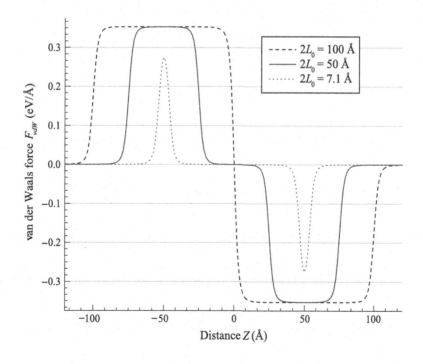

FIG. 6.15

Force distribution of $(5, 5)$ nanotube of length $2L_0$ oscillating inside sixfold symmetry bundle comprising $(5, 5)$ nanotubes of length $2L = 100\,\text{Å}$.

at $Z = 0$ inside the bundle. By pulling the tube away from the minimum energy configuration in either direction, the van der Waals force tends to propel the nanotube back towards the centre of the bundle. Accordingly, we obtain an oscillatory motion of the nanotube inside the bundle. We also observe that when $L_0 \ll L$, we obtain peak-like forces which are similar to those obtained in Section 6.2 for a C_{60} oscillating inside a single-walled nanotube.

6.4.2 OSCILLATORY BEHAVIOUR

By adopting the simplified model of motion for oscillating nanotubes presented in Section 6.3, we can show that when $L \geqslant L_0$, the van der Waals force F_{vdW} experienced by the oscillating nanotube can be approximated by

$$F_{dvW} = W\,[H(Z + L + L_0) - H(Z + L - L_0) - H(Z - L + L_0)$$
$$+ H(Z - L - L_0)]\,,$$

where $H(z)$ is the Heaviside unit-step function and W is the suction energy per unit length, which here we call the suction force, as given in Eq. (5.26). From this model, it can be shown that, assuming the nanotube is initially at rest and extruded by a distance d out of the nanotube bundle, in a similar manner to Eq. (6.24) the resulting oscillatory frequency f is given by

$$f = \frac{1}{4}\sqrt{\frac{2W}{M}} \left(\frac{\sqrt{d}}{2d + (L - L_0)} \right), \tag{6.31}$$

where M is the mass of the oscillating nanotube, which is given by $M = 4\pi r_0 L_0 \eta m_0$, and m_0 is the mass of a single carbon atom.

As found in Section 6.3, the maximum frequency occurs when the extrusion distance satisfies the relationship $d = (L - L_0)/2$. In this case the maximum frequency f_{max} is given by

$$f_{max} = \frac{1}{8}\sqrt{\frac{W}{M(L - L_0)}}. \tag{6.32}$$

However, there are certain limitations on those oscillators which can attain this frequency. Firstly, the extrusion distance d must be less than the length of the oscillating nanotube $2L_0$. This leads to an upper limit on the ratio of the bundle length L to the oscillator length L_0, that is $L < 5L_0$. Bundle oscillators longer than this will not be able to attain the theoretical maximum frequency f_{max} given in Eq. (6.32). Secondly, a sensible lower limit is required on the extrusion distance. The reason for this is that when $L = L_0$, then extremely high frequencies are theoretically achieved by choosing very small extrusion distances d. However, for an oscillator to be practical, the total displacement of the oscillating nanotube needs to be measurable, and this represents a practical limit on the design of such oscillators. We denote this practical limitation on the minimum extrusion distance as d_{min}, and therefore we conclude that Eq. (6.32) only applies when the bundle length lies within the limits

$$L_0 + 2d_{min} < L < 5L_0. \tag{6.33}$$

We note that the constraints (6.33) also apply equally well to double-walled oscillators.

A comparison between results of the present model and that of the molecular dynamics study by Kang et al. (2006) shows reasonable overall agreement considering the assumptions of the model presented here. By using this model, we investigate in Worked Example 6.6 the oscillation of a $(5, 5)$ nanotube initially at rest in a sixfold nanotube bundle also comprising of $(5, 5)$ nanotubes. As an example, for the case when both and the oscillating nanotube have the same half-length $L = L_0 = 50\,\text{Å}$ and R is assumed to be $9.9283\,\text{Å}$, we obtain $W = 0.3998\,\text{eV}\,\text{Å}^{-1}$ which for an extrusion distance of $50\,\text{Å}$ produces a frequency of $f = 15.72\,\text{GHz}$.

WORKED EXAMPLE 6.6

Show the frequency profiles for the oscillating $(5, 5)$ carbon nanotube in a sixfold nanotube bundle comprised of $(5, 5)$ carbon nanotubes versus the extrusion distance d, where the bundle length is assumed to be $2L = 100\,\text{Å}$ and the oscillating tube lengths are assumed to be $2L_0 = 100, 50$, and $7.1\,\text{Å}$.

Solution

The relation between the oscillation frequency and the extrusion distance for the nanobundles is given by Eq. (6.31). In this case, we have the suction energy $W = -NE_{tt}$, where N denotes the number of nanotubes in the bundle and E_{tt} is the interaction potential per unit length for two parallel carbon nanotubes, which is given by Eq. (4.35). The algebraic package MAPLE is utilised to plot such a relation:

```
> restart;
> A:=15.2:
> B:=24100:
> eta:=0.3812:
> M:= (r,L0) -> evalf(L0*4*Pi*r*eta*1.993E-26):
> F2:=(a,b1,b2,c1,c2,x,y,N)->add(evalf(pochhammer(a,r)
*pochhammer(b1,r)*pochhammer(b2,r)/(r!*pochhammer(c1,r)
*pochhammer(c2,r))*x^r*y^r*hypergeom([a+r,b1+r],[c1+r],x)
*hypergeom([a+r,b2+r],[c2+r],y)),r=0..N):

> Ett:=(R,r,N)->evalf(3/2*eta^2*r^2*Pi^3*(R^2-2*r*R)^(-5/2)
*(-A*F2(5/2,-3/2,1/2,1,1,-r^2/(R^2-2*r*R),-4*r/(R-2*r),N)
+21/32*(R^2-2*r*R)^(-3)*B*F2(11/2,-9/2,1/2,1,1,-r^2/(R^2-2*r*R),
-4*r/(R-2*r),N))):
```

We need to change the unit of the suction energy, which is $\text{eV}\,\text{Å}^{-1}$, to SI units, where $1\,\text{eV} = 1.602 \times 10^{-19}\,\text{kg}\,\text{m}^2\,\text{s}^{-2}$. Moreover, we present the frequency in the gigahertz range.

```
> W:= (R,r,N) -> -N*Ett(R,r,N)*1.602*10^(-9):
> f:= (R,r,N,L,L0,d) -> 10^(-9)*10^5*sqrt(2*W(R,r,N)/M(r,L0))
*(sqrt(d)/(2*d+L-L0))/4:
> plot([f(9.9283,3.392,6,100,100,d),f(9.9283,3.392,6,100,50,d),
f(9.9283,3.392,6,100,3.55,d)],d=2..50);
```

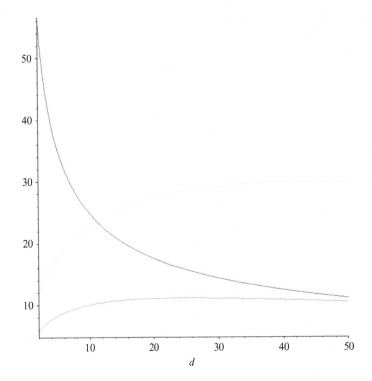

EXERCISES

6.1. Derive the total axial van der Waals interaction force between the C_{60} molecule and the carbon nanotube as given in Eq. (6.3).

6.2. Plot the relation between the total axial van der Waals interaction force and the position distance Z for the C_{60} molecule oscillating inside both a $(9,9)$ $(b = 6.106\,\text{Å})$ and an $(11,11)$ $(b = 7.463\,\text{Å})$ carbon nanotube with lengths of $100\,\text{Å}$.

6.3. For a C_{60} molecule oscillating inside a carbon nanotube, plot the relationship between the oscillation frequency and the radius b for any carbon nanotube with the length of $2L = 100\,\text{Å}$ on assuming the system is frictionless and that the C_{60} molecule is initially at rest. Find the radius b which gives the highest oscillation frequency.

6.4. Show that for a $(9,9)$ nanotube $(b = 6.106\,\text{Å})$ with a length of $2L = 100\,\text{Å}$, the initial velocity v_0 needs to be approximately $1152\,\text{ms}^{-1}$ for the C_{60} molecule to penetrate into the tube.

6.5. From the frequency equation for the double-walled carbon nanotubes Eq. (6.27), show that the maximum frequency occurs where $d = (\alpha^2 \ell_4 - 2v_0^2)/(2\alpha^2)$. Assuming that $v_0 \equiv 0$, show that the maximum frequency occurs at $d = \ell_4/2$.

6.6. Find the relationship between the oscillation frequency and the inner tube length $2L_1$ for double-walled carbon nanotubes where the tube is initially at rest with an extrusion distance of $d = (L_2 - L_1)/2$ and where the outer tube length assumed to be $2L_2 = 100$ Å.

6.7. Determine the oscillatory behaviour for a carbon atom inside a carbon nanotube.

6.8. Determine the oscillatory behaviour for a C_{60} fullerene inside a nanotube bundle.

MECHANICS OF MORE COMPLICATED STRUCTURES: NANOPEAPODS AND SPHEROIDAL FULLERENES

7

7.1 INTRODUCTION

The discovery of carbon nanostructures has led to the possible creation of many nanodevices, for example, Baughman et al. (1999, 2002), Bianco et al. (2005), and Shenderova et al. (2002). Carbon nanotubes in particular have attracted much attention due to their mechanical and electronic properties. One further aspect of carbon nanotubes which is of particular interest is their accessible internal storage capacity via their open ends, which can be filled with other molecular structures. It was first observed by Smith et al. (1998) that hollow nanotubes may be filled with C_{60} fullerene molecules, with the resulting structures being usually referred to as nanopeapods. While earlier attention was given to C_{60} molecules inside the nanotubes, Hodak and Girifalco (2003) report different types of fullerene nanopeapods, including C_{70}, C_{78}, and C_{80}. These nanopeapods possess potential applications as superconducting nanowires. Nanopeapods are superior to empty carbon nanotubes as a superconducting nanowire because the charge can travel not only along the tube wall, but also along the chain of fullerenes inside the nanotube.

For C_{60} peapods the interwall spacing between a fullerene and the nanotube depends only on the radii of the tube and the fullerene. However, for peapods with ellipsoidal shaped fullerenes, such as C_{70} and C_{80}, which are referred to as spheroidal fullerenes, the interwall spacing is also dependent on the orientation of the fullerenes. Evidently, due to their nonspherical structure, spheroidal fullerenes can pose different orientations inside the nanotubes, unlike spherical fullerenes C_{60}. This also leads to a possible advantage that spheroidal fullerene peapods may have compared to the C_{60} peapods, which is the capacity to control the electronic properties of the system by simply controlling the orientations of the fullerenes.

There are two main sections in this chapter. We begin in Section 7.2 by considering nanopeapods comprising only C_{60} molecules inside a single-walled carbon nanotube. Due to the complicated analysis arising from the antisymmetric structures of the C_{70} and C_{80} fullerenes, only the potential expressions are determined; they are given in Section 7.3.

Modelling and Mechanics of Carbon-based Nanostructured Materials. http://dx.doi.org/10.1016/B978-0-12-812463-5.00007-8

7.2 NANOPEAPODS

Carbon nanostructures, such as carbon nanotubes and C_{60} fullerenes, have received considerable attention because of their underlying unique mechanical properties arising from the van der Waals interaction force and their electronic properties arising from the large surface to volume ratio. The combination of a single-walled carbon nanotube and a C_{60} fullerene chain, a so-called nanopeapod, also embodies such properties as a new hybrid nanostructure. Nanopeapods were originally observed by Smith et al. (1998), as shown in Fig. 7.1. Nanopeapods can be thought of as the prototype nanocarrier for drug delivery, where the carbon nanotube is the nanocontainer and the C_{60} molecular chain can be considered as the drug molecules.

While the investigation of the packing of C_{60} molecules inside a carbon nanotube is usually achieved through either experimentation or large-scale computation, this chapter again adopts the themes of elementary mechanical principles and classical applied mathematical modelling techniques to formulate explicit analytical criteria for such encapsulations. In particular, the Lennard-Jones potential and the continuum approximation are employed to determine three encapsulation mechanisms for a C_{60} fullerene entering a tube: (i) directly through the tubes' open end (termed 'head-on'), (ii) around the edge of the tubes' open end, and (iii) through a defect opening on the tube wall; these mechanisms are presented in Sections 7.2.1, 7.2.2, and 7.2.3, respectively. All configurations are assumed to be in a vacuum, and the C_{60} fullerene is assumed to be initially at rest. Double integrals are performed to determine the energy of the system, and analytical expressions are obtained in terms of hypergeometric functions. Moreover, the packing of C_{60} fullerene chains inside a single-walled carbon nanotube is investigated, again by utilising the Lennard-Jones potential function and the continuum approximation. Both zigzag and spiral chain configurations are examined. Analytical expressions in terms of hypergeometric functions for the potential energy for such configurations are obtained, and these configurations are presented in Sections 7.2.4 and 7.2.5, respectively.

We note that in order to determine the interaction energy between a spherical fullerene and a carbon nanotube for a typical point on the carbon nanotube, the surface integral of the Lennard-Jones potential over the sphere is first performed, which is detailed in Section 3.3.2. The interaction energy between a C_{60} molecule and a nanotube is subsequently obtained by performing another surface integral over the cylindrical tube.

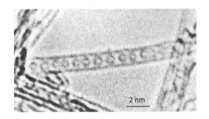

FIG. 7.1

A single-walled carbon nanotube containing a row of closed carbon shells concentric with the tubule axis.

From Smith, B.W., Monthioux, M., Luzzi, D.E., 1998. Encapsulated C_{60} in carbon nanotubes. Nature 396, 323–324.

7.2.1 ENCAPSULATION OF C$_{60}$ HEAD-ON AT AN OPEN END

In this section, we determine the potential for the encapsulation of a C$_{60}$ molecule into a single-walled carbon nanotube head-on into the tube's open end, as shown in Fig. 7.2. The C$_{60}$ fullerene is assumed to be initially at rest and remain on the tube axis. With reference to a rectangular Cartesian coordinate system (x, y, z), with the origin located on the tube axis and at the tube end, a typical point on the surface of the tube has the coordinates $(b \cos \theta, b \sin \theta, z)$, where b is the radius of the semi-infinite tube. Similarly, with reference to the same rectangular Cartesian coordinate system (x, y, z), the centre of the C$_{60}$ molecule has coordinates $(0, 0, Z)$, where Z is the distance in the z-direction, which can be either positive or negative. Thus the distance δ between the centre of the C$_{60}$ fullerene and a typical point on the tube is given by

$$\delta^2 = b^2 + (z - Z)^2. \tag{7.1}$$

Using the Lennard-Jones potential function (3.1) together with the continuum approximation, the total potential can be written as

$$E = b\eta_g \int_{-\pi}^{\pi} \int_0^{\infty} E^*(\delta) \, dz \, d\theta,$$

which is the same as Eq. (4.12) but the lower limit for z is changed from $-\infty$ to 0. Note that η_g represents the mean atomic surface density of the carbon nanotube, δ is given by Eq. (7.1) and $E^*(\delta)$ denotes the interaction energy between an atom and a C$_{60}$ fullerene, which is given by Eq. (3.15) and is detailed in Section 3.3.2. For clarity the interaction energy $E^*(\delta)$ is restated here:

$$E^*(\delta) = \frac{\pi a \eta_f}{\delta} \left[\frac{A}{2} \left(\frac{1}{(\delta + a)^4} - \frac{1}{(\delta - a)^4} \right) - \frac{B}{5} \left(\frac{1}{(\delta + a)^{10}} - \frac{1}{(\delta - a)^{10}} \right) \right], \tag{7.2}$$

where in this case the variable η in Eq. (3.15) is replaced by $\eta = \eta_f$, and we note that this is identical to Eq. (5.5) used in Section 5.2.1. By expanding the denominators and reducing to fractions in terms of powers of $(\delta^2 - a^2)$, as shown in Eqs (5.8) and (5.9), the integrals for the total potential energy E which need to be evaluated are of the form

$$G_n = \int_{-\pi}^{\pi} \int_0^{\infty} \frac{1}{(\delta^2 - a^2)^n} \, dz \, d\theta = \int_{-\pi}^{\pi} \int_0^{\infty} \frac{1}{[b^2 - a^2 + (z - Z)^2]^n} \, dz \, d\theta, \tag{7.3}$$

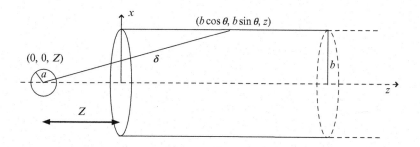

FIG. 7.2

C$_{60}$ fullerene encapsulated in a carbon nanotube head-on at an open end.

where n is a certain positive integer; we refer the reader to Eqs (5.8) and (5.9) for the expansion of Eq. (7.2). It is clear that Eq. (7.3) is independent of θ so that

$$G_n = 2\pi \int_0^\infty \frac{1}{[b^2 - a^2 + (z - Z)^2]^n} dz. \tag{7.4}$$

The details of the analytical evaluation of Eq. (7.4) are presented in Worked Example 7.1, and the resulting numerical solution follows.

WORKED EXAMPLE 7.1

Determine an analytical expression for G_n, as defined by Eq. (7.4).

Solution

We consider the integral G_n, which is defined in Eq. (7.4). Letting $\lambda^2 = b^2 - a^2$ and making the substitution $x = z - Z$, it can be deduced

$$G_n = 2\pi \int_{-Z}^\infty \frac{dx}{(\lambda^2 + x^2)^n},$$

where n is a certain positive integer. The substitution $x = \lambda \tan \psi$ yields

$$G_n = 2\pi \int_{-\tan^{-1}(Z/\lambda)}^{\pi/2} \frac{\lambda \sec^2 \psi}{\lambda^{2n} \sec^{2n} \psi} d\psi = \frac{2\pi}{\lambda^{2n-1}} \int_{-\tan^{-1}(Z/\lambda)}^{\pi/2} \cos^{2(n-1)} \psi \, d\psi. \tag{7.5}$$

The evaluation for Eq. (7.5) can be found in Gradshteyn and Ryzhik (2000) (p. 149, No. 2.513 3) from which it can be deduced:

$$\int \cos^{2(n-1)} \psi \, d\psi = \frac{1}{2^{2(n-1)}} \left[\binom{2(n-1)}{n-1} \psi + \sum_{k=0}^{n-2} \binom{2(n-1)}{k} \frac{\sin[(2n - 2k - 2)\psi]}{n - k - 1} \right], \tag{7.6}$$

where $\binom{n}{m}$ is the binomial coefficient given by Eq. (2.10). By evaluating Eq. (7.6) at $\psi = \pi/2$ and $\psi = -\tan^{-1}(Z/\lambda)$, the following analytical expression for G_n can be obtained:

$$G_n = \frac{4\pi}{(2\lambda)^{2n-1}} \left[\binom{2n - 2}{n - 1} \left(\frac{\pi}{2} + \tan^{-1} \frac{Z}{\lambda} \right) + \sum_{k=0}^{n-2} \binom{2n - 2}{k} \frac{\sin\left[(2n - 2k - 2)\tan^{-1} \frac{Z}{\lambda} \right]}{n - k - 1} \right].$$

Using the parameter values from Tables 1.5–1.7, the relationship between the potential energy and the distance Z for the C_{60} molecule encapsulated head-on into the $(10, 10)$, $(16, 16)$, and $(20, 20)$ carbon nanotubes is shown in Fig. 7.3. The energetically favourable location for the C_{60} fullerene of all three cases is inside the tube, which is the positive direction of Z. Furthermore, the binding energies, which are defined as the energy required to separate the two bodies, are 3.222, 0.326, and 0.109 eV for the $(10, 10)$, $(16, 16)$, and $(20, 20)$ carbon nanotubes, respectively. The lowest potential energy is observed to occur for the $(10, 10)$ tube, because the preferred location of the C_{60} molecule is on the tube axis. As a result, offset locations from the tube axis for the $(16, 16)$ and $(20, 20)$ tubes give rise to the most stable configurations. We refer the reader to Section 4.4 for the calculation details for the preferred location of a C_{60} molecule inside a carbon nanotube.

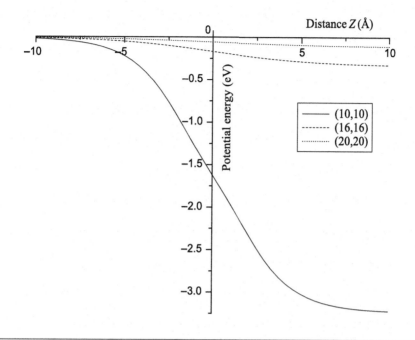

FIG. 7.3

Energy profile for C_{60} encapsulated by head-on at open end.

7.2.2 ENCAPSULATION OF C_{60} AROUND THE EDGE OF AN OPEN END

In this section the energy for a C_{60} molecule encapsulated into a carbon nanotube by entering the tube around the tube edge at the open end is investigated. With reference to the same rectangular Cartesian coordinate system (x, y, z), a typical point on the surface of the tube has the coordinates $(b \cos \theta, b \sin \theta, z)$, where b is the radius of the semi-infinite tube. Similarly, with reference to the rectangular Cartesian coordinate system (x, y, z), the centre of the C_{60} molecule has the coordinates $(x, 0, Z)$, where Z is the distance in the z-direction, which can be either positive or negative. The distance Z and the coordinate x can also be described in terms of an angle ϕ and the distance r in the radial direction, $Z = r \cos \phi$ and $x = r \sin \phi + b$, as illustrated in Fig. 7.4. Thus the distance δ between the centre of the C_{60} fullerene and a typical point on the tube is given by

$$\delta^2 = (b \cos \theta - x)^2 + b^2 \sin^2 \theta + (z - Z)^2$$
$$= (b - x)^2 + 4bx \sin^2(\theta/2) + (z - Z)^2. \tag{7.7}$$

The total potential energy is obtained by integrating $E^*(\delta)$, which is defined by Eq. (7.2), over the tube length and the angle θ. Thus, there is only one form of the integral for the total potential energy E which needs to be evaluated, and this is given by

$$H_n = \int_{-\pi}^{\pi} \int_0^{\infty} \frac{1}{(\delta^2 - a^2)^n} dz d\theta, \tag{7.8}$$

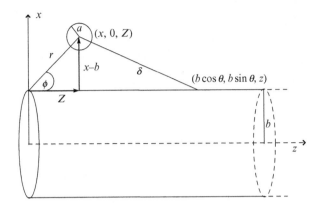

FIG. 7.4

C_{60} fullerene encapsulated in carbon nanotube around the edge of an open end.

where δ is given by Eq. (7.7). Furthermore, there are three possible expressions arising from Eq. (7.8), and these details are presented in Worked Example 7.2. Although the analytical details for Eq. (7.8) are clearly complicated, numerical values may be readily evaluated using the algebraic computer package MAPLE. Note that the total potential energy in terms of the distance r and the angle ϕ can be obtained by replacing $Z = r\cos\phi$ and $x = r\sin\phi + b$.

WORKED EXAMPLE 7.2

Determine an analytical expression for the integral H_n, as defined by Eq. (7.8).

Solution

The integral H_n is defined in the form of Eq. (7.8), where n is a certain positive integer. Letting $\lambda^2 = (b-x)^2 + 4bx\sin^2(\theta/2) - a^2$, it can be deduced

$$H_n = \int_{-\pi}^{\pi}\int_0^{\infty} \frac{1}{[\lambda^2 + (z-Z)^2]^n}\, dz\, d\theta.$$

Making the substitution $u = z - Z$, H_n becomes

$$H_n = \int_{-\pi}^{\pi}\int_{-Z}^{\infty} \frac{1}{(\lambda^2 + u^2)^n}\, du\, d\theta = \int_{-\pi}^{\pi}\int_{-\tan^{-1}(Z/\lambda)}^{\pi/2} \frac{\lambda\sec^2\psi}{\lambda^{2n}\sec^{2n}\psi}\, d\psi\, d\theta,$$

where the final line is obtained by substituting $x = \lambda\tan\psi$. Finally, H_n simplifies to become

$$H_n = \int_{-\pi}^{\pi}\int_{-\tan^{-1}(Z/\lambda)}^{\pi/2} \frac{1}{\lambda^{2n-1}}\cos^{2(n-1)}\psi\, d\psi\, d\theta,$$

which gives the same result as that given in Worked Example 7.1, namely Eq. (7.6). In evaluating the above equation at $\psi = \pi/2$ and $\psi = -\tan^{-1}(Z/\lambda)$, there are three forms of the integral for $\theta \in (0, \pi/2)$ which need to be determined, and these are given by

$$J_s = \int_0^{\pi/2} \frac{dv}{\lambda^s}, \quad K_{s,t} = \int_0^{\pi/2} \frac{dv}{\lambda^t(\lambda^2 + Z^2)^s}, \quad L_s = \int_0^{\pi/2} \frac{1}{\lambda^s}\tan^{-1}\left(\frac{Z}{\lambda}\right) dv, \tag{7.9}$$

where now λ is given by $\lambda^2 = (b-x)^2 + 4bx \sin^2(v/2) - a^2$. These integrals are evaluated in Baowan et al. (2007), and the detailed integrations can also be found in Worked Examples 5.2, 5.3, and 5.4. Explicitly, the integrals in Eq. (7.9) can be performed to yield

$$J_s = \frac{\pi}{2(b+x)^s} F\left(\frac{s}{2}, \frac{1}{2}; 1; \frac{4bx}{(b+x)^2}\right),$$

$$K_{s,t} = \frac{\pi}{2(b+x)^t \left[(b+x)^2 + Z^2\right]^s} \sum_{i=0}^{\infty} \frac{(1/2)_i (t/2)_i}{(i!)^2} F\left(\frac{1}{2} + i, s; 1 + i; \frac{4bx}{(b+x)^2 + Z^2}\right)$$
$$\times \left[\frac{4bx}{(b+x)^2}\right]^i,$$

$$L_s = \frac{\pi}{2(b+x)^s} \sum_{k=0}^{\infty} \sum_{i=0}^{\infty} \frac{Z^{2k+1}(2k)!}{2^{2k}(k!)^2(2k+1)\left[(b+x)^2 + Z^2\right]^{k+1/2}} \frac{(1/2)_i (s/2)_i}{(i!)^2}$$
$$\times F\left(\frac{1}{2} + i, \frac{1}{2} + k; 1 + i; \frac{4bx}{(b+x)^2 + Z^2}\right) \left[\frac{4bx}{(b+x)^2}\right]^i,$$

where $F(a, b; c; z)$ denotes the standard hypergeometric function.

To confirm our results the numerical evaluation for the encapsulation of the C_{60} molecule around the edge of the tube end is determined using both polar and Cartesian coordinate systems. Using $x = r \sin\phi + b$ and $Z = r \cos\phi$, the relation between the binding energy and the equilibrium distance for different angles ϕ is presented in Table 7.1. The lowest binding energy is observed to occur at $\phi \approx 165°$ for all three cases due to the edge effect. Consequently, this value of ϕ is the critical value which determines whether the C_{60} molecule is encapsulated into the tube. The equilibrium distance (E_0), which is the distance between the tube edge and the centre of the fullerene at equilibrium, are obtained as 6.775, 6.540, and 6.550 Å for $\phi = 270°$ and for each of the (10, 10), (16, 16), and (20, 20) tubes, respectively. These values are equivalent to 0.009, 4.306, and 7.007 Å, respectively, for the offset location, which is the distance away from the tube axis to the centre of the C_{60} fullerene in the x-direction.

Using the Cartesian coordinate system (x, y, z), the potential energy of the system depends on both distances in the x- and z-directions. An example of the potential energy versus the distance Z for the encapsulation of the C_{60} fullerene into the (10, 10) tube is presented. Primarily, our interest is in the positive z-direction, where the C_{60} molecule is located above the tube. As shown in Fig. 7.5, the C_{60} fullerene will not be encapsulated into the tube if its location is too far from the edge of the tube. This is because of the lower energy level at that position and the high energy peak near the tube end. If the value of x is greater than 13.034 Å, then the C_{60} fullerene has no chance of being sucked into the carbon nanotube because the global minimum energy position is located further along the tube in the positive z-direction. We note that if the C_{60} molecule overcomes the energy barrier and becomes positioned on the negative z-axis, then the analysis for the suction by head-on applies for the encapsulation.

Table 7.1 Numerical Values for Binding Energy (BE) in eV and the Equilibrium Distance (E_0) in Å for a C_{60} Fullerene Encapsulated in a Carbon Nanotube Around the Tube Edge at the Open End for Different Angles ϕ

	(10, 10)		(16, 16)		(20, 20)	
ϕ	BE	E_0	BE	E_0	BE	E_0
15°	0.53424	25.16055	0.58315	25.14903	0.60398	25.14903
30°	0.53026	13.01349	0.57883	13.00486	0.59953	13.00486
45°	0.51050	9.18529	0.55756	9.20351	0.57737	9.18603
60°	0.45467	7.53991	0.49675	7.56151	0.51479	7.54922
75°	0.35970	6.77549	0.39343	6.79463	0.40775	6.81484
90°	0.26722	6.47536	0.29169	6.51166	0.30211	6.51166
105°	0.20322	6.32529	0.22146	6.36167	0.22892	6.35916
120°	0.16640	6.30654	0.17929	6.32301	0.18476	6.32103
135°	0.14647	6.26761	0.15413	6.21275	0.15967	6.28291
150°	0.13894	6.26761	0.14378	6.21275	0.14722	6.28291
165°	0.14169	6.26761	0.14250	6.21275	0.14412	6.28291
180°	0.15563	6.26761	0.14998	6.21275	0.14929	6.28291
195°	0.18511	6.26761	0.16666	6.21275	0.16318	6.28291
210°	0.24079	6.30654	0.19779	6.30216	0.18779	6.28291
225°	0.34809	6.34427	0.24633	6.30216	0.22694	6.32103
240°	0.56623	6.44215	0.32206	6.34055	0.28746	6.32103
255°	1.01175	6.66655	0.43751	6.42551	0.38012	6.40225
270°	1.62119	6.77519	0.60665	6.53999	0.51827	6.55048

7.2.3 ENCAPSULATION OF C_{60} AT A DEFECT ON THE TUBE WALL

In this section, we determine the potential energy for a C_{60} fullerene encapsulated into a carbon nanotube at a rectangular-shaped defect opening on the tube wall, which is centrally located midway along the tube length. In terms of the interaction for the C_{60} entering the tube through the defect, the Lennard-Jones potential is only effective at short range, and therefore the carbon nanotube may be assumed to be infinite in length. The total potential energy of the system is obtained by subtracting the total energy of the C_{60} fullerene interacting with the defect pad from the total potential energy of the C_{60} fullerene interacting with the entire infinite carbon nanotube, as illustrated in Fig. 7.6.

Again with reference to the rectangular Cartesian coordinate system (x, y, z), a typical point on the surface of the tube has the coordinates $(b \cos \theta, b \sin \theta, z)$, where b is the radius of the infinite tube. Similarly, with reference to the rectangular Cartesian coordinate system (x, y, z) with the origin located at the centre of the tube, the centre of the C_{60} molecule is assumed to have coordinates $(x, 0, Z)$, where Z is the distance in the z-direction and can be either positive or negative. Thus the distance δ between the centre of the C_{60} fullerene and a typical point on the tube is again given by Eq. (7.7). The total potential energy for the entire tube interacting with the C_{60} fullerene is given by

$$E_{tube} = b \eta_g \int_{-\pi}^{\pi} \int_{-\infty}^{\infty} E^*(\delta) dz d\theta, \tag{7.10}$$

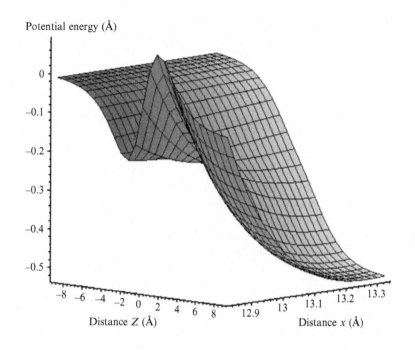

FIG. 7.5

Energy profile for C_{60} encapsulated into (10, 10) tube.

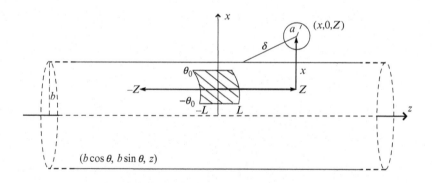

FIG. 7.6

C_{60} fullerene encapsulated in carbon nanotube through a defect opening on the tube wall.

where η_g denotes the mean atomic surface density of the carbon nanotube, $E^*(\delta)$ is defined by Eq. (7.2) and δ is given in Eq. (7.7). Note that Eq. (7.10) is the same equation as Eq. (4.12). The defect pad is assumed to occupy the region $Z \in (-L, L)$ and $\theta \in (-\theta_0, \theta_0)$ so that the interaction energy between the C_{60} molecule and the defect pad is given by

$$E_{pad} = b\eta_g \int_{-\theta_0}^{\theta_0} \int_{-L}^{L} E^*(\delta)dzd\theta, \tag{7.11}$$

where δ is again given by Eq. (7.7). Thus the total potential energy for the C_{60} fullerene encapsulated in the carbon nanotube at the defect opening on the tube wall is obtained from

$$E = b\eta_g \left(\int_{-\pi}^{\pi} \int_{-\infty}^{\infty} E^*(\delta)dzd\theta - \int_{-\theta_0}^{\theta_0} \int_{-L}^{L} E^*(\delta)dzd\theta \right). \tag{7.12}$$

By precisely the same analytical method as shown in Section 7.2.2, Eqs (7.10) and (7.11) are separately determined, and the total potential energy (7.12) is numerically calculated for the system.

The defect pad is arbitrarily chosen to be a square such that the length L is the radius a of the C_{60} fullerene plus the equilibrium interspacing between the C_{60} fullerene and the graphene, which is 3.25 Å so that $L = a + 3.25 = 6.8$ Å. Using the arc length formula $s = b\theta$, the limit of the integration θ_0 is adopted to be determined from $L = b\theta_0$. Note that varying θ_0 has only a minor effect on the energy profile and that the overall properties of the system remain the same when L is greater than the critical value 6.8 Å.

The relation between the potential energy and the distance Z for different values of x, which is the interspacing distance between the C_{60} molecule and the tube wall, is examined, and all cases have a similar behaviour. An example for the energy profile for the interaction between the C_{60} molecule and a (10, 10) tube is shown in Fig. 7.7. In terms of the binding energy, this energy is concentrated at both

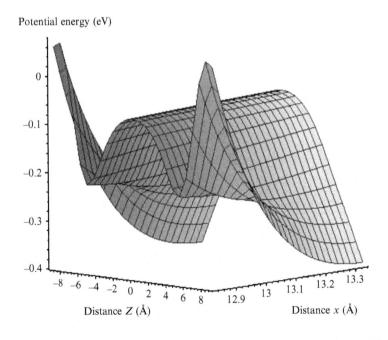

FIG. 7.7

Potential energy profile along the (10, 10) tube.

edges of the defect pad and arises from the point-force singularities operating at the defect edges. In this case an approximate value of 0.225 eV is obtained for the binding energy from both edges of the defect pad. Moreover, two potential energy peaks near the edges of the defect pad for $x \leq 13.034$ Å are observed, so that if the C_{60} molecule is located outside the region of the pad, an initial energy is required for the C_{60} fullerene to be absorbed into the nanotube. However, the C_{60} molecule is spontaneously sucked in through the defect opening when its position is directly above the defect. Furthermore, if the value of x is greater than 13.034 Å, the global minimum energy position is always located outside the region of the pad along the tube in the z-position. Subsequently, the C_{60} fullerene will not be adsorbed through the pad, and a nanopeapod cannot be formed. We note that because we assume a constant temperature $T = 0°$ K throughout the book, any thermal fluctuations would tend to elevate the global minimum energy.

WORKED EXAMPLE 7.3

Show that if $x = 13.034$ Å, a C_{60} molecule has no chance of being encapsulated into a $(10, 10)$ carbon nanotube.

Solution

This Worked Example uses MAPLE to plot the relationship between the total potential energy versus the distance Z. We begin by defining the parameters, which can be found in Tables 1.5–1.7.

```
> restart:
> a:= 3.55:
> nf:= 0.3789: ng:= 0.3812:
> A:= 17.4: B:= 29000:
> L:=(a+3.25):
> t:= b -> L/b:
```

The distance between the centre of the C_{60} molecule and the surface element of a carbon nanotube is given by Eq. (7.7).

```
> d:= (b,x) -> sqrt((b-x)^2 + 4*b*x*sin(theta/2)^2+(z-Z)^2):
```

We utilise the expanded expression of Eq. (7.2) for the interaction energy between an atom and a C_{60} molecule. Subsequently, the numerical integration command Int() is employed to determine the total potential energy of the system.

```
> P:= r -> 4*Pi*a^2*nf*((B*( 5/(r^2-a^2)^6 + 80*a^2/(r^2-a^2)^7
  + 336*a^4/(r^2-a^2)^8 + 512*a^6/(r^2-a^2)^9 + 256*a^8/(r^2-a^2)
  ^(10))/5 - A*(1/(r^2-a^2)^3 + 2*a^2/(r^2-a^2)^4 ))):
```

The total energy of the C_{60} molecule interacting with the defect is obtained as

```
> E1:= (b,x) -> b*ng*Int(Int(P(d(b,x)),theta=-Pi..Pi),
  z=-infinity..infinity);
```

The total potential energy of the C_{60} molecule interacting with the infinite carbon nanotube is obtained as

```
> E2:= (b,x) -> b*ng*Int(Int(P(d(b,x)),theta=-t(b)..t(b)),z=-L..L);
```

The total potential energy of the system is obtained by subtracting E_1 from E_2.

```
> Energy:= (b,x) -> E1(b,x)-E2(b,x):
```

At $x = 13.034$ Å and $b = 6.784$ Å, the relation between the total potential and the distance Z can be illustrated as follows.

```
> plot(Energy(6.784,13.034),Z=-10..10);
```

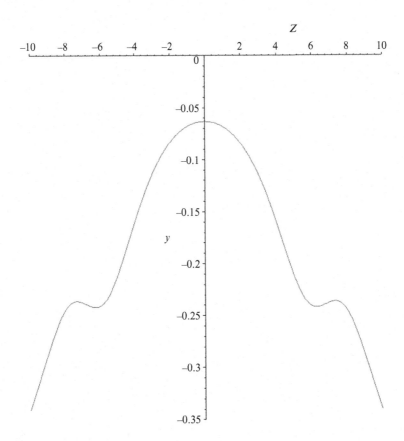

From this graph, we can see that the global minimum energy position is always located outside the region of the pad along the tube in the z-position, and therefore, the C_{60} fullerene will not be encapsulated through the pad. However, the energy barrier for this case is a very small fraction of 1 eV, so the C_{60} molecule may be accepted into the tube if an initial force is given.

7.2.4 ZIGZAG NANOPEAPODS COMPRISING $(2K + 1)$ C_{60} MOLECULES

In this subsection and the next, we use the continuum approximation and the Lennard-Jones potential function to determine the potential energy of a nanopeapod, which is assumed to form either a zigzag or a spiral configuration. The analysis for zigzag nanopeapods comprising $(2k + 1)$ C_{60} molecules is presented here. The investigation for nanopeapods with a spiral configuration and comprising k C_{60}

Linear chain

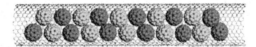

Zigzag configuration

Spiral configuration

FIG. 7.8

Three configurations for C_{60}-nanopeapods.

Adapted with permission from Troche et al., 2005. Copyright 2005 American Chemical Society.

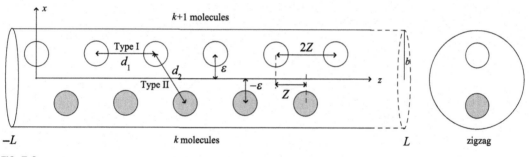

FIG. 7.9

Zigzag configuration for $(2k + 1)$ C_{60} fullerenes inside a carbon nanotube.

molecules is presented in Section 7.2.5. In Fig. 7.8, we show three possible configurations for C_{60} peapods as obtained from the molecular dynamics simulations of Troche et al. (2005). We note that here we simply assume the general nature of these configurations, and then deduce the numerical values of the spacing and the offset parameters to be those for which the interaction energy is minimised. It would be a far more arduous investigation to determine the configurations and well beyond the scope of this book. Rather than employing such large-scale computations, here we utilise classical mathematical modelling to determine such configurations.

We assume that the preferred pattern for a chain of C_{60} fullerenes inside a single-walled carbon nanotube is a zigzag pattern and that the interaction energy is determined in the following manner: A configuration, as shown in Fig. 7.9, is assumed to comprise $(2k + 1)$ C_{60} molecules located as indicated, and the total energy of the system is assumed to comprise:

(i) $(2k + 1)$ C_{60} fullerenes each interacting with all the carbon atoms of the carbon nanotube,

(ii) $2(2k − 1)$ type I interactions, comprising two for each $j = 2, 3, \ldots, k$ and $i = 2, 3, \ldots, k − 1$ and one for each of $j = 1, k + 1$ and $i = 1, k$; thus $2(k − 1 + k − 2) + 4 = 2(2k − 1)$,

(iii) $4k$ type II interactions, comprising two for each $j = 2, 3, \ldots, k$ and $i = 1, 2, \ldots, k$ and one for each of $j = 1, k + 1$; thus $2(k − 1 + k) + 2 = 4k$,

where the nearest neighbour interactions of types I and II are shown in Fig. 7.9. Furthermore, all other non-nearest neighbour interactions are assumed to be sufficiently small such that their contributions to the total energy can be neglected. We note that the van der Waals force is a short-range force so that, for example, for two interacting C_{60} fullerenes, the nearest neighbour approximation involves a distance between fullerene centres of at least 10.0550 Å, and therefore only the nearest neighbour interactions need to be considered. The distance in the z direction between centres of adjacent molecules is assumed to be Z, so that the centre of the upper jth C_{60} molecule ($j = 1, 2, \ldots, k + 1$) is located at position $2(j − 1)Z$, while the centre of the lower ith C_{60} molecule ($i = 1, 2, \ldots, k$) is located at the position $Z + 2(i − 1)Z = (2i − 1)Z$.

With reference to the rectangular Cartesian coordinate system (x, y, z) located as indicated in Fig. 7.9, a typical point on the surface of the tube has coordinates $(b \cos \theta, b \sin \theta, z)$, where b is the radius of the tube.

The length of the tube is assumed to be $2L$, where L may tend to infinity. Again, with reference to the rectangular Cartesian coordinate system (x, y, z) with the origin located on the tube axis and through the centre of the leftmost C_{60} molecule, the centre of the upper jth C_{60} molecule has coordinates $(\varepsilon, 0, 2Z(j − 1))$ ($j = 1, 2, \ldots, (k + 1)$) and the centre of the lower ith C_{60} molecule has coordinates $(−\varepsilon, 0, Z(2i − 1))$ ($i = 1, 2, \ldots, k$), where Z is the distance between centres of adjacent C_{60} fullerenes and ε is the offset position from the centre of the tube to the centre of the C_{60} fullerene in the x direction, as illustrated in Fig. 7.9.

In the case of a many-body problem, the potential energy of the system is the total energy between each pair of molecules, which is called the pair potential approximation, and is given by

$$E^{tot} = \frac{1}{2} \sum_{i,j=1, i \neq j}^{N} E(\rho_{ij}), \tag{7.13}$$

where ρ_{ij} denotes the distance between a surface element on molecule i and a surface element on molecule j, and $1/2$ denotes the elimination of double counting. Thus from Eq. (7.13), the total potential energy for the zigzag configuration is obtained by

$$E^{tot} = \sum_{i=1}^{k} E_i(\delta_i) + \sum_{j=1}^{k+1} E_j(\delta_j) + (2k − 1)E^{**}(d_1) + 2kE^{**}(d_2),$$

where d_1 and d_2 are the distances between the centres of C_{60} fullerenes, as shown in Fig. 7.9, and $d_1^2 = 4Z^2$ and $d_2^2 = 4\varepsilon^2 + Z^2$. The potential functions $E^{**}(d_1)$ and $E^{**}(d_2)$ arise from type I and type II interactions between a pair of C_{60} molecules, respectively, and are defined by

$$E^{**}(d) = −AP_6 + BP_{12}, \tag{7.14}$$

where

$$P_n = \frac{4\pi^2 a^2 \eta_f^2}{d(2-n)(3-n)} \left(\frac{1}{(2a+d)^{n-3}} - \frac{1}{d^{n-3}} - \frac{1}{(2a-d)^{n-3}} + \frac{1}{(-d)^{n-3}} \right), \tag{7.15}$$

as derived in Chapter 3.

The potential functions E_i and E_j represent the energy of a C_{60} fullerene interacting with the carbon nanotube, which is obtained from

$$E_m = b\eta_g \int_{-\pi}^{\pi} \int_{-L}^{L} E^*(\delta_m) dz d\theta \quad (m = i, j) \tag{7.16}$$

where η_g is the mean atomic surface density for the carbon nanotube and the length L is subsequently taken to be infinite. The potential function E^* is defined by Eq. (7.2) and δ_m ($m = i$ and j) are given by

$$\delta_i^2 = (b+\varepsilon)^2 - 4b\varepsilon \sin^2(\theta/2) + [z - Z(2i-1)]^2,$$
$$\delta_j^2 = (b-\varepsilon)^2 + 4b\varepsilon \sin^2(\theta/2) + [z - 2Z(j-1)]^2.$$

According to Eqs (5.8), (5.9), and (7.16), the following integral needs to be evaluated

$$I_n = \int_{-\pi}^{\pi} \int_{-\infty}^{\infty} \frac{1}{(\delta_m^2 - a^2)^n} dz d\theta, \tag{7.17}$$

where n is an integer. The details for evaluating Eq. (7.17) are presented in Worked Example 7.4, where an analytical expression is determined, which can be written as

$$I_n = \frac{\pi^2}{2^{2n-3}(\alpha_m + \beta_m)^{n-1/2}} \binom{2(n-1)}{n-1} F\left(n - \frac{1}{2}, \frac{1}{2}; 1; 1 - \gamma_m\right),$$

where $F(a, b; c; z)$ denotes the usual hypergeometric function, $\binom{x}{y}$ represents the usual binomial coefficient, $\gamma_m = \alpha_m/(\alpha_m + \beta_m)$ ($m = i$ and j), $\alpha_i = (b+\varepsilon)^2 - a^2$, $\beta_i = -4b\varepsilon$, $\alpha_j = (b-\varepsilon)^2 - a^2$, and $\beta_j = 4b\varepsilon$.

WORKED EXAMPLE 7.4

Determine an analytical expression for the integral I_n, as defined by Eq. (7.17).

Solution

The integral I_n is given by Eq. (7.17), where $m = i$ and j. By letting $\lambda_i^2 = (b+\varepsilon)^2 - 4b\varepsilon \sin^2(\theta/2) - a^2$ and $\lambda_j^2 = (b-\varepsilon)^2 + 4b\varepsilon \sin^2(\theta/2) - a^2$, I_n becomes

$$I_n = \int_{\pi}^{-\pi} \int_{-L}^{L} \frac{1}{[\lambda_m^2 + (z + Z_m)^2]^n} \, dz \, d\theta,$$

where $Z_i = Z(2i - 1)$ $(i = 1, 2, \ldots, k)$ and $Z_j = 2Z(j - 1)$ $(j = 1, 2, \ldots, k + 1)$. Upon making the substitution $x_m = z + Z_m$, it can be deduced that

$$I_n = \int_{\pi}^{-\pi} \int_{Z_m - L}^{Z_m + L} \frac{1}{(\lambda_m^2 + x_m^2)^n} \, dx_m \, d\theta = \int_{-\pi}^{\pi} \int_{-\pi/2}^{\pi/2} \frac{\lambda_m \sec^2 \psi}{\lambda_m^{2n} \sec^{2n} \psi} \, d\psi \, d\theta,$$

where the final line is obtained by substituting $x_m = \lambda_m \tan \psi$ and letting L tend to infinity. Finally, I_n simplifies to become

$$I_n = \int_{-\pi}^{\pi} \int_{-\pi/2}^{\pi/2} \frac{1}{\lambda_m^{2p+1}} \cos^{2p} \psi \, d\psi \, d\theta, \tag{7.18}$$

where $p = n - 1$. The evaluation of Eq. (7.18) can be found in Gradshteyn and Ryzhik (2000) (p. 149, No. 2.513 3) and is given by Eq. (7.6). By evaluating Eq. (7.6) at $\psi = \pi/2$ and $\psi = -\pi/2$ and by the fact that $\sin 2x = 2 \sin x \cos x$, it may be deduced that

$$I_n = \frac{\pi}{2^{2p}} \binom{2p}{p} \int_{-\pi}^{\pi} \frac{1}{\lambda_m^{2p+1}} \, d\theta = \frac{4\pi}{2^{2p}} \binom{2p}{p} \int_0^{\pi/2} \frac{1}{\lambda_m^{2p+1}} \, dx,$$

where $x = \theta/2$ and $\lambda_i^2 = (b + \varepsilon)^2 - 4b\varepsilon \sin^2 x - a^2$ and $\lambda_j^2 = (b - \varepsilon)^2 + 4b\varepsilon \sin^2 x - a^2$. The analytical evaluation of this integral can be found in Worked Example 5.2.

7.2.4.1 Numerical solutions for zigzag nanopeapods

By minimising the total energy of the system, the offset location ε from the centre of the tube to the centre of the C_{60} fullerene, and the equilibrium distance Z between centres of a pair of C_{60} molecules for zigzag nanopeapods are determined. The total potential energy consists of two nearest neighbour interactions of two C_{60} fullerenes and one interaction between the C_{60} fullerene and the carbon nanotube. Infinite length nanopeapods comprising $(2k + 1)$ C_{60} molecules inside $(10, 10)$, $(16, 16)$, and $(20, 20)$ carbon nanotubes are examined. Using the algebraic computer package MAPLE together with the parameter values in Tables 1.5–1.7, the numerical values for the offset location ε, the equilibrium distance Z and the total potential energy E^{tot} are presented in Table 7.2. Note that the global minimum energy location of the system is first plotted to ensure a genuine global minimum, and the optimisation package in MAPLE is then utilised to find the optimum values for each parameter at this location.

In the case of the $(10, 10)$ carbon nanotube an offset position is obtained of $\varepsilon = 0$, which is equivalent to a distance of $3.234 \, \text{Å}$ from the tube wall to the nearest atom on the C_{60} molecule. The equilibrium distance is shown to be $Z = 10.0550 \, \text{Å}$ for three C_{60} molecules inside the tube. As a result, all C_{60} fullerenes inside the $(10, 10)$ tube are likely to align and form a linear chain along the tube axis. The equilibrium distance decreases slightly as the number of the C_{60} molecules is increased due to the packing of the molecules. Moreover, the C_{60} fullerenes move closer to the wall as the radius of the tube increases. The offset positions of $\varepsilon = 4.2977 \, \text{Å}$ and $\varepsilon = 7.0213 \, \text{Å}$ are obtained; these are equivalent to the equilibrium distances of $Z = 5.02176 \, \text{Å}$ and $Z = 5.0267 \, \text{Å}$ for the $(16, 16)$ and $(20, 20)$ carbon nanotubes, respectively. For these two cases the zigzag pattern is more clearly evident along the tube. However, for the three C_{60} fullerenes inside the $(20, 20)$ carbon nanotube, the

Table 7.2 Equilibrium Distance Z (Å), Offset Location ε (Å) and Total Potential Energy of the System E^{tot} (eV) for Each Pair of C_{60} Fullerenes in a Zigzag Configuration Nanopeapod Comprising $(2k+1)$ C_{60} Molecules

k	(10, 10)			(16, 16)			(20, 20)		
	Z	ε	E^{tot}	Z	ε	E^{tot}	Z	ε	E^{tot}
1	10.0550	0	−6.7632	5.2176	4.2977	−2.7048	0	7.0213	−2.0941
2	10.0543	0	−13.8074	5.0390	4.3216	−6.2354	5.0267	7.0220	−4.7420
3	10.0542	0	−20.8516	5.0366	4.3232	−9.7692	5.0269	7.0217	−7.3973
4	10.0542	0	−27.8958	5.0358	4.3239	−13.3031	5.0269	7.0215	−10.0526
5	10.0541	0	−34.9400	5.0354	4.3244	−16.8370	5.0269	7.0214	−12.7079
10	10.0541	0	−70.1612	5.0347	4.3251	−34.5067	5.0270	7.0212	−25.9845
15	10.0540	0	−105.3823	5.0345	4.3255	−52.1764	5.0270	7.0211	−39.2611
20	10.0540	0	−140.6034	5.0344	4.3255	−69.8460	5.0270	7.0211	−52.5377
25	10.0540	0	−175.8245	5.0344	4.3256	−87.5157	5.0270	7.0210	−65.8143
50	10.0540	0	−351.9301	5.0343	4.3257	−175.8641	5.0270	7.0210	−132.1973
100	10.0540	0	−704.1413	5.0342	4.3258	−352.5610	5.0270	7.0210	−274.9634

equilibrium distance is obtained as $Z = 0$, which means that although a zigzag pattern exists, all three of the C_{60} molecules are in the same plane. This is because there is a sufficient amount of space for the three C_{60} molecules to align themselves along the diameter due to the large circumference of the tube.

The offset locations for all three nanopeapod configurations found in this investigation are in strong agreement with the case of a single C_{60} fullerene inside a single-walled carbon nanotube (see Section 4.4). Moreover, the interaction energy between the C_{60} fullerenes is observed to have more influence on the chain formation than the interaction energy between the tube and the C_{60} fullerene. For example, an equilibrium distance of 10.054 Å is obtained for a (10, 10) nanopeapod, which is comparable to the equilibrium distance between two C_{60} molecules. Furthermore, the number of C_{60} molecules in the system makes only a minor contribution to the alignment of the molecules, as shown in Table 7.2, and we might expect that the equilibrium spacing for two C_{60} molecules to be almost identical to that for three C_{60} molecules, as we assume that the intermolecular interactions are dominated by pairwise interactions.

7.2.5 SPIRAL NANOPEAPOD COMPRISING K C_{60} MOLECULES

In this section, a spiral configuration is assumed for k C_{60} fullerenes, which are located inside a single-walled carbon nanotube, as shown in Fig. 7.10. More precisely the fullerene centres are assumed to lie on the three-dimensional helix $(\varepsilon \cos \alpha t, \varepsilon \sin \alpha t, \beta t)$, and the energy of the system is minimised such that the values of the angular spacing α, the longitudinal spacing β and the offset location ε are determined so that the total potential energy is minimised, namely $\partial E^{tot}/\partial \alpha = \partial E^{tot}/\partial \beta = \partial E^{tot}/\partial \varepsilon = 0$. The total potential energy of the system is assumed to comprise:

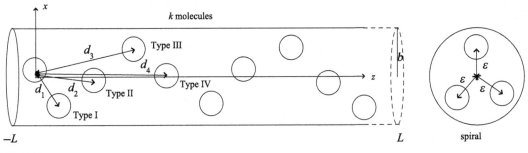

FIG. 7.10

Spiral configuration for k C_{60} molecules inside a carbon nanotube.

(i) k C_{60} fullerenes each interacting with all the carbon atoms of the carbon nanotube,

(ii) $2(k-1)$ type I interactions, comprising two for each $i = 2, 3, \ldots, k-1$ and one for each of $i = 1$ and $i = k$; thus $2(k-2) + 2 = 2(k-1)$,

(iii) $2(k-2)$ type II interactions, comprising two for each $i = 3, 4, \ldots, k-2$ and one for each of $i = 1, 2$ and $i = k-1, k$; thus $2(k-4) + 4 = 2(k-2)$,

(iv) $2(k-3)$ type III interactions, comprising two for each $i = 4, 5, \ldots, k-3$ and one for each of $i = 1, 2, 3$ and $i = k-2, k-1, k$; thus $2(k-6) + 6 = 2(k-3)$,

(v) $2(k-4)$ type IV interactions, comprising two for each $i = 5, 6, \ldots, k-4$ and one for each of $i = 1, 2, 3, 4$ and $i = k-3, k-2, k-1, k$; thus $2(k-8) + 8 = 2(k-4)$,

where the four nearest neighbour interactions of types I, II, III and IV are as shown in Fig. 7.10.

With reference to the rectangular Cartesian coordinate system (x, y, z) shown in Fig. 7.10, a typical point on the surface of the tube has the coordinates $(b \cos \theta, b \sin \theta, z)$, where b is the radius of the tube. The length of the tube is assumed to be $2L$, where L can tend to infinity. Similarly, with reference to the rectangular Cartesian coordinate system (x, y, z) with the origin located on the tube axis and through the centre of the left most C_{60} molecule, the centres of C_{60} molecules are assumed to have coordinates $(\varepsilon \cos \alpha i, \varepsilon \sin \alpha i, \beta i)$ $(i = 1, 2, \ldots, k)$, where ε represents the offset location, and α and β denote the angular and the longitudinal spacings for the helix, respectively. We note that $\alpha = \pi$, which gives rise to the special case of the zigzag pattern previously considered. From the potential energy for a many-body system (7.13), the total potential energy is given by

$$E^{tot} = \sum_{i=1}^{k} E_i(\delta_i) + (k-1)E^{**}(d_1) + (k-2)E^{**}(d_2) + (k-3)E^{**}(d_3)$$
$$+ (k-4)E^{**}(d_4), \tag{7.19}$$

where d_ℓ ($\ell = 1, 2, 3, 4$) are the distances between centres of C_{60} fullerenes, as shown in Fig. 7.10, and

$$d_\ell^2 = 4\varepsilon^2 \sin^2(\ell \alpha/2) + (\ell \beta)^2. \tag{7.20}$$

The potential function $E^{**}(d_\ell)$ represents types I, II, III, and IV interactions, which are the potential energies between a pair of C_{60} fullerenes defined by Eq. (7.14). The potential function E_i ($i = 1, 2, \ldots, k$) represents the energy of a C_{60} fullerene interacting with the carbon nanotube, which is obtained from Eq. (7.16), where in this case $m = i$ and $i = 1, 2, \ldots, k$, and the function E^* is defined by Eq. (7.2), where δ_i ($i = 1, 2, \ldots, k$) is given by

$$\delta_i^2 = (b - \varepsilon)^2 + 4b\varepsilon \sin^2[(\theta - \alpha i)/2] + (z - \beta i)^2.$$

Due to the assumed symmetry of the tube, the term αi has no effect for the integral in Eq. (7.16), and so assuming $\alpha i \equiv 0$, δ_i^2 simplifies to obtain

$$\delta_i^2 = (b - \varepsilon)^2 + 4b\varepsilon \sin^2(\theta/2) + (z - \beta i)^2.$$

In the limit as the length of the tube L tends to infinity, the following integral needs to be evaluated

$$I_n = \int_{-\pi}^{\pi} \int_{-\infty}^{\infty} \frac{1}{(\delta_i^2 - a^2)^n} \, dz \, d\theta, \tag{7.21}$$

where n is an integer. Using precisely the same method as the derivation of Eq. (7.17), it can be shown that the evaluation of Eq. (7.21) is again in the form of the hypergeometric function and can be written as

$$I_n = \frac{\pi^2}{2^{2n-3}(\alpha_i + \beta_i)^{n-1/2}} \binom{2(n-1)}{n-1} F\left(n - \frac{1}{2}, \frac{1}{2}; 1; 1 - \gamma_i\right),$$

whereas before $F(a, b; c; z)$ denotes the usual hypergeometric function, $\binom{x}{y}$ represents the usual binomial coefficient, $\alpha_i = (b - \varepsilon)^2 - a^2$, $\beta_i = 4b\varepsilon$, and $\gamma_i = [(b - \varepsilon)^2 - a^2]/[(b + \varepsilon)^2 - a^2]$.

7.2.5.1 Numerical solutions for spiral nanopeapods

The energy minimisation technique is employed here to determine the stable configurations of a spiral chain of C_{60} fullerenes inside a single-walled carbon nanotube. Nanopeapods comprising k C_{60} molecules inside infinite (10, 10), (16, 16), and (20, 20) carbon nanotubes with four possible nearest neighbour interactions for two C_{60} molecules and one interaction between the C_{60} molecule and all the atoms of the carbon nanotube are considered. Again, using the algebraic computer package MAPLE and the parameter values in Tables 1.5–1.7, numerical values for the angular spacing α, the longitudinal spacing β, the offset location ε and the total potential energy E^{tot} for such a chain are obtained and are shown in Table 7.3. Note that β is analogous to the equilibrium distance Z for the zigzag configuration.

For the (10, 10) carbon nanotube, the offset location is also obtained as $\varepsilon = 0$. Moreover, from Eq. (7.20), the angular spacing α has little effect on this configuration, and the longitudinal spacing β is found to be 10.0545 Å. Subsequently, the C_{60} fullerenes form a linear chain along the tube axis. These three parameters, α, β, and ε, change slightly as the number of C_{60} fullerenes in the tube increases. The angular spacing is $\alpha \simeq \pi$ for the (16, 16) tube, which corresponds to the zigzag

Table 7.3 Angular Spacing α, Longitudinal Spacing β, Offset Location ε in Å and Total Potential Energy of the System E^{tot} (eV) for Each Pair of C_{60} Fullerenes in a Spiral Configuration Nanopeapod Comprising k C_{60} Molecules

tube	k	α	β	ε	E^{tot}
(10, 10)	3	0	10.0545	0	−10.2852
(10, 10)	4	0	10.0543	0	−13.8075
(10, 10)	5	0	10.0543	0	−17.3297
(10, 10)	10	0	10.0541	0	−34.9409
(10, 10)	15	0	10.0540	0	−52.5521
(10, 10)	20	0	10.0540	0	−70.1633
(10, 10)	25	0	10.0540	0	−87.7745
(10, 10)	50	0	10.0540	0	−175.8306
(10, 10)	100	0	10.0539	0	−351.9426
(16, 16)	4	3.1416	5.0385	4.3216	−6.2395
(16, 16)	5	3.1416	5.0366	4.3226	−8.0120
(16, 16)	10	3.1416	5.0341	4.3244	−16.8749
(16, 16)	15	3.1416	5.0335	4.3249	−25.7379
(16, 16)	20	3.1416	5.0333	4.3251	−34.6029
(16, 16)	25	3.1416	5.0331	4.3253	−43.4640
(16, 16)	50	3.1416	5.0329	4.3256	−87.7791
(16, 16)	100	3.1416	5.0327	4.3259	−176.4093
(20, 20)	3	1.3490	4.9083	7.0251	−3.6709
(20, 20)	4	1.5566	0.8831	7.0239	−5.2978
(20, 20)	5	1.5436	2.4984	7.0057	−6.6966
(20, 20)	10	1.5565	2.5063	6.9840	−14.9446
(20, 20)	15	1.5649	2.5078	6.9709	−23.1989
(20, 20)	20	1.7194	2.2931	6.9181	−33.5703
(20, 20)	25	1.7194	2.2934	6.9163	−42.5048
(20, 20)	50	1.7194	2.2941	6.9126	−87.1785
(20, 20)	100	1.7194	2.2944	6.9108	−176.5269

configuration and is close to $\pi/2$ for the (20, 20) tube. For $k = 100$, $\beta = 5.0327\,\text{Å}$, $\varepsilon = 4.3259\,\text{Å}$, and $\beta = 2.2944\,\text{Å}$, $\varepsilon = 6.9108\,\text{Å}$ for the (16, 16) and the (20, 20) tubes, respectively. Consequently, spiral patterns for C_{60} fullerenes in both the (16, 16) and the (20, 20) nanotubes are clearly observed.

In particular, the zigzag configuration can be thought of as a special case of the spiral configuration with angular spacing $\alpha = \pi$. The comparable numerical values for the offset location ε and the longitudinal spacing β for all sizes of the tubes are obtained, and an example is shown for the case of a (16, 16) carbon nanotube in Table 7.3. Moreover, in the case of a (20, 20) tube, at least four C_{60}

molecules are required to form a stable spiral configuration. This observation compares well with the findings of Hodak and Girifalco (2003) in the sense that four molecules are required on each layer within the carbon nanotube with the radius lying in the range 13.5–14.05 Å.

WORKED EXAMPLE 7.5

For a spiral configuration of a $(10, 10)$ carbon nanotube, show that the longitudinal spacing β is obtained as $\beta \approx 10.05$ Å, where from Table 7.3 we assume that $\alpha = 0$ and $\varepsilon = 0$.

Solution

We utilise MAPLE to plot the relation between the longitudinal spacing β and the total potential energy E^{tot} as given in Eq. (7.19). Again, we employ the command Int() to determine E^{tot}.

```
> restart;
> a:= 3.55:
> nf:= 0.3789: ng:= 0.3812:
> Agc:= 17.4: Bgc:= 29000:

> Acc:= 20: Bcc:= 34800:
```

Firstly, we consider the four types of the interaction energies between two C_{60} molecules. The distance between the centres of C_{60} fullerenes is defined as a function of l, where $l = 1, 2, 3, 4$ corresponds to the four types of the interactions.

```
> d:= l -> sqrt(4*e^2*(sin(l*alpha/2))^2 + (l*beta)^2):
```

The interaction energy between a pair of C_{60} molecule is given by Eq. (7.14).

```
> P_ball:= (r,n) -> 4*Pi^2*a^2*nf^2/(r*(2-n)*(3-n))*
  (1/(2*a+r)^(n-3)- 1/(2*a-r)^(n-3) - 2/r^(n-3)):
> Un:= r -> -Acc*P_ball(r,6)+Bcc*P_ball(r,12):
```

The total interaction energy for four interactions of a C_{60} molecules is obtained as

```
> E_ball:= k -> (k-1)*Un(d(1))+(k-2)*Un(d(2))+(k-3)*Un(d(3))
  +(k-4)*Un(d(4)) :
```

Next, we consider the interaction between a C_{60} fullerene and a carbon nanotube. We utilise the expanded expression of Eq. (7.2) for the interaction energy between an atom and a C_{60} molecule.

```
> P_tube:= r -> 4*Pi*a^2*nf*((B*( 5/(r^2-a^2)^6 + 80*a^2/(r^2-a^2)^7
  + 336*a^4/(r^2-a^2)^8 + 512*a^6/(r^2-a^2)^9 + 256*a^8/(r^2-a^2)
  ^(10))/5 - A*(1/(r^2-a^2)^3 + 2*a^2/(r^2-a^2)^4 ))): Then we input the distance δ.
> delta_i:= (b,i) -> sqrt((b-e)^2+4*b*e*sin(theta/2)^2+(z-beta*i)^2):
```

The interaction energy between a C_{60} fullerene and a carbon nanotube is given by

```
> E_tube:= (b,k) -> sum(2*b*ng*Int(Int(P_tube(delta_i(b,i)),
  theta=-Pi..Pi),z=0..infinity),i=1..k): Thus the total energy of the system is
> E:= (b,k) -> E_ball(k) + E_tube(b,k):
```

Assuming $\alpha = 0$ and $\varepsilon = 0$, we use the command plot() to illustrate the relation between the total potential energy and the longitudinal spacing β. We note that the radius of the $(10, 10)$ carbon nanotube is 6.784 Å.

```
> e:=0:alpha:=0:
> plot(E(6.784,1),beta=8..16);
```

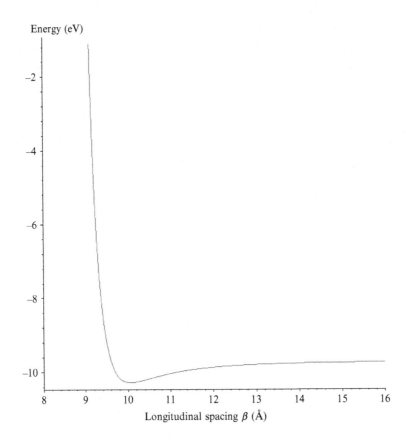

From the graph, the minimum energy occurs at $\beta \approx 10$ Å, which is in agreement with Table 7.3. Moreover, we can use the command fsolve() to find the exact minimum point which we obtain as $\beta \approx 10.0545$ Å.

```
> fsolve(diff(E(6.784,3),beta),beta=10);
```
$$10.05451843$$

7.3 SPHEROIDAL FULLERENES

A spheroid refers to an ellipsoid with two equal semiaxis, and in Section 7.3.1 the expression for the potential energy between a spheroid and a semi-infinite cylinder is derived by an analytical integration of the angular variables and the length variable along the length of the cylinder. For the potential energy, the semi-infinite integrals are evaluated in terms of Appell's F_1 hypergeometric functions, which can be transformed into a series of standard hypergeometric functions and further transformed into a series of incomplete beta functions. In Section 7.3.2 the expression derived in Section 7.3.1 is evaluated using numerical integration of the azimuthal variable where necessary to calculate the acceptance and suction energy for spherical and spheroidal fullerenes interacting with various configurations of carbon nanotubes. The procedure is validated by comparison with the previously derived result for the sphere.

7.3.1 POTENTIAL ENERGY OF A SPHEROIDAL FULLERENE AND A SINGLE-WALLED CARBON NANOTUBE OF SEMI-INFINITE EXTENT

Firstly, a rectangular coordinate system (x, y, z), as shown in Fig. 7.11, is established and used as a reference system. Next, a cylindrical polar coordinate system (r, ψ, z) is introduced with the same origin, such that $x = r \cos \psi$ and $y = r \sin \psi$. The carbon nanotube is assumed to be a semi-infinite cylinder of radius b with centre on the z-axis, which extends from the origin in the positive z direction. The parametric form of the surface of the cylinder in terms of the rectangular coordinates (x_c, y_c, z_c) is

$$x_c = b \cos \psi, \quad y_c = b \sin \psi, \quad z_c = z,$$

where $\psi \in [0, 2\pi]$ and $z \in [0, \infty)$.

Next, as shown in Fig. 7.11, a spheroid is introduced with the axis of rotation collinear with the z-axis and centred at the point $(0, 0, Z)$ with equatorial semiaxes of length a and a polar semiaxis (along the z axis) of length c. Furthermore, a system of spheroidal coordinates (ξ, ϕ, θ) is defined with an origin located at the centre of the spheroid. The relation between the spheroidal coordinates and the rectangular coordinates is given by

$$x = \alpha \sinh \xi \sin \phi \cos \theta, \quad y = \alpha \sinh \xi \sin \phi \sin \theta, \quad z = \alpha \cosh \xi \cos \phi + Z,$$

where α is the parameter of the coordinate system, $\xi \in [0, \infty)$, $\phi \in [0, \pi]$, and $\theta \in [0, 2\pi]$. Now α and ξ are fixed such that the equatorial semiaxes of the spheroid are $a = \alpha \sinh \xi$ and the polar semiaxis of the spheroid is $c = \alpha \cosh \xi$. The parametric form of the surface of the spheroid can then be expressed in rectangular Cartesian coordinates (x_s, y_s, z_s) by the relations

$$x_s = a \sin \phi \cos \theta, \quad y_s = a \sin \phi \sin \theta, \quad z_s = c \cos \phi + Z,$$

as shown in Fig. 7.11.

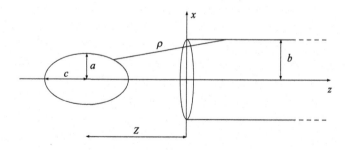

FIG. 7.11

Geometry for a spheroidal fullerene interacting with a semiinfinite carbon nanotube.

Using the Lennard-Jones potential $\Phi(\rho)$ given by Eq. (3.1), where ρ is the distance between typical surface elements on the spheroid and the cylinder, the total potential E, between a spheroidal and a cylindrical surface is given by

$$E = \eta_c \eta_s \int \int_{S_s} \int \int_{S_c} \Phi(\rho) \, dS_c \, dS_s,$$

where S_s and S_c are the surfaces of the spheroid and cylinder respectively, η_c and η_s are their atomic densities per unit area and dS_s and dS_c are their surface elements. From the parametric forms of the surfaces, it may be shown that

$$\rho^2 = (a \sin\phi \cos\theta - b \cos\psi)^2 + (a \sin\phi \sin\theta - b \sin\psi)^2 + (c \cos\phi + Z - z)^2, \tag{7.22}$$

and furthermore, the surface elements are given by

$$dS_c = b \, d\phi \, dz, \quad dS_s = a \sin\phi \sqrt{a^2 \cos^2\phi + c^2 \sin^2\phi} \, d\phi \, d\theta.$$

Thus the problem of determining the total potential E consists of evaluating the following integral

$$E = ab\eta_c \eta_s \int_0^{2\pi} \int_0^{\pi} \int_0^{\infty} \int_0^{2\pi} \left(-\frac{A}{\rho^6} + \frac{B}{\rho^{12}} \right)$$
$$\times \sin\phi \sqrt{a^2 \cos^2\phi + c^2 \sin^2\phi} \, d\psi \, dz \, d\phi \, d\theta. \tag{7.23}$$

WORKED EXAMPLE 7.6

For a spheroid with coordinates

$$x = a \sin\phi \cos\theta, \quad y = a \sin\phi \sin\theta, \quad z = c \cos\phi,$$

show that its surface element is given by $dS_s = a \sin\phi \sqrt{a^2 \cos^2\phi + c^2 \sin^2\phi} \, d\phi \, d\theta$.

Solution
We note that

$$\iint_S f(x, y, z) dS = \int_\theta \int_\phi f(\mathbf{r}(\theta, \phi)) \left| \frac{\partial \mathbf{r}}{\partial\theta} \times \frac{\partial \mathbf{r}}{\partial\phi} \right| d\theta d\phi, \tag{7.24}$$

where $\mathbf{r} = (a \sin\phi \cos\theta, a \sin\phi \sin\theta, c \cos\phi)$. We have that

$$\frac{\partial \mathbf{r}}{\partial\theta} = (-a \sin\phi \sin\theta, a \sin\phi \cos\theta, 0), \quad \frac{\partial \mathbf{r}}{\partial\phi} = (a \cos\phi \cos\theta, a \cos\phi \sin\theta, -c \sin\phi),$$

for which

$$\frac{\partial \mathbf{r}}{\partial\theta} \times \frac{\partial \mathbf{r}}{\partial\phi} = \begin{vmatrix} \hat{i} & \hat{j} & \hat{k} \\ -a \sin\phi \sin\theta & a \sin\phi \cos\theta & 0 \\ a \cos\phi \cos\theta & a \cos\phi \sin\theta & -c \sin\phi \end{vmatrix}$$
$$= -ac \sin^2\phi \cos\theta \hat{i} - ac \sin^2\phi \sin\theta \hat{j} - a^2 \sin\phi \cos\phi \hat{k}.$$

Thus

$$\left| \frac{\partial \mathbf{r}}{\partial\theta} \times \frac{\partial \mathbf{r}}{\partial\phi} \right|^2 = a^2 c^2 \sin^4\phi \cos^2\theta + a^2 c^2 \sin^4\phi \sin^2\theta + a^4 \sin^2\phi \cos^2\phi$$

$$= a^2 \sin^2 \phi \left(c^2 \sin^2 \phi \cos^2 \theta + c^2 \sin^2 \phi \sin^2 \theta + a^2 \cos^2 \phi \right)$$

$$= a^2 \sin^2 \phi \left(a^2 \cos^2 \phi + c^2 \sin^2 \phi \right).$$

As a result, we obtain

$$\left| \frac{\partial \mathbf{r}}{\partial \theta} \times \frac{\partial \mathbf{r}}{\partial \phi} \right| = a \sin \phi \sqrt{a^2 \cos^2 \phi + c^2 \sin^2 \phi},$$

which upon substituting into Eq. (7.24), we find that the surface element of the spheroid is given by $dS_s = a \sin \phi \sqrt{a^2 \cos^2 \phi + c^2 \sin^2 \phi}\, d\phi d\theta$.

7.3.1.1 Rotational symmetry: θ or ψ integration

As the problem possesses rotational symmetry around the z-axis, it is possible, without loss of generality, to remove the dependence on one of either θ or ψ by choosing a particular value and then performing the trivial integration of that variable immediately. Mathematically, this is because the integrand depends explicitly on an even function of $\theta - \psi$, namely $\cos(\theta - \psi)$. By choosing to perform this by adopting the value $\psi = 0$ and then by evaluating that integration, there remains

$$E = 2\pi a b \eta_c \eta_s \int_0^{2\pi} \int_0^{\pi} \int_0^{\infty} \left(-\frac{A}{\rho_\star^6} + \frac{B}{\rho_\star^{12}} \right) \sin \phi \sqrt{a^2 \cos^2 \phi + c^2 \sin^2 \phi}\, dz\, d\phi\, d\theta, \tag{7.25}$$

where

$$\rho_\star^2 = (a \sin \phi \cos \theta - b)^2 + (a \sin \phi \sin \theta)^2 + (c \cos \phi + Z - z)^2. \tag{7.26}$$

Alternatively, one may choose $\theta = 0$ and likewise obtain

$$E = 2\pi a b \eta_c \eta_s \int_0^{2\pi} \int_0^{\pi} \int_0^{\infty} \left(-\frac{A}{\rho_\diamond^6} + \frac{B}{\rho_\diamond^{12}} \right) \sin \phi \sqrt{a^2 \cos^2 \phi + c^2 \sin^2 \phi}\, dz\, d\phi\, d\psi, \tag{7.27}$$

where

$$\rho_\diamond^2 = (a \sin \phi - b \cos \psi)^2 + (b \sin \psi)^2 + (c \cos \phi + Z - z)^2. \tag{7.28}$$

The two forms of E from Eqs (7.25) and (7.27) are equivalent and if θ is substituted for ψ, then it can be shown that $\rho_\star = \rho_\diamond$, so that either form may be used as the starting point for the integration of the second axial variable.

WORKED EXAMPLE 7.7

Show that $\rho_\star = \rho_\diamond$ if θ is substituted for ψ.

Solution

From Eq. (7.26), we have

$$\rho_\star^2 = (a \sin \phi \cos \theta - b)^2 + (a \sin \phi \sin \theta)^2 + (c \cos \phi + Z - z)^2$$

$$= (a \sin \phi \cos \theta)^2 - 2ab \sin \phi \cos \theta + b^2 + (a \sin \phi \sin \theta)^2 + (c \cos \phi + Z - z)^2$$

$$= (a \sin \phi)^2 (\cos^2 \theta + \sin^2 \theta) - 2ab \sin \phi \cos \theta + b^2 + (c \cos \phi + Z - z)^2$$
$$= (a \sin \phi)^2 - 2ab \sin \phi \cos \theta + b^2 (\cos^2 \theta + \sin^2 \theta) + (c \cos \phi + Z - z)^2$$
$$= (a \sin \phi - b \cos \theta)^2 + (b \sin \theta)^2 + (c \cos \phi + Z - z)^2$$
$$= \rho_\diamond^2,$$

on replacing θ by ψ.

7.3.1.2 Integration along the length of the cylinder

Here, the case of a semi-infinite carbon nanotube ($0 \leqslant z < \infty$) is considered. The potential E for an ellipsoidal fullerene interacting with the carbon nanotube is given by Eq. (7.23) and ρ is defined in Eq. (7.22). In this derivation the infinite extent of z in one direction makes it possible to effect this integration first. To this end, we make the substitution

$$\kappa^2 = (a \sin \phi \cos \theta - b \cos \psi)^2 + (a \sin \phi \sin \theta - b \sin \psi)^2,$$
$$= \beta_1^2 - (\beta_1^2 - \beta_2^2) \sin^2((\theta - \psi)/2), \tag{7.29}$$

where $\beta_1 = b - a \sin \phi$ and $\beta_2 = b + a \sin \phi$, and then we consider I_z, namely

$$I_z = \int_{-\infty}^{\infty} \left\{ -A[\kappa^2 + (z - c \cos \phi - Z)^2]^{-3} + B[\kappa^2 + (z - c \cos \phi - Z)^2]^{-6} \right\} dz.$$

Making the further substitution $\gamma = z - c \cos \phi - Z$ produces

$$I_z = \int_{\gamma_0}^{\infty} \left[-A(\kappa^2 + \gamma^2)^{-3} + B(\kappa^2 + \gamma^2)^{-6} \right] d\gamma,$$

where $\gamma_0 = -c \cos \phi - Z$ and the substitution $\gamma = \kappa \tan \tau$ gives

$$I_z = \int_{\tan^{-1}(\gamma_0/\kappa)}^{\pi/2} \left(-A\kappa^{-5} \cos^4 \tau + B\kappa^{-11} \cos^{10} \tau \right) d\tau,$$

whereupon on integrating we obtain

$$I_z = -A \left\{ 3 \left[\pi/2 - \tan^{-1}(\gamma_0/\kappa) - \kappa\gamma_0(\kappa^2 + \gamma_0^2)^{-1} \right] - 2\kappa^3 \gamma_0(\kappa^2 + \gamma_0^2)^{-2} \right\} / 8\kappa^5$$
$$+ B \left\{ 315 \left[\pi/2 - \tan^{-1}(\gamma_0/\kappa) - \kappa\gamma_0(\kappa^2 + \gamma_0^2)^{-1} \right] - 210\kappa^3 \gamma_0(\kappa^2 + \gamma_0^2)^{-2} \right.$$
$$- 168\kappa^5 \gamma_0(\kappa^2 + \gamma_0^2)^{-3} - 144\kappa^7 \gamma_0(\kappa^2 + \gamma_0^2)^{-4}$$
$$\left. - 128\kappa^9 \gamma_0(\kappa^2 + \gamma_0^2)^{-5} \right\} / 1280\kappa^{11},$$

noting that as in Eq. (7.29), κ^2 can be written as $\kappa^2 = \beta_1^2 - (\beta_1^2 - \beta_2^2) \sin^2((\theta - \psi)/2)$. Next, the integral of I_z is evaluated over the angles θ and ψ, namely

$$I_{\theta,\psi} = \int_0^{2\pi} \int_0^{2\pi} I_z \, d\theta \, d\psi.$$

This double integral can be reduced by appealing to the same arguments of rotational symmetry presented in Section 7.3.1.1, yielding

$$I_{\theta,\psi} = 2\pi \int_0^{2\pi} I_z|_{\psi=0}\, d\theta. \tag{7.30}$$

Substituting $x = \theta/2$ and bisecting the interval, Eq. (7.30) becomes

$$\begin{aligned}
I_{\theta,\psi} = \pi \int_0^{\pi/2} \Big\{ &-A \Big[3\left(\pi/2 - \tan^{-1}(\gamma_0 p^{-1/2}) \right) p^{-5/2} \\
&- \gamma_0 \left(3p^{-2}q^{-1} + 2p^{-1}q^{-2} \right) \Big] + B \Big[315\left(\pi/2 - \tan^{-1}(\gamma_0 p^{-1/2}) \right) p^{-11/2} \\
&- \gamma_0 \left(315p^{-5}q^{-1} + 210p^{-4}q^{-2} + 168p^{-3}q^{-3} + 144p^{-2}q^{-4} \right. \\
&\left. + 128p^{-1}q^{-5} \right) \Big] / 160 \Big\}\, dx,
\end{aligned} \tag{7.31}$$

where $p = \beta_1^2 - (\beta_1^2 - \beta_2^2)\sin^2 x$ and $q = p + \gamma_0^2$. Noting that the integrals in Eq. (7.31) involving q are all of the form

$$J_{m,n} = \int_0^{\pi/2} p^{-m} q^{-n} dx,$$

where $m + n = 3$ or 6. Making the substitution $u = \sin^2 x$ gives

$$\begin{aligned}
J_{m,n} = \frac{1}{2\beta_1^{2m}(\beta_1^2 + \gamma_0^2)^n} \int_0^1 &u^{-\frac{1}{2}}(1-u)^{-\frac{1}{2}} \left[1 + u\left(1 - \frac{\beta_2^2}{\beta_1^2} \right) \right]^{-m} \\
&\times \left[1 + u\left(\frac{\beta_1^2 - \beta_2^2}{\beta_1^2 + \gamma_0^2} \right) \right]^{-n} du,
\end{aligned}$$

where $\beta_1 = b - a\sin\phi$, $\beta_2 = b + a\sin\phi$, and $\gamma_0 = -c\cos\phi - Z$ and which, from Section 5.8.2(5) of Erdélyi et al. (1953), can be written as

$$J_{m,n} = (\pi/2)\beta_1^{-2m}(\beta_1^2 + \gamma_0^2)^{-n} F_1\left(\frac{1}{2}, m, n, 1\,; \zeta_1, \zeta_2 \right), \tag{7.32}$$

where $\zeta_1 = 1 - \beta_2^2/\beta_1^2$, $\zeta_2 = (\beta_1^2 - \beta_2^2)/(\beta_1^2 + \gamma_0^2)$, and $F_1(a, b, b', c; z, z')$ is the Appell hypergeometric function of the first kind, which is detailed in Section 2.7.

Furthermore, the integrals in Eq. (7.31), which do not involve q terms need to be evaluated, namely

$$K_m = \int_0^{\pi/2} p^{-m-\frac{1}{2}}\, dx,$$

for $m = 2$ and 5. Again, using the substitution of $u = \sin^2 x$ produces

$$K_m = \frac{1}{2\beta_1^{2m+1}} \int_0^1 u^{-\frac{1}{2}}(1-u)^{-\frac{1}{2}}(1 - \zeta_1 u)^{-m-\frac{1}{2}}\, du,$$

which is the fundamental integral for the standard hypergeometric function, thus

$$K_m = \frac{\pi}{2}\beta_1^{-2m-1} F\left(m + \frac{1}{2}, \frac{1}{2}\,; 1\,; \zeta_1 \right).$$

From Sections 2.9(4) and 3.2(28) of Erdélyi et al. (1953), one can show that

$$F\left(a, 1/2 ; 1 ; 1 - \beta_2^2/\beta_1^2\right) = (\beta_1/\beta_2)^a \, P_{a-1}(\epsilon),$$ (7.33)

where $\epsilon = (\beta_1/\beta_2 + \beta_2/\beta_1)/2$, and therefore K_m can be expressed as

$$K_m = \pi(\beta_1\beta_2)^{-m-\frac{1}{2}} P_{m-\frac{1}{2}}(\epsilon)/2,$$ (7.34)

where $P_n(z)$ is the Legendre function of the first kind.

Finally, the terms of Eq. (7.31) which have not been integrated need to be considered, namely

$$L_m = \int_0^{\pi/2} p^{-m-\frac{1}{2}} \tan^{-1}(\gamma_0 p^{-\frac{1}{2}}) \, dx,$$ (7.35)

where $m = 2$ and 5. Here, two different series representations of the arctangent are used, depending on the value of the argument. One is given in Section 1.644(1) of Gradshteyn and Ryzhik (2000), and this is

$$\tan^{-1}(x) = \frac{x}{\sqrt{1+x^2}} \sum_{k=0}^{\infty} \frac{(2k)!}{2^{2k}(k!)^2(2k+1)} \left(\frac{x^2}{1+x^2}\right)^k,$$ (7.36)

valid for any real x. The other series representation is also from Section 1.644(2) of Gradshteyn and Ryzhik (2000), and this is

$$\tan^{-1}(x) = \text{sgn}(x)\frac{\pi}{2} - \sum_{k=0}^{\infty}(-1)^k \frac{1}{(2k+1)x^{2k+1}},$$ (7.37)

valid for real $|x| > 1$ and $\text{sgn}(x)$ is ± 1 according as $x > 0$ or $x < 0$. An investigation of the convergence indicates that 10 terms are sufficient for five significant digits of accuracy, provided that Eq. (7.36) is used when $|x| < 1.34$ and which we designate $L_m^\star$; Eq. (7.37) is used with $|x| \geqslant 1.34$, and which we designate $L_m^\diamond$. First considering Eq. (7.36), it follows that

$$p^{-m-\frac{1}{2}} \tan^{-1}(\gamma_0 p^{-\frac{1}{2}}) = \frac{\gamma_0}{p^{m+\frac{1}{2}}q} \sum_{k=0}^{\infty} \frac{(2k)!}{2^{2k}(k!)^2(2k+1)} \left(\frac{\gamma_0^2}{q}\right)^k,$$

and finally, by substitution in Eq. (7.35),

$$L_m^\star = \sum_{k=0}^{\infty} \frac{(2k)!\gamma_0^{2k+1}}{2^{2k}(k!)^2(2k+1)} \int_0^{\pi/2} p^{-(m+\frac{1}{2})} q^{-(k+1)} \, dx,$$

upon which one observes that the integral is of the form previously defined as $J_{m+\frac{1}{2},k+1}$, and therefore

$$L_m^\star = \frac{\pi}{2\gamma_0\beta_1^{2m+1}} \sum_{k=0}^{\infty} \frac{(2k)!}{2^{2k}(k!)^2(2k+1)} \left(\frac{\gamma_0^2}{\beta_1^2 + \gamma_0^2}\right)^{k+1}$$
$$\times F_1\left(\frac{1}{2}, m + \frac{1}{2}, k + 1, 1 ; \zeta_1, \zeta_2\right).$$ (7.38)

Considering the second series expansion for $\tan^{-1} x$, Eq. (7.37) gives

$$p^{-m-\frac{1}{2}} \tan^{-1} \left(\frac{\gamma_0}{\sqrt{p}} \right) = p^{-m-\frac{1}{2}} \left[\mathrm{sgn} \left(\frac{\gamma_0}{\sqrt{p}} \right) \frac{\pi}{2} - \sum_{k=0}^{\infty} \frac{(-1)^k}{2k+1} \left(\frac{\gamma_0}{\sqrt{p}} \right)^{-2k-1} \right],$$

and as $p > 0$ and the remaining integrals only involve p, which have been considered previously, giving

$$L_m^{\diamond} = \mathrm{sgn}(\gamma_0) \frac{\pi}{2} K_m - \sum_{k=0}^{\infty} \frac{(-1)^k}{(2k+1)\gamma_0^{2k+1}} K_{m+k+1/2},$$

and therefore

$$L_m^{\diamond} = \frac{\pi}{2(\beta_1 \beta_2)^{m+1/2}} \left[\mathrm{sgn}(\gamma_0) \frac{\pi}{2} P_{m-1/2}(\epsilon) - \sum_{k=0}^{\infty} \frac{(-1)^k P_{m+k}(\epsilon)}{(2k+1)(\gamma_0^2 \beta_1 \beta_2)^{k+1/2}} \right]. \tag{7.39}$$

By using either Eq. (7.38) or Eq. (7.39), one is able to calculate a rapidly converging value of L_m, and therefore additionally using Eqs (7.32) and (7.34) a value for $I_{\theta,\psi}$ can be determined, as shown in Eq. (7.31) as

$$I_{\theta,\psi} = \pi \left\{ -A \left[3 \left(K_2 \pi/2 - L_2 \right) - \gamma_0 \left(3J_{2,1} + 2J_{1,2} \right) \right] + B \left[315 \left(K_5 \pi/2 - L_5 \right) \right. \right.$$
$$\left. \left. - \gamma_0 \left(315 J_{5,1} + 210 J_{4,2} + 168 J_{3,3} + 144 J_{2,4} + 128 J_{1,5} \right) \right] / 160 \right\}.$$

From Bailey (1972) and Eq. (2.33), Appell hypergeometric function of two variables $F_1(1/2, m, n, 1; \zeta_1, \zeta_2)$ contained in $J_{m,n}$ and $L_m^{\star}$ can be written in terms of the standard hypergeometric functions, thus

$$F_1(\tfrac{1}{2}, m, n, 1; \zeta_1, \zeta_2) = \sum_{i=0}^{\infty} \frac{(\tfrac{1}{2})_i (m)_i}{(i!)^2} \zeta_1^i F(\tfrac{1}{2} + i, n; 1 + i; \zeta_2),$$

where $(a)_i$ is the Pochhammer symbol. Finally, the potential energy E can then be obtained from

$$E = ab\eta_c \eta_s \int_0^{\pi} I_{\theta,\psi} \sin\phi \sqrt{a^2 \cos^2 \phi + c^2 \sin^2 \phi} \, d\phi, \tag{7.40}$$

and the final ϕ integration must be undertaken numerically.

7.3.2 INTERACTION OF SPHEROIDAL MOLECULES LOCATED ON THE AXIS OF A SINGLE-WALLED CARBON NANOTUBE

Following the analysis in Chapter 5 the acceptance and suction energies are calculated for spheroidal molecules interacting with carbon nanotubes of various configurations. The acceptance energy calculation addresses the issue of whether van der Waals forces alone are sufficient for the molecule to be accepted into the interior of the carbon nanotube. The suction energy is the magnitude of the total energy acquired by the molecule from atomic interactions upon entering the carbon nanotube. The analysis in Chapter 5 demonstrates that the suction energy is positive for certain configurations of spherical fullerenes and carbon nanotubes, indicating that the inside of the nanotube is energetically favourable, though the acceptance energy is negative. This situation indicates that the energetically favourable position in the interior of the nanotube is unreachable due to an energetically unfavourable barrier at the nanotube opening unless some additional energy, perhaps in the form of kinetic energy, is imparted to the fullerene by means of a mechanical impulse. It is quite likely that analogous configurations also exist for spheroidal fullerenes and more general surfaces of revolution; therefore the analysis here includes a calculation for both the acceptance and the suction energies. In the analysis below, it is assumed that the Lennard-Jones constants A and B are those that apply for graphitic carbon interactions for C_{60} graphene, as given in Table 1.7, and the other constants used in the calculations are as shown in Tables 1.5 and 1.6.

7.3.2.1 Interaction of a spherical C_{60} fullerene molecule

In this section the energy expression derived in Section 7.3.1 is used to calculate the acceptance and suction energies for a spherical C_{60} fullerene and carbon nanotubes of various radii. The purpose of this is to validate the expressions derived in this chapter for the special case of spherical particle interactions in comparison to Section 5.2.

The Lennard-Jones potential E, as calculated in Section 7.3.1 for a spherical C_{60} fullerene and various sized carbon nanotubes of radius b, is shown in Fig. 7.12, where the final ϕ integration (7.40) is performed using Simpson's rule with 20 intervals. Following the analysis for the mechanics of fullerenes and carbon nanotube interactions in Chapter 5, the graph shows that for carbon nanotubes

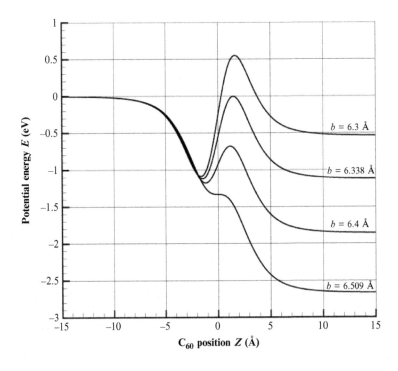

FIG. 7.12

Potential energy for C_{60} fullerene at varying position Z for carbon nanotubes of varying radius b.

where the radius is less than some critical value, the energy has a local maximum in the neighbourhood of the tube extremity; the difference between the potential energy well outside of the nanotube and this local maximum is termed the acceptance energy in Chapter 5. Fig. 7.12 shows that the critical value of nanotube radius where the acceptance energy is zero is $b \approx 6.338$ Å. Furthermore, the value of the nanotube radius b, where the force experienced by the fullerene is not negative at any stage, and therefore the local maximum ceases to be a feature of the energy graph, is in excess of 6.509 Å. Both these calculations show excellent agreement with those in Worked Example 5.1 in Chapter 5. It can be seen from the plot that a nanotube with radius 6.3 Å would not accept a C_{60} fullerene from rest because there exists an energy barrier at the tube opening. The magnitude of this is approximately 0.548 eV, which again is in excellent agreement with the results in Chapter 5.

The difference between the potential energy well outside the nanotube and well inside the nanotube is termed suction energy in Chapter 5. In Fig. 7.12, this is shown as the asymptotic limit, which the potential energy is rapidly approaching as Z increases. Using an arbitrarily larger value of Z leads to an even closer agreement, and the calculation repeated for a value of $Z = 100$ Å gives a relative error that is less than 0.0001 for the four nanotube radii shown.

7.3.2.2 Interaction of spheroidal fullerene molecules

With the method validated by the analysis in the previous section, the analysis now proceeds to the interaction between carbon nanotubes and spheroidal fullerenes. In particular the C_{70} fullerene and C_{80} fullerene are investigated and modelled as prolate spheroidal surfaces of uniform atomic density. The major and minor axes dimensions are taken from Nakao et al. (1994), and the atomic densities are calculated by dividing the number of atoms by the surface area of the idealised spheroid; these values are shown in Table 7.4. The method used is detailed in Section 7.3.1, and as in the previous section, the ϕ integration in Eq. (7.40) is performed numerically using Simpson's Rule with 20 partitions.

The results for the calculation for a C_{70} fullerene and carbon nanotubes of various radii are shown in Fig. 7.13, which indicates that an increase in the radius of the nanotube is necessary to accommodate the C_{70} molecule, as compared to the case of the C_{60} fullerene.

The radius of the nanotube which leads to a zero acceptance energy is $b \approx 6.383$ Å, which is a small increase on that of a C_{60} fullerene. Fig. 7.14 shows the potential energy for a C_{80} fullerene and carbon nanotubes of various radii.

Table 7.4 Constants Used for Spheroidal Fullerenes		
Mean surface density for graphene	η_g =	0.3812 Å^{-2}
C_{70} equatorial semiaxis length	a =	3.59 Å
C_{70} polar semiaxis length	c =	4.17 Å
Mean surface density for fullerene C_{70}	η_{70} =	0.3896 Å^{-2}
C_{80} equatorial semiaxis length	a =	3.58 Å
C_{80} polar semiaxis length	c =	4.73 Å
Mean surface density for fullerene C_{80}	η_{80} =	0.4072 Å^{-2}
Attractive constant	A =	17.4 eV Å^6
Repulsive constant	B =	29×10^3 eV Å^{12}

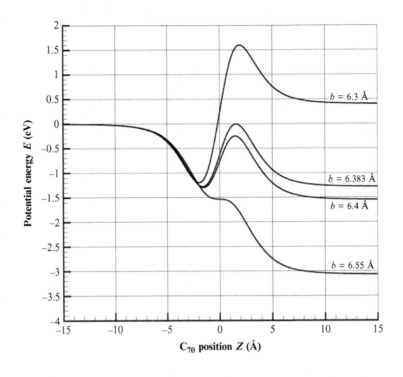

FIG. 7.13

Potential energy for C_{70} fullerene at varying position Z for carbon nanotubes of varying radius b.

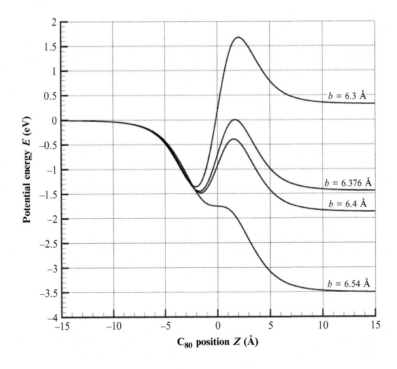

FIG. 7.14

Potential energy for C_{80} fullerene at varying position Z for carbon nanotubes of varying radius b.

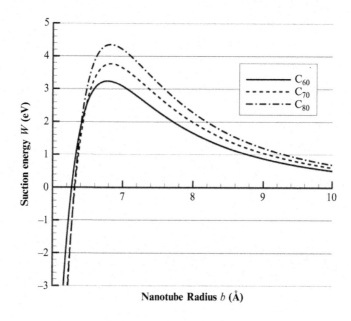

FIG. 7.15

Suction energy W for fullerenes C_{60}, C_{70}, and C_{80} for carbon nanotubes of varying radius b.

It shows that C_{80} fullerenes are accepted into slightly smaller tubes, which demonstrates that the equatorial semiaxes length a is the major determining factor in prescribing an appropriate nanotube size. In other words, the polar semiaxis length (or longitudinal dimension) c, which is 0.56 Å longer for a C_{80} than for a C_{70}, does not increase the size of tube necessary to accept the fullerene. However, the 0.01 Å decrease in the equatorial semiaxes length a does reduce the required nanotube radius b. The nanotube radius which leads to a zero acceptance energy is $b \approx 6.376$ Å, which is a decrease of approximately the same magnitude as the decrease in equatorial semiaxes length as compared to the C_{70} fullerene.

To conclude this section, the suction energies W for C_{60}, C_{70}, and C_{80} fullerenes are calculated and shown in Fig. 7.15. The nanotube radius which maximises the suction energy for a C_{70} fullerene is $b \approx 6.83$ Å, which is just a small increase in the radius when compared to the findings for the C_{60} molecule. The maximum value of the suction energy W is increased in comparison to that of the C_{60} case, but there is no practical increase in the energy per atom; this leads to the same final velocity for a molecule being sucked into the nanotube. Therefore, it can be concluded that there is no configuration for a C_{70} fullerene and a single-walled carbon nanotube which leads to higher frequencies of oscillation, as compared to a similar configuration with a C_{60} molecule. The nanotube radius, whose maximum suction energy W for a C_{80} fullerene is $b \approx 6.82$ Å, is again very similar to the value for a C_{70} fullerene, and as before, the maximum potential energy per atom in the fullerene is very similar. Therefore in terms of producing a maximal oscillator, the C_{80} would not appear to offer any advantages in terms of van der Waals suction energy per unit mass.

EXERCISES

7.1. In the encapsulation of a C_{60} molecule around the edge of a tube open end process, show that if $x = 13.034$ Å, the C_{60} molecule has no chance to be encapsulated into the $(10, 10)$ carbon nanotube.

7.2. Using MAPLE to show that the distance Z between centres of adjacent C_{60} molecules in the zigzag pattern of a $(10, 10)$ nanopeapod is obtained as $Z \approx 10.055$ Å.

7.3. Following the determination given in Section 7.3.1.2, determine the potential energy of a spheroidal fullerene inside a single-walled carbon nanotube of infinite extent.

NANOTUBES AS DRUG DELIVERY VEHICLES

8.1 INTRODUCTION

The prospect that nanocapsules may realise the 'magic bullet' concept, first proposed at the beginning of the 20th century by the Nobel Prize winner Paul Ehrlich (1854–1915), has generated immense interest in their development. The ideal drug carrier, or 'magic bullet', is envisaged as a transporter of drugs or other molecular cargo to a specific site in the body which then unloads the cargo in a controlled manner. Although this notion sounds like science fiction, the advent of nanotechnology means that it is rapidly becoming scientific fact. Despite the prominence of carbon nanotubes in the broader area of nanotechnology, the field of nanotube biotechnology is in its infancy, and there is still much work that needs to occur before specific products can be marketed. Drug delivery is one of the most promising biomedical applications of nanotechnology, and as stated by Hillebrenner et al. (2006) in a review of template synthesised nanotubes for biomedical delivery applications: 'The future challenges for nanotubes as drug delivery vehicles are substantial but not insurmountable'.

The ability of new materials, such as nanocapsules and nanotubes, to serve as biocompatible transporters has received relatively little attention. However, the biocompatibility, and unique physical, electrical, optical, and mechanical properties of single-walled nanotubes provide an impressive foundation for a new class of materials that may be exploited for drug, protein, and gene delivery applications, and which will provide the next generation of medical therapeutics. Furthermore, nanocapsules and nanotubes offer the possibility of drug delivery hybrids that, when used in conjunction with magnetic resonance imaging and inductive heating, will play a key role in new and powerful cancer treatment regimes. Their ability to be functionalised and visualised in biological environments using simple fluorescence microscopy has further enhanced their nanomedical potential.

The major advantage of targeted drug delivery is that it enables a stronger drug to be used at a smaller dosage and has fewer side effects than current delivery methods, such as chemotherapy, which is limited by the inadequate and nonspecific delivery of therapeutic concentrations to target the tumour tissue. In a review on the emerging field of nanotube biotechnology, Martin and Kohli (2003) were one of the first to propose the use of nanotubes for drug delivery because of their capacity to be functionalised and filled with a drug payload. In addition, nanotubes offer the perfect isolated environment for the drug until it reaches the target site, both from degradation and reaction with healthy cells. Nanotubes also provide a means of delivery without the need to include solvents, which often are the cause of as much damage to healthy cells as the drug itself. This technique has potential advantages

Modelling and Mechanics of Carbon-based Nanostructured Materials. http://dx.doi.org/10.1016/B978-0-12-812463-5.00008-X

not only for cancer treatment, but also for infections, metabolic diseases, autoimmune diseases, and pain treatment, and may also be used in gene therapy.

Typically, nanoparticles have been used in the areas of drug delivery and cosmetics, since they are spherical and therefore easier to make than other shapes. In addition, nanoparticles can be manufactured from a wide range of materials. Nanoparticles with an iron-platinum core, called a yolk-shell nanocrystal, achieve a more effective delivery with a smaller dosage than the parent drug cisplatin.

Several nanocarrier formulations, such as Doxil and Abraxane, have already been approved for clinical use in cancer chemotherapy. Doxil is an example of a successful application of nanoparticles and comprises a long-circulating, doxorubicin-loaded, polyethylene glycol-coated liposome. The Doxil formulation has shown a sixfold improvement in effectiveness and in reducing cardiotoxicity in comparison to the pure drug. One disadvantage of liposomal technology is that there is a fast elimination from the blood, leading to low drug efficiency and high drug leakage. A good nanocarrier is considered to be one that circulates in the blood for long periods of time and delivers the drugs with minimal side effects. The practical implication of persistent circulation is an increase in the exposure of the drugs to cancer cells or other specific cells.

The main advantage of nanoparticles and nanotubes as drug carriers is their multifunctionality, which means they can incorporate multiple therapeutic, diagnostic, targeting, and barrier-avoiding agents, as illustrated in Fig. 8.1. These additional features may be engineered either onto their surface or incorporated within their interiors so that the nanocapsules carry one or more functional groups. For example, these may include: the ability to cause cell apoptosis (cell death) via heat using near-infrared agents, or acting as contrast agents (optical, magnetic or other) to enable a means of tracking the treatment, and recognition signals for targeting. However, functionalising nanoparticles for directed delivery has associated problems and payload capacity is also difficult. Therefore although liposomal and nanoparticle drug delivery methods offer considerable improvements on existing procedures, it is possible to obtain even better results using nanotubes.

Nanotubes offer a promising alternative with a number of advantages such as:

- a large inner volume that can be filled with the desired chemicals, ranging in size from small molecules to proteins,
- distinct inner and outer surfaces,
- a large aspect ratio, which provides multiple attachment sites for various functionalisation along the nanotube axis,
- open ends that make the inner volume and surfaces accessible,
- the ability, like nanoparticles, to be readily taken up by cells in addition to being able to reach the cell nuclei, which suggests the possibility of gene therapy.

On the other hand, the major disadvantages or limitations of use of nanotubes for drug delivery are:

- their toxicity when they are not functionalised. Pristine carbon nanotubes are harmful to cells and can accumulate in various organs, such as the liver and lungs,
- the difficulty in solubilising nanotubes as part of their biocompatibility,
- the need to have *in vitro* and *in vivo*, short- and long-term effects accurately determined,
- the lack of understanding and consensus, as to the cellular uptake mechanism,
- the need to balance the small scale of the nanocapsule or nanotube with the quantities of drug that are clinically necessary.

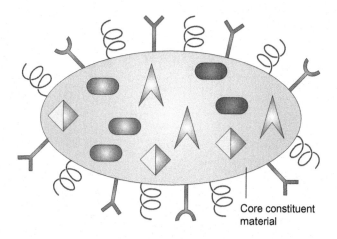

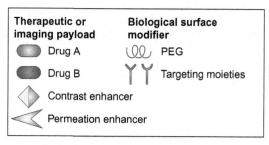

FIG. 8.1

A schematic of a multifunctional nanoparticle.

Source: Ferrari, M., 2005. Cancer nanotechnology: opportunities and challenges. Nat. Rev. Cancer 5, 161–171.

Hillebrenner et al. (2006) investigate a nano test tube-style drug carrier, using the nanotube as a test tube with one open end, which may be filled and subsequently capped, to provide a convenient filling method. Assuming a nano test tube style delivery vehicle, an overview of the general process is shown in Fig. 8.2 and comprises:

(a) The nanotube surface is functionalised with a chemical receptor, such as the folate targeting of tumours that overexpress folic acid. The chemistry of carbon nanotubes offers the possibility of introducing more than one function on the same tube; for example, both functionalising for biocompatibility and targeting specific cells. The drug is then encapsulated within the nanotube interior by either simply filling it with solution or by suction of the drug molecules or by simply attaching drug molecules to the tube sidewalls,

(b) Once the drug is encapsulated or attached, the open end is capped, which may be biodegradable or chemically removable,

(c) The nanocapsule is then inserted into the body either orally or by intravenous injection. Due to the functionalised surface, the nanocapsule arrives at the designated site in the body,

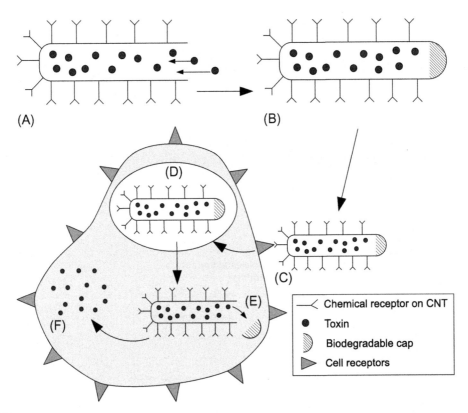

FIG. 8.2

Outline of the proposed drug delivery process: (A) The Nanotube surface is functionalised with a chemical receptor and the drug molecules are encapsulated, (B) open end is capped, (C) nanocapsule is ingested and locates to target site due to the functionalised surface, (D) cell internalises the capsule; for example by receptor mediated endocytosis, (E) cap is removed or it biodegrades inside cell, and (F) drug molecules are released.

Source: Hilder, T.A., Hill, J.M., 2009. Modeling the loading and unloading of drugs into nanotubes. Small 5(3), 300–308.

(d) The cell then internalises the nanocapsule, either by receptor-mediated endocytosis or by needle-like passive penetration,

(e) The cap is then removed or it biodegrades inside the cell, which may be caused by some chemical or external trigger, such as a change in pH, near-infrared light, or a chemically removable cap that 'uncorks' when triggered in the body to release its contents,

(f) The drug molecules are then released into the cell by a cleavable bond or by a change in the local environment.

Most of the research to date has been through experimental and molecular dynamics simulations investigating toxicity, solubility, biocompatibility, uptake, and functionalisation. Recent experimental and theoretical investigations look at the loading and unloading of molecular cargo. For example, it can be shown theoretically that carbon nanotubes can be readily filled with particles, and they obtain reasonable agreement with experiments. Similarly, by using molecular dynamics the spontaneous insertion of a DNA molecule into a carbon nanotube can be demonstrated. As mentioned in Chapter 1, Ferrari (2005) states that:

> Novel mathematical models are needed, in order to secure the full import of nanotechnology into oncology.

In the next section, we outline the underlying mathematics used throughout this chapter. Following this, we investigate the encapsulation of drug molecules into nanotubes, and in particular the anticancer drug cisplatin.

8.2 UNDERLYING MATHEMATICS
8.2.1 HYBRID DISCRETE-CONTINUUM APPROXIMATION

As mentioned in Section 1.3, in the interest of modelling irregularly shaped molecules, such as drugs, an alternative hybrid discrete-continuum approximation can be used to determine the interaction energy. The hybrid model, which is represented by elements of both Eqs (1.11) and (1.12), is given by

$$E = \sum_{\gamma} \eta \int \Phi(\rho_\gamma)\, dS, \qquad (8.1)$$

where η is the surface density of atoms on the molecule, which is considered continuous, ρ is the distance between a typical surface element dS on the continuously modelled molecule and an atom γ in the molecule, which is modelled as a summation over each of its atoms. Again, $\Phi(\rho_\gamma)$ is the potential function, and the interaction energy is obtained by summing over all atoms in the drug molecule.

A general method is developed that can be used to model any shaped molecule. More specifically, each atom γ of a given molecule is defined as an offset atom located at a distance z_γ from the end of the carbon nanotube and at a radius r_γ from the centre of mass (CoM) of the molecule, as shown in Fig. 8.3.

More specifically, each individual atom in the drug molecule is defined in terms of the Cartesian coordinates (x_1, x_2, x_3), which is related to r_γ and z_γ through

$$r_\gamma = \sqrt{x_1^2 + x_2^2}, \quad z_\gamma = Z + x_3$$

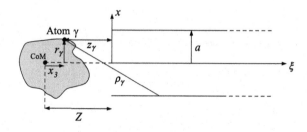

FIG. 8.3

General offset atom and a carbon nanotube.

where Z is the distance between the centre of mass of the drug molecule and the end of the nanotube, as shown in Fig. 8.3, since the tube is assumed to be semiinfinite in length.

It is assumed that the centre of mass of the entering drug molecule is located on the axis of the nanotube at a distance Z from the nanotube end. With a cylindrical polar coordinate system (r, θ, z) with the origin located on the tube axis and at the tube end, the offset atom γ is located at $(r_\gamma, 0, z_\gamma)$, while a typical point on the nanotube is defined by $(a \cos \theta, a \sin \theta, \xi)$, where the nanotube is assumed to be semiinfinite, as illustrated in Fig. 8.3.

The distance between the offset atom γ and a typical point on the nanotube is given by

$$\rho^2 = (x_2 - x_1)^2 + (y_2 - y_1)^2 + (z_2 - z_1)^2$$
$$= (a + r_\gamma)^2 - 4ar_\gamma \cos^2(\theta/2) + (\xi - z_\gamma)^2. \tag{8.2}$$

Substituting Eq. (8.2) into Eq. (8.1) and the Lennard-Jones potential function (3.1), we obtain the energy of one offset atom interacting with the carbon nanotube, namely

$$E = \eta a \sum_{\gamma=1}^{N_a} \int_0^\infty \int_{-\pi}^\pi \left(\frac{-A}{[(a + r_\gamma)^2 + (\xi - z_\gamma)^2 - 4ar_\gamma \cos^2(\theta/2)]^3} \right.$$
$$\left. + \frac{B}{[(a + r_\gamma)^2 + (\xi - z_\gamma)^2 - 4ar_\gamma \cos^2(\theta/2)]^6} \right) d\theta \, d\xi,$$

where η is the surface density of carbon atoms on the nanotube, a is the radius of the nanotube and N_a is the number of atoms in the discrete molecule.

Letting $\phi = \theta/2$ obtains the general θ integration

$$I_n = 2 \int_{-\pi/2}^{\pi/2} \frac{d\phi}{[\alpha_1^2 + (\alpha_2^2 - \alpha_1^2) \cos^2 \phi]^n}, \tag{8.3}$$

where $\alpha_1^2 = (a + r_\gamma)^2 + (\xi - z_\gamma)^2$, $\alpha_2^2 = (a - r_\gamma)^2 + (\xi - z_\gamma)^2$, $\alpha_2^2 - \alpha_1^2 = -4ar_\gamma$, and of primary interest are the values $n = 3$ and $n = 6$. The interaction energy E is then simply

$$E = \eta a \sum_{\gamma=1}^{N_a} \int_0^\infty (-AI_3 + BI_6) \, d\xi.$$

WORKED EXAMPLE 8.1

Evaluate the integral I_n given by Eq. (8.3).

Solution

Given the symmetry of the integral and making the substitution $t = \cos^2 \phi$, we obtain

$$I_n = 4 \int_0^{\pi/2} \frac{d\phi}{[\alpha_1^2 + (\alpha_2^2 - \alpha_1^2) \cos^2 \phi]^n} = \frac{2}{\alpha_1^{2n}} \int_0^1 \frac{t^{-1/2}(1 - t)^{-1/2}}{[1 - (1 - (\alpha_2/\alpha_1)^2)t]^n} \, dt.$$

Using the relation given by Eq. (2.29) enables the integral to be expressed in terms of the ordinary hypergeometric function, $F(a, b; c; z)$, namely

$$I_n = \frac{2}{\alpha_1^{2n}} B\left(\frac{1}{2}, \frac{1}{2}\right) F\left(n, \frac{1}{2}; 1; 1 - \left(\frac{\alpha_2}{\alpha_1}\right)^2\right) = \frac{2\pi}{\alpha_1^{2n}} F\left(n, \frac{1}{2}; 1; 1 - \left(\frac{\alpha_2}{\alpha_1}\right)^2\right).$$

From Erdélyi et al. (1953) (pp. 64, 69) it can be shown that the hypergeometric function given above is quadratic, because two of the numbers $1 - c$, $\pm(a - b)$, $\pm(c - a - b)$ are equal, as well as degenerate because at least one of the numbers a, b, $c - a$, $c - b$ is an integer. Using the transformation given by Eq. (2.25) and restated here

$$F(a, b; c; x) = (1 - x)^{c-a-b} F(c - a, c - b; c; x),$$

we obtain two degenerate functions, given $n = 3$ and $n = 6$, corresponding to

$$F(3, 1/2; 1; x) = \left(\frac{\alpha_1}{\alpha_2}\right)^5 \left(1 - x + \frac{3x^2}{8}\right),$$

$$F(6, 1/2; 1; x) = \left(\frac{\alpha_1}{\alpha_2}\right)^{11} \left(1 - \frac{5x}{2} + \frac{15x^2}{4} - \frac{25x^3}{8} + \frac{175x^4}{128} - \frac{63x^5}{256}\right),$$

where $x = 1 - (\alpha_2/\alpha_1)^2$.

From Worked Example 8.1, the interaction energy is given by

$$E = 2\pi \eta a \sum_{\gamma=1}^{N_a} \int_0^\infty \left[-\frac{A}{\alpha_1^6} F(3, 1/2; 1; x) + \frac{B}{\alpha_1^{12}} F(6, 1/2; 1; x) \right] d\xi. \tag{8.4}$$

By expanding these results and substituting into the original interaction energy equation, we get

$$E = 2\pi \eta a \sum_{\gamma=1}^{N_a} \left[-A\left(\frac{1}{4}J_{3,3} + \frac{3}{8}(J_{1,5} + J_{5,1})\right) \right.$$

$$\left. +B\left(\frac{63}{256}(J_{1,11} + J_{11,1}) + \frac{35}{256}(J_{3,9} + J_{9,3}) + \frac{30}{256}(J_{5,7} + J_{7,5})\right) \right],$$

where

$$J_{m,n} = \int_0^\infty \frac{d\xi}{\alpha_1^m \alpha_2^n}. \tag{8.5}$$

By expanding Eq. (8.5), the integral becomes

$$J_{m,n} = \int_0^\infty \frac{d\xi}{[(a + r_\gamma)^2 + (\xi - z_\gamma)^2]^p [(a - r_\gamma)^2 + (\xi - z_\gamma)^2]^q}, \tag{8.6}$$

where $m = 2p$ and $n = 2q$. The integral given by Eq. (8.6) may be evaluated using both Appell's hypergeometric functions, as illustrated in Worked Example 8.2, and elliptic functions. Details for the evaluation of Eq. (8.6) using elliptic functions is given in Hilder and Hill (2007).

WORKED EXAMPLE 8.2

In this example, we evaluate Eq. (8.6) using Appell's hypergeometric functions:

Solution

Making the substitution $\xi - z_\gamma = (a + r_\gamma) \tan \omega$ into Eq. (8.6) yields

$$J_{2p,2q} = \int_{\omega_1}^{\pi/2} \frac{(a + r_\gamma) \sec^2 \omega \, d\omega}{[(a + r_\gamma)^2 + (a + r_\gamma)^2 \tan^2 \omega]^p [(a - r_\gamma)^2 + (a + r_\gamma)^2 \tan^2 \omega]^q}$$

$$= (a + r_\gamma)^{1-2p} \int_{\omega_1}^{\pi/2} \frac{(\cos^2 \omega)^{p-1} \cos^{2q} \omega \, d\omega}{[(a - r_\gamma)^2 \cos^2 \omega + (a + r_\gamma)^2 \sin^2 \omega]^q}$$

$$= (a + r_\gamma)^{1-2p-2q} \int_{\omega_1}^{\pi/2} \frac{(\cos^2 \omega)^{p+q-1} \, d\omega}{[1 - (1 - v^2) \cos^2 \omega]^q},$$

where $\omega_1 = \tan^{-1}(-z_\gamma/(a + r_\gamma))$ and $v = (a - r_\gamma)/(a + r_\gamma)$. By making the further substitutions $t = \cos^2 \omega$ and $t = t_1 u$, we obtain

$$J_{2p,2q} = \frac{(a + r_\gamma)^{1-2p-2q}}{2} \int_0^{t_1} t^{p+q-1} t^{-1/2} (1 - t)^{-1/2} (1 - (1 - v^2)t)^{-q} \, dt$$

$$= \frac{[(a + r_\gamma)^2 + z_\gamma^2]^{1/2-p-q}}{2} \int_0^1 u^{p+q-3/2} (1 - t_1 u)^{-1/2} [1 - (1 - v^2)t_1 u]^{-q} \, du,$$

where $t_1 = (a + r_\gamma)^2 / [(a + r_\gamma)^2 + z_\gamma^2]$. Using the relation (2.32), namely

$$\int_0^1 u^{\alpha-1}(1 - u)^{\gamma-\alpha-1}(1 - ux)^{-\beta_1}(1 - uy)^{-\beta_2} \, du = \frac{\Gamma(\alpha)\Gamma(\gamma - \alpha)}{\Gamma(\gamma)} F_1(\alpha; \beta_1, \beta_2; \gamma; x, y),$$

and writing $J_{2p,2q}$ in terms of Appell's hypergeometric function of two variables, gives

$$J_{m,n} = \frac{[(a + r_\gamma)^2 + z_\gamma^2]^{-M}}{2} \frac{\Gamma(M)}{\Gamma(N)} F_1\left(M; \frac{n}{2}, \frac{1}{2}; N; (1 - v^2)t_1, t_1\right),$$

where $M = (m + n - 1)/2$ and $N = (m + n + 1)/2$.

The interaction energy is given by

$$E = 2\pi \eta a \sum_{\gamma=1}^{N_a} \left[-A \left(\frac{1}{4} J_{3,3} + \frac{3}{8}(J_{1,5} + J_{5,1}) \right) \right.$$

$$\left. + B \left(\frac{63}{256}(J_{1,11} + J_{11,1}) + \frac{35}{256}(J_{3,9} + J_{9,3}) + \frac{30}{256}(J_{5,7} + J_{7,5}) \right) \right], \tag{8.7}$$

where, in terms of Appell's hypergeometric functions (see Bailey, 1972), $J_{m,n}$ is given by

$$J_{m,n} = \frac{[(a + r_\gamma)^2 + z_\gamma^2]^{-M}}{2} \frac{\Gamma(M)}{\Gamma(N)} F_1\left(M; \frac{n}{2}, \frac{1}{2}; N; (1 - v^2)t_1, t_1\right),$$

for $M = (m + n - 1)/2$, $N = (m + n + 1)/2$, $v = (a - r_\gamma)/(a + r_\gamma)$, $t_1 = (a + r_\gamma)^2/[(a + r_\gamma)^2 + z_\gamma^2]$, and $F_1(a; b_1, b_1; c; x; y)$ is an Appell hypergeometric function. This offset atom formula may be used to determine both the force and energy for any shaped molecule, such as those presented in the following sections.

Given the symmetry of the problem, the axial force is of primary interest so that on noting $F(\xi) = -dE/d\xi$ and using Eq. (8.4), the interaction force is given by

$$F = 2\pi\eta a \sum_{\gamma=1}^{N_a} \left[\frac{A}{\alpha_1^6} F(3, 1/2; 1; 1 - \alpha_2^2/\alpha_1^2) - \frac{B}{\alpha_1^{12}} F(6, 1/2; 1; 1 - \alpha_2^2/\alpha_1^2) \right], \tag{8.8}$$

where $\alpha_1^2 = (a + r_\gamma)^2 + (\xi - z_\gamma)^2$, $\alpha_2^2 = (a - r_\gamma)^2 + (\xi - z_\gamma)^2$. Note that as usual, $F(a, b; c; x)$ denotes the ordinary hypergeometric function.

The hybrid discrete-continuum approximation has been shown to compare well with the typically used Eqs (1.11) and (1.12) approaches. In particular, the hybrid model is within 2% of the discrete summation given by Eq. (1.11), and generally within 10% of the continuum formulation given by Eq. (1.12). In the following Worked Example 8.3, we compare the continuum approximation with the hybrid discrete-continuum formulation for one specific molecular interaction. We refer the reader to Hilder and Hill (2007) for further details regarding the hybrid discrete-continuum model and its comparison to other approximations used.

WORKED EXAMPLE 8.3

The interaction between a C_{60} fullerene and a carbon nanotube may be examined using the continuous approach in Section 5.2 or the discrete-continuum approach given by Eq. (8.1). In the first instance the atoms of the fullerene are assumed to be uniformly distributed over the surface of a sphere, and in the latter case the atoms are given by distinct coordinates. Using the algebraic package MAPLE and the equation for force for the continuous approach Eq. (5.7), and hybrid discrete-continuous approach Eq. (8.8), we may illustrate the difference between these two models.

Solution

Refresh the memory and define relevant constants.

```
> restart; with(plots):
> n := evalf(4*sqrt(3)/(9*(1.42)^2)): A := 17.4: B := 29*10^3:
> nb := (b) -> 60/(4*Pi*b^2):
```

Define the force for the C_{60} fullerene using the continuum model (TF1).

```
> lambda := (a,b,Z) -> (a^2-b^2+Z^2)/b^2:
> FT1 := (a,b,Z) -> 8*Pi^2*n*nb(b)*a*(A*(1+2/lambda(a,b,Z))
 - B*(5+80/lambda(a,b,Z)+336/lambda(a,b,Z)^2
 + 512/lambda(a,b,Z)^3+256/lambda(a,b,Z)^4)
 /(5*b^6*lambda(a,b,Z)^3))/(b^4*lambda(a,b,Z)^3):
```

Include the geom3d tool. Note that this tool includes a truncated icosahedron which can be used to approximate a C_{60} fullerene, as discussed in Section 1.2.3.

```
> with(geom3d):
```

Define the object which will be the C_{60} fullerene. Note we use a radius of 3.55 Å as given in Table 1.5.

```
> TruncatedIcosahedron(bb,point(o,0,0,0),3.55):
```

Define the force for one atom for the hybrid discrete-continuum model. Note that ξ has been set to zero, because we are interested in the force at the end of the nanotube.

```
> alpha1 := (a,epsilon,z) -> sqrt((a+epsilon)^2 + (z)^2):
> alpha2 := (a,epsilon,z) -> sqrt((a-epsilon)^2 + (z)^2):
> Force := (a,epsilon,z) -> evalf(2*Pi*n*a*((-A/alpha1(a,epsilon,z)^6)
  *hypergeom([3,1/2],[1],(alpha1(a,epsilon,z)^2
  -alpha2(a,epsilon,z)^2)/alpha1(a,epsilon,z)^2)
  + (B/alpha1(a,epsilon,z)^(12))
  *hypergeom([6,1/2],[1],(alpha1(a,epsilon,z)^2
  -alpha2(a,epsilon,z)^2)/alpha1(a,epsilon,z)^2))):
```

Define the force for the fullerene for the hybrid discrete-continuum model (FT2).

```
> FT2 := proc(a,Z,0)
      local t, x:
      t := 0:
      for x in vertices(0) do
        t := t - Force(a,sqrt(x[1]^2+x[2]^2),Z+x[3]):
      end do:
      return evalf(t):
    end proc:
```

Plot both cases for a radius of 6.4 Å.

```
> plot([FT1(6.4,3.55,Z),FT2(6.4,Z,bb)],Z=-15..15,linestyle=[SOLID,DASH]);
```

Fig. 8.4 illustrates the interaction force for the C_{60} fullerene and a carbon nanotube for both approaches for a radius of 6.4 Å.

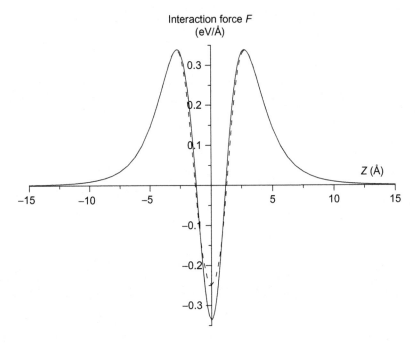

FIG. 8.4

C_{60} force ($a = 6.4$ Å) for both hybrid discrete-continuous (- -) and continuous (—).

8.2.2 **LENNARD-JONES PARAMETERS—MIXING RULES**

In some cases the force constants for the interaction between two atoms is unavailable, as is generally the case in this chapter. When information regarding the force constants is lacking, it is possible to use the empirical combining laws or mixing rules previously discussed in Section 4.2 and detailed in Worked Example 4.1, thus

$$\epsilon_{12} = (\epsilon_1 \epsilon_2)^{1/2}, \quad \sigma_{12} = (\sigma_1 + \sigma_2)/2, \tag{8.9}$$

where 1 and 2 refer to the respective individual atoms. Roughly speaking the assumptions underlying Eq. (8.9) are that the dispersion forces dominate the interactions (Israelachvili, 1992). Furthermore, the calculation of σ_{12} from Eq. (8.9) is exact for rigid spherical molecules, and the calculation of ϵ_{12} follows the simple interpretation of the dispersion forces in terms of the polarisabilities of the individual molecules. The force constants derived from these laws seem to give reasonably good results in calculations involving mixtures, where their accuracy depends on the accuracy of the experimentally derived individual-like molecule force constants. Note that the attractive and repulsive constants A and B in Eq. (3.1) are determined simply by $A = 4\epsilon\sigma^6$ and $B = 4\epsilon\sigma^{12}$, respectively Eq. (4.6), and we refer the reader to Hirschfelder et al. (1954) for further details.

8.2.3 **ENCAPSULATION**

The aim of this section is to understand the encapsulation behaviour of certain drug molecules entering nanotubes. To do this, we use the hybrid discrete-continuum model outlined in Section 8.2.1 and the concepts of an acceptance condition and the suction energy, which were described in Sections 5.1.1 and 5.1.2, respectively. To remind the reader, the acceptance condition is given by $W_a > 0$, or more specifically

$$\int_{-\infty}^{Z_0} F(Z)dZ > 0,$$

and this is the formal condition necessary for a molecule to enter a nanotube. Similarly, the suction energy is the total energy acquired by the molecule as it enters the nanotube and is given by

$$W = \int_{-\infty}^{\infty} F(Z)dZ,$$

where the interaction force $F(Z)$ is given by Eq. (8.8).

8.2.4 **ASSUMPTIONS**

The encapsulation of drug molecules into nanotubes is a complicated process. As a first approximation and in order to make possible a mathematical model, a number of assumptions are made:

Assumption 1
The encapsulation occurs within a vacuum and under isothermal conditions.

Often, drug molecules are administered to patients in a solvent medium. For example, the anticancer drug cisplatin is administered intravenously. It is supplied as a powder containing sodium chloride, mannitol, and hydrochloric acid for a pH adjustment, and it is then reconstituted with 10mL of sterile water for injection. The use of nanotubes may avoid the need for solvents because unaltered cisplatin can cross cell membranes and react within the cell. For the case where the drug molecule is contained in a solvent medium, the dispersion contribution still dominates even though the van der Waals force is much reduced. This may be accounted for by reducing the interactions by a factor of the dielectric constant of the solvent. Similar techniques are used for the interaction of highly polar water molecules with carbon nanotubes to mimic changes in the local polarity and solvent conditions.

We hope that by using nanotubes for drug delivery we may avoid the need for a solvent medium by providing a protected environment for the drug molecule; for example, from chemical reactivity prior to reaching the intended destination. It is important to avoid the use of solvents as they can cause as many adverse side effects to the patient as the drug itself. Therefore in this chapter the inclusion of a carrier solvent is not specifically addressed.

Assumption 2

Each atom is clearly defined by a coordinate position, and this position does not change throughout encapsulation.

In reality the position of each atom in the drug molecule will change according to environmental changes, such as temperature or pH. However, as a result of Assumption 1 we can assume that the conditions during encapsulation are constant, and therefore changes to the atom coordinates is negligible.

Assumption 3

There are no electrostatic effects generated between the drug molecule and the nanotube.

The atomic interactions (nonbonded interactions) between cisplatin, a highly polar molecule, and a carbon nanotube (nonpolar) consist of dispersion, induction, and orientation (electrostatic) forces. In the interaction between dissimilar molecules of which one is nonpolar, as is the case in Section 8.3, the interaction energy is almost completely dominated by the dispersion contribution. As such, the electrostatic interactions are assumed to be in this investigation. Here, the dispersion contribution is represented by the Lennard-Jones potential, which is believed to be applicable to nonpolar interactions. However, the drug molecule cisplatin is highly polar. Therefore, to more thoroughly represent the interaction between a carbon nanotube and a drug such as cisplatin, an electrostatic potential would also need to be incorporated (Israelachvili, 1992).

In some cases the solvent used to deliver the drug may also be highly polar, and in this situation the carbon nanotube may become polarised. In this case an additional electrostatic interaction energy must be included. This effect would be more pronounced in a metallic carbon nanotube, which may increase the overall interaction. However, in this model, as a first modelling attempt, the effect of polarisation is not specifically addressed. The nonbonded interaction is evaluated using the Lennard-Jones potential given by Eq. (3.1). The Lennard-Jones potential is typically used for nonpolar interactions, and as such this model provides only a first approximation, because the drug molecule may also generate electrostatic interactions with the carbon nanotube.

Assumption 4

The nanotube is empty prior to filling, and the drugs enter from one end only. The other end of the nanotube is capped and far enough away from encapsulation that it has no effect on encapsulation.

In other words, since the interaction force is a short-range force, the nanotube end is assumed to be sufficiently far away such that its contribution to the overall energy is small and does not affect encapsulation of the drug molecule.

As mentioned in Assumption 1, should the nanotube contain a buffer solution prior to filling, this is accommodated by reducing the interactions by division by the appropriate dielectric constant. However, it is assumed that the nanotube is empty of both buffer solution and other drug molecules prior to filling.

Assumption 5

Encapsulation of subsequent molecules into the nanotube is independent of molecules which are already contained within the nanotube interior.

In other words, since the interaction force is a short-range force, the already-encapsulated molecule is assumed to be sufficiently far away, such that its contribution to the overall energy is small and does not affect the encapsulation of the subsequent molecule. As the nanotube becomes filled with the encapsulated molecule, this assumption may no longer be valid.

Therefore the encapsulation of subsequent drug molecules follows the same procedure described herein and will stack inside the tube over time with the distance apart determined by their van der Waals diameter similar to the stacking in a nanopeapod, described in Section 7.2. Thus the nanotube length may be varied so as to control the dosage amount.

Assumption 6

The drug molecule does not react with the nanotube wall.

In other words, the encapsulation is such that the two molecules (drug and nanotube) remain nonbonded throughout encapsulation.

Assumption 7

The centre of mass lies on the nanotube axis.

In order for the drug molecule to be taken up within the carbon nanotube, it must lie within the general vicinity of the tube end; for definiteness, we assume that its centre of mass lies on the nanotube axis. In this way, we are able to discern the major qualitative mechanical features of the uptake process of the drug into the tube. In principle, we may undertake a similar but more complicated calculation, assuming that the centre of mass is slightly displaced from the tube axis, but we may anticipate qualitatively similar results. In practice, for larger radii tubes, the centre of mass may not always lie on the nanotube axis or the drug molecule may shift off axis, similar to that found for C_{60} fullerenes inside nanotubes. However, for small radii (i.e. minimum radius of acceptance) the assumption of the centre of mass lying on the tube axis is an entirely reasonable assumption in the first instance.

As mentioned, the encapsulation of drug molecules into nanotubes is a complicated process, thus in order to construct a mathematical model it is necessary to make the above assumptions. However, in practical terms the encapsulation of drug molecules into nanotubes may result in a number of these assumption being violated. For example, Assumption 2 assumes that the drug molecule's atomic positions remain constant throughout encapsulation. In reality, these atom positions may change as a result of atomic vibrations from temperature or pH changes. Thus all of the above assumptions will introduce an element of error to the solution, but as with any complicated physical process making assumptions allows the mathematician to solve a less complicated problem which may provide insight for the original.

8.3 ENCAPSULATION OF CISPLATIN INTO A CARBON NANOTUBE

As an example, this section specifically investigates the interaction of cisplatin, an anticancer agent, with a carbon nanotube. Cisplatin, one of the most frequently used anticancer drugs, is a platinum-based chemotherapy drug used to treat various types of cancer. It is most widely used for tumours of the testis, ovary, head, neck, lung, and bladder. Side effects of cisplatin include kidney damage, anaphylactic reactions, nerve damage, hearing loss, nausea, and vomiting. Two isomers exist termed *cis* and *trans*, where the *cis* form is used here, as this is the more effective against cancer. Cisplatin was chosen for its relatively simple structure, $Pt(NH_3)_2Cl_2$ and because it has a small number of atoms. The hybrid discrete-continuum method is easily extended to more complicated drugs of many atoms, provided the atom coordinates are known.

Table 8.1 outlines the numerical values of the well depth ϵ and van der Waals diameter σ for the various atoms studied in this chapter. As there is interest in the interaction of a drug with a carbon nanotube, the constants given in Table 8.1 must be modified accordingly using the empirical combining rules given by Eq. (8.9). By using the mixing formulae 8.9 the Lennard-Jones force constants for the interaction of the drug molecule cisplatin and a carbon nanotube are generated and are shown in Table 8.2.

Table 8.1 Well Depth ϵ and van der Waals Diameter σ for Carbon (C), Chlorine (Cl), Hydrogen (H), Nitrogen (N), and Platinum (Pt)

	ϵ (eV×10^{-2})	σ (Å)
C-C	0.2864	3.469
Cl-Cl	2.724	4.217
H-H	0.5146	2.827
N-N	0.6155	3.798
Pt-Pt	1.9833	3.92

Table 8.2 Approximate Modified Lennard-Jones Constants

	A (eVÅ^6)	B (eVÅ^{12})
C-Cl	113.81	366,602
C-H	14.94	14,544
C-N	38.65	88,933
C-Pt	76.66	194,943

Table 8.3 Average Dimensions of Cisplatin	
Bonded Pair	**Bond Length (Å)**
Pt-Cl	2.33 ± 0.01
Pt-N	2.01 ± 0.04
N-H	1.05
$\angle$Cl-Pt-Cl	$91.9 \pm 0.4°$

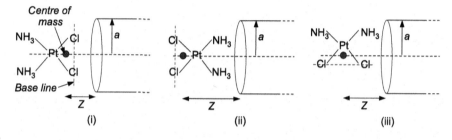

FIG. 8.5

Schematic of cisplatin entering a carbon nanotube with orientation (i), (ii), and (iii) and the location of the centre of mass.

The bond lengths and angles for cisplatin are given in Table 8.3. For the NH_3, one hydrogen atom is nearly in plane with the remaining atoms and the other two lie symmetrically above and below with N-H bonds of approximately 1.05 Å. For simplicity, it is assumed that all bond angles around the platinum atom (Pt) are 90° because $\angle$Cl-Pt-Cl is the only angle which slightly deviates from this value.

As mentioned, it is assumed that the centre of mass of the drug molecule is located on the central axis of the carbon nanotube. It is therefore important to determine the location of the centre of mass for cisplatin. The centre of mass R_M, is given by $R_M = \sum m_\gamma p_\gamma / M$, where M is the total mass, m_γ is the mass of atom γ and p_γ is the position of atom γ. The position of each atom is defined by taking a base line as the line made between both chlorine atoms, as shown in Fig. 8.5(i), and using the bond lengths given in Table 8.3. The atomic mass of each atom comprising the molecule cisplatin is obtained, and the total atomic mass of cisplatin is calculated in Worked Example 8.4 as 300.068amu, where amu stands for atomic mass unit and is equal to $1.66053873 \times 10^{-27}$kg. The centre of mass, shown in Fig. 8.5(i), is subsequently evaluated to be 1.42 Å from the base line, as shown in Worked Example 8.4. In this chapter, three different orientations of cisplatin are considered as it enters the carbon nanotube, each of which has the centre of mass located on the nanotube axis. Orientation (i), (ii), and (iii) are shown in Fig. 8.5.

WORKED EXAMPLE 8.4

Determine the location of the centre of mass for the cisplatin molecule.

Solution

The mass of each atom comprising cisplatin is given in Table 8.4. Summing up these masses for cisplatin obtains a total mass M of

$$M = 2 \times 35.45 + 6 \times 1.008 + 2 \times 14.01 + 195.1 = 300.068 \text{ amu.}$$

Table 8.4 Atomic Mass of Atoms Comprising Cisplatin

Atom Symbol	Mass (amu)
Cl	35.45
H	1.008
N	14.01
Pt	195.1

We define the base line as the line made between both the chlorine atoms as illustrated in Fig. 8.5(i). Note that the cisplatin molecule is symmetrical about the line which crosses the platinum atom and runs between two points which are equidistant between the chlorine and nitrogen atoms. Therefore the centre of mass must lie on this mirror line so that the only coordinates of interest in terms of calculating the centre of mass is the distance from the base line, which we term p. Using the bond lengths given in Table 8.3, the atom coordinates are then calculated relative to this base line, thus

$$p_{Cl} = 0, \quad p_{Pt} = 2.33/\sqrt{2}, \quad p_{NH_3} = 4.34/\sqrt{2},$$

so that the centre of mass is located a distance

$$\begin{aligned} R_M &= \frac{1}{M} \sum m_\gamma p_\gamma \\ &= \frac{0 \times 2 \times 35.45 + 2.33 \times 195.1/\sqrt{2} + 4.34 \times (2 \times 14.01 + 6 \times 1.008)/\sqrt{2}}{300.068} \\ &= 1.42 \text{ Å.} \end{aligned}$$

Thus the centre of mass of cisplatin is located on the mirror line at a distance 1.42 Å from the base line.

8.3.1 ENCAPSULATION BEHAVIOUR

Using the algebraic software package MAPLE, the interaction force F_z (Eq. 8.8) and the interaction energy E (Eq. 8.7) are plotted for cisplatin and a carbon nanotube of varying radii and chirality (n, m) for each orientation against the distance Z from the edge of the open end of the nanotube. Negative and positive Z relate to the cisplatin molecule being either outside or inside the nanotube, respectively, and the tube end is located at $Z = 0$. The interaction force for orientations (i) and (ii) are shown in Figs 8.6 and 8.7, respectively. Orientations (i) and (ii) are reflections of each other with respect to an axis passing through the platinum atom and perpendicular to the tube axis, and Figs 8.6 and 8.7 clearly

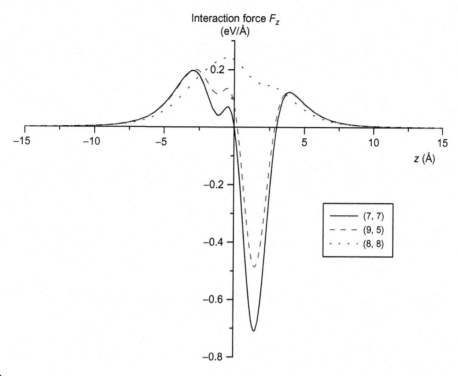

FIG. 8.6

Interaction force for cisplatin entering a carbon nanotube with orientation (i), for varying nanotube radii a.

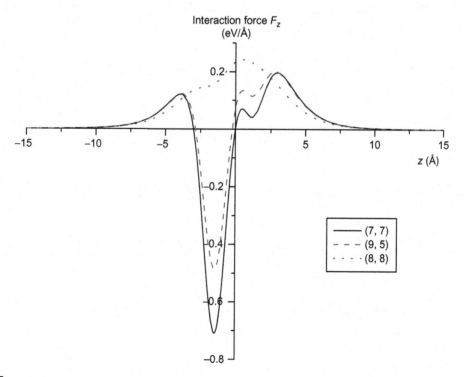

FIG. 8.7

Interaction force for cisplatin entering a carbon nanotube with orientation (ii), for varying nanotube radii a.

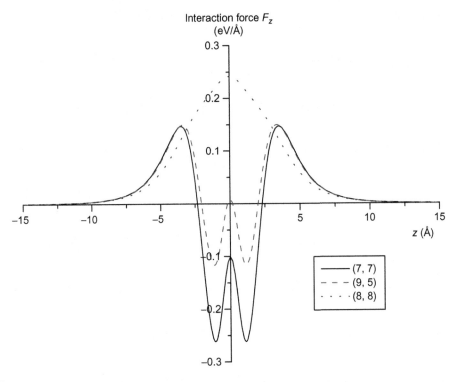

FIG. 8.8

Interaction force for cisplatin entering a carbon nanotube with orientation (iii), for varying nanotube radii *a*.

reflect this symmetry. For example, a $(9, 5)$ carbon nanotube with radius 4.81 Å will accept cisplatin by van der Waals forces alone for orientation (i) but not for orientation (ii). A $(8, 8)$ carbon nanotube with radius 5.43 Å will accept both orientations, as shown in Figs 8.6 and 8.7, since in both cases the force is positive.

Similarly for orientation (iii), Fig. 8.8 illustrates the interaction force of cisplatin with a carbon nanotube for different nanotube radii.

In this case the drug molecule is symmetric with respect to the axis passing through the platinum atom and perpendicular to the tube axis, and the interaction force for this orientation reflects this symmetry. In Fig. 8.8, a $(7, 7)$ carbon nanotube of radius 4.747 Å does not accept cisplatin because the force is negative, however the other two nanotubes $(9, 5)$ and $(8, 8)$ with radii 4.81 Å and 5.43 Å, respectively will accept cisplatin into their interior.

Alternatively, the suction characteristics of a particular carbon nanotube can be illustrated using the interaction energy. Figs 8.9–8.11 illustrate the interaction energy for varying nanotube radii for orientations (i) to (iii), respectively. In these figures, it is easy to see those carbon nanotubes which accept the cisplatin molecule and those which do not. If the energy inside the nanotube is less than the energy outside and there is no energy barrier at the tube end then the molecule will be accepted.

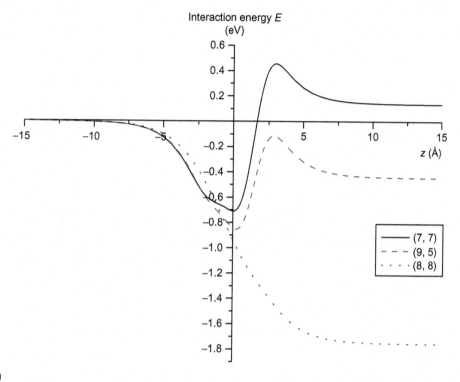

FIG. 8.9

Interaction energy for cisplatin entering a carbon nanotube with orientation (i), for varying nanotube radii *a*.

For example, in Figs 8.9–8.11 the $(8, 8)$ nanotube easily accepts cisplatin into its interior because the energy profiles decrease smoothly.

It is possible to determine the condition for which the cisplatin molecule will be accepted into the carbon nanotube using a similar procedure to that described in Section 5.2 for a C_{60} fullerene and a carbon nanotube. The formal condition for a drug molecule to be accepted into a carbon nanotube is given by Eq. (5.1) and outlined in Section 5.1.1. Fig. 8.12 illustrates the acceptance energy, and the point where the curve crosses the horizontal axis is the minimum nanotube radius which will accept cisplatin into its interior.

Consequently, for cisplatin with orientation (i), (ii), and (iii) to be accepted into a carbon nanotube the nanotube must have a radius of at least 4.795 Å, 4.895 Å, and 4.785 Å, respectively. Therefore for all three orientations to be accepted, the nanotube must have a radius of at least 4.895 Å, which is approximately equivalent to a $(10, 4)$ nanotube. Note that although orientation (i) and (ii) are mirror images of each other, they have quite different acceptance profiles.

Similarly, the suction energy W is given by Eq. (5.3) and outlined in Section 5.1.2. More specifically, it is defined as the energy that is acquired by the drug molecule from the carbon nanotube, which acts to suck the drug into the nanotube's interior. Fig. 8.13 illustrates the suction energy of cisplatin for the three orientations. As expected, the suction characteristics of orientation (i) and (ii) are identical, since it represents an integration of the entire force which were previously shown to be mirror images of each

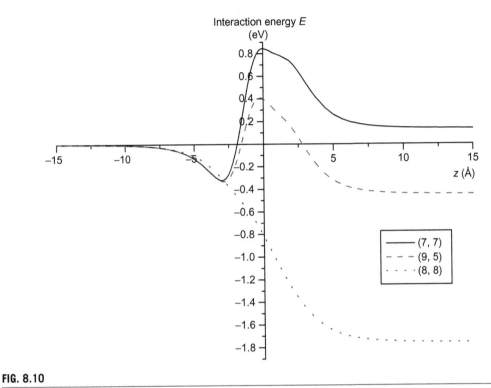

FIG. 8.10

Interaction energy for cisplatin entering a carbon nanotube with orientation (ii), for varying nanotube radii a.

other. Maximum suction energy occurs when the carbon nanotube radius is $5.34\,\text{Å}$ for orientations (i) and (ii), and $5.27\,\text{Å}$ for orientation (iii), which is approximately equal to a $(11, 4)$ nanotube. Note that as a quick estimate, it is possible to obtain an optimal radius of $5.1\,\text{Å}$ by adding half the length between the two chlorine atoms, plus the van der Waals radii of chlorine and carbon.

The same method might be used for a subsequent molecule to enter the nanotube. Eventually, a chain of drug molecules will form in the nanotube, and depending on the diameter of the tube, they will stack inside the tube accordingly, similar to the packing in nanotube peapods, Section 7.2. For the maximum suction radius obtained above, the drug molecules will form a single-file chain, and the nanotube length can be used to tune the number of drug molecules to be encapsulated, such that the molecules situate themselves a certain distance apart, which is equivalent to their van der Waals diameter.

The point in Fig. 8.13 where the curve crosses the horizontal axis is the minimum radius of nanotube, where it is energetically favourable for cisplatin to enter its interior. For orientation (i) and (ii), this occurs at $a = 4.76\,\text{Å}$ and for orientation (iii) at $a = 4.74\,\text{Å}$. For the case where the carbon nanotube radius is between this energetically favourable value and the acceptance condition, the suction energy is not sufficient to overcome the energy barrier at the nanotube end. However, in this range, it is energetically favourable for cisplatin to be inside the carbon nanotube, thus cisplatin will be accepted

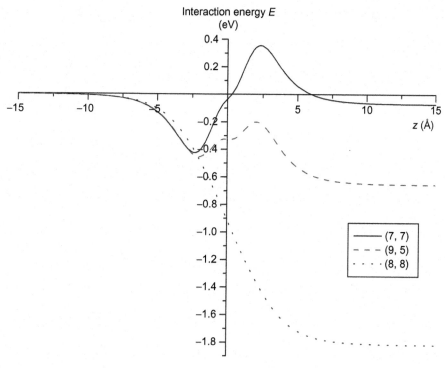

FIG. 8.11

Interaction energy for cisplatin entering a carbon nanotube with orientation (iii), for varying nanotube radii a.

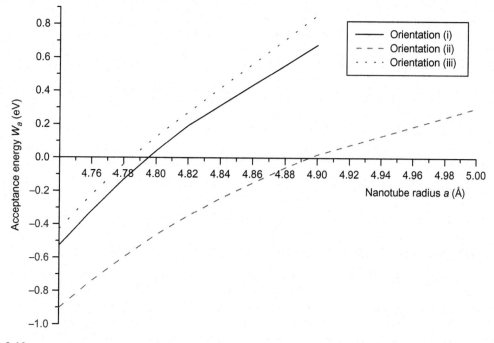

FIG. 8.12

Acceptance energy for cisplatin entering a carbon nanotube for the three orientations.

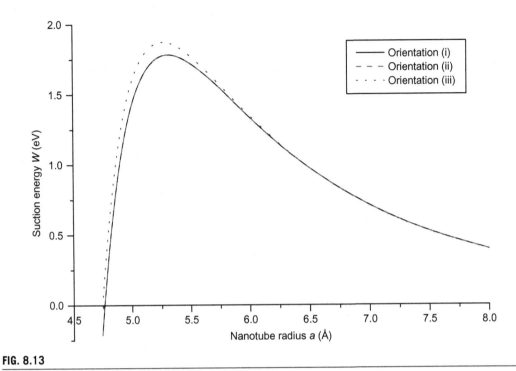

FIG. 8.13

Suction energy of cisplatin into a carbon nanotube for the three orientations.

if an additional energy is applied to cisplatin by some external force. For example, for orientation (ii), this range is $4.76 < a < 4.895\,\text{Å}$. This is further illustrated in Fig. 8.10 for orientation (ii) for the $(9, 5)$ nanotube with radius $4.81\,\text{Å}$, where there is a clear energy barrier which needs to be overcome for cisplatin to enter the nanotube, and the energy inside the tube is less than that outside.

As the nanotube radius gets larger, there is no longer enough energy to encapsulate the drug molecule into the carbon nanotube by interatomic forces alone, and as such as the nanotube radius a increases, the suction energy approaches zero. However, it is still energetically favourable for the molecule to enter the nanotube, and in principle, additional energy may be applied to the cisplatin molecule, but of course this might well involve some practical challenges.

8.4 ALTERNATIVE NANOTUBE MATERIALS

Single-walled nanotubes may be fabricated from a range of alternative materials, such as boron nitride, silicon, and boron carbide, and the list of possible materials which can form nanotubes is constantly expanding. Some of these alternative materials may be more biocompatible or biodegradable and may offer significant opportunities for potential nanobiotechnology tools and devices, such as those already demonstrated for carbon nanotubes. It is therefore important to understand the advantages and disadvantages of these alternative materials.

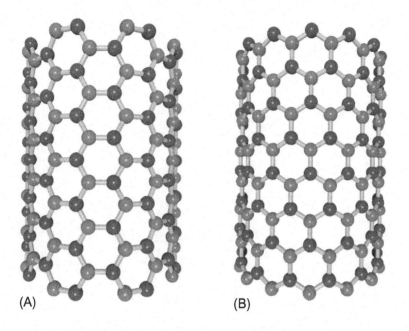

(A)   (B)

FIG. 8.14

Atomic structure of (A) armchair (8, 8), and (B) zigzag (14, 0) boron nitride nanotubes.

Boron nitride nanotubes (BNNT) have many of the excellent properties of carbon nanotubes (CNT) because they share the same electronic structure and are therefore expected to be just as desirable for applications as carbon nanotubes (Ishigami et al., 2003). Fig. 8.14 illustrates the structure of both a (8, 8) and (14, 0) boron nitride nanotube (Rubio et al., 1994). Compared to carbon nanotubes, boron nitride nanotubes offer improved performance; for example, through their high chemical stability and high resistance to oxidation at high temperatures. Boron nitride tubes have already shown improvement over carbon as high frequency oscillators in molecular dynamics simulations, where boron nitride based nanoscale oscillators give rise to higher frequencies (Lee, 2006). However, in terms of the filling of nanotubes, it has been suggested that capillary-induced filling of boron nitride tubes may be more difficult than for carbon tubes, since it has limited intercalation chemistry. General cytotoxicity studies of ceramics, including boron nitride, show no cytotoxic effect to cells, therefore suggesting that they may be a suitable candidate for medical devices where biocompatibility is vital. In recent experiments by Zhi et al. (2005) boron nitride nanotubes have shown a natural affinity to protein, so that boron nitride nanotubes may be particularly suitable for biological applications. However, the usefulness of boron nitride nanotubes for nanomedical applications has to date not been extensively researched to date.

On the other hand, silicon materials including single-crystal silicon, porous silicon, and silicon nanowires, have been widely used in the development of biomedical devices, such as neural prostheses, controlled drug delivery systems, biochips, and chemical/biological sensors. Silicon has been widely recognised as the most important material of the 20th century with high stability towards oxidation. Most of the published papers on silicon nanotubes (SiNT) are on the theoretical creation and existence of the silicon nanotube structure. Certain classes of silicon-based tubular nanostructures are stable and

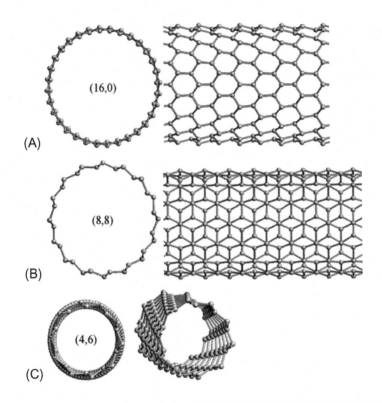

FIG. 8.15

Optimised structures of (A) zigzag (16, 0), (B) armchair (8, 8), and (C) chiral (4, 6) silicon nanotubes.

Source: J. Phys. Chem. B, 109:8605–8612 (2005).

energetically viable. As long as the dangling bonds on a silicon nanotube are properly terminated, a silicon nanotube can in principle be formed. However, the resulting energy minimised silicon nanotube has a slightly puckered structure with a corrugated surface. Consequently, the dominant structure for single-walled silicon nanotubes contains alternating sp^3- and sp^2-like bonds, which exhibit a 'gearlike' structure, and some researchers suggest that this structure is more stable than a smooth-walled tube. Fig. 8.15 illustrates the structure of three possible silicon nanotubes, exhibiting a slightly 'gearlike' structure. However, others have theoretically shown that the same amount of energy is necessary to roll a graphite-like sheet of silicon into a nanotube as with the equivalent carbon sheet, suggesting that a smooth-walled silicon tube may be formed.

Silicon nanotubes (SiNT) have not appeared in nature yet, largely due to silicon preferring sp^3 bonding. For carbon, sp^2 bonding is the most stable, whereas sp^3 bonding is the most stable for silicon. This is reflected in graphite being the most stable structure for carbon, and the diamond structure being the most stable for silicon. As a result, silicon nanowires with sp^3 bonding have been observed much more readily than the silicon nanotube, with sp^2 bonding. However, recently Sha et al. (2002) synthesised silicon nanotubes using a chemical vapour deposition process, but there are no current

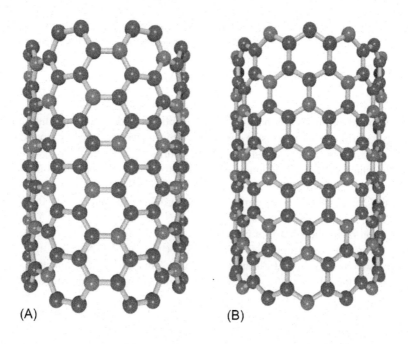

(A)   (B)

FIG. 8.16

Atomic structure of (A) armchair (8, 8), and (B) zigzag (14, 0) boron carbide nanotubes.

studies on their use in nanomedicine. One experimental report by De Crescenzi et al. (2005) provides evidence of synthesised clean single-walled silicon nanotubes, with diameters ranging from 20 to 350 Å. The biocompatibility of silicon nanotubes has been proven experimentally by Mu et al. (2007) and therefore silicon nanotubes are also a good candidate for nanomedical applications.

Another material which can theoretically form graphite-like tubules is boron carbide. Electron micrographs show that boron carbide is a homogeneous product with a sheet-like character and graphite-like symmetry. The relative stability of boron carbide nanotubes (BCNT), obtained from rolling hexagonal boron carbide sheets, is comparable to that of carbon tubules (Miyamoto et al., 1994). Fig. 8.16 illustrates an armchair and zigzag structure of a boron carbide nanotube. In particular, the strain energies of the boron carbide tubules are smaller than those of the carbon tubules and comparable to those of the boron nitride tubules. Therefore it has been predicted that the boron carbide tubules are likely to form as well as the carbon and boron nitride tubules. Similarly, from computational studies using density functional theory, narrow boron carbide nanotubes have been shown to be more energetically stable than the corresponding strips obtained from boron carbide hexagonal sheets. In addition, all the boron carbide tubes preserve their hollow shape and hexagonal boron carbide network. These theoretical predictions have been confirmed by Weng-Sieh et al. (1995) through the successful synthesis of boron carbide nanotubes. Boron–carbon graphites offer the possibility of greater electrical conductivity relative to graphite and may provide advantages in other areas such as nanomedicine. However, the use of boron-carbide nanotubes in nanomedicine has not yet been investigated.

Boron nitride, silicon, and boron carbide can all form graphite-like nanotubes and may perform better as drug delivery vehicles. In particular, some advantages which alternative materials offer are:

- improved biocompatibility,
- improved chemical stability,
- cheaper synthesis costs.

Thus in this section the encapsulation behaviour of the anticancer drug cisplatin into a boron nitride, silicon, and boron carbide nanotube is investigated. This technique may be applied to any drug molecule, however cisplatin was chosen in the interest of drawing a direct comparison with the results of Section 8.3, in which the encapsulation of cisplatin into a carbon nanotube is investigated. In the following subsection the encapsulation behaviour of the three alternative materials is provided and compared to that of carbon. First, we determine the Lennard-Jones parameters for the three alternative materials.

Similar to carbon nanotubes, boron nitride tubes can be thought of as a hexagonal lattice sheet, with alternating boron (B) and nitrogen (N) atoms, as shown in Fig. 8.17A, rolled up to form a tubular structure with a boron–nitrogen bond distance of 1.45 Å. Thus the surface density η for both boron and nitrogen atoms on the tube is $0.122\,\text{Å}^{-2}$. A graphite-like sheet of silicon (Si) rolled to form a tube has a bond distance of 2.245 Å and an atom surface density of $0.1527\,\text{Å}^{-2}$. Again, similar to carbon nanotubes, boron carbide tubes can be thought of as a hexagonal lattice sheet with boron (B) and carbon (C) atoms arranged in an array composed of a ratio of 1:3 boron–carbon atoms (BC_3), as shown in Fig. 8.17B, rolled into a tubular structure. Bond lengths for a boron carbide tube are carbon–carbon of 1.42 Å and boron–carbon of 1.55 Å. It is important to note that for the boron nitride and boron carbide nanotubes, the surface integral is performed over the tube for each atom type defined by their respective densities. For example, the boron carbide tube has a different surface density for its boron and carbon atoms, i.e. a hexagonal and triangular array, respectively, as shown in Fig. 8.17B.

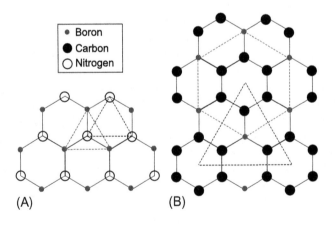

FIG. 8.17

Atomic arrangement for a sheet of (A) boron nitride, and (B) boron carbide.

WORKED EXAMPLE 8.5

Determine the surface density of boron and nitrogen atoms on the boron nitride nanotube.

Solution

As shown in Fig. 8.17A, the boron and nitrogen atoms both exhibit a triangular array of atoms on the sheet. Given that the distance between boron and nitrogen atoms in the nanotube is 1.45 Å throughout, both atoms will possess the same surface density.

In the triangular array the distance between boron atoms x or nitrogen atoms is given by

$$x = 2 \times 1.45 \sin 60° = 2.51 \text{ Å}.$$

Thus the surface density of boron, or nitrogen atoms on the boron nitride nanotube is

$$\eta = \frac{4\sqrt{3}}{9\sigma^2} = 0.122 \text{ atoms/Å}^2.$$

Note that the equation above is simply the equation for the surface density of atoms on a graphene sheet or carbon nanotube, as discussed in Worked Example 1.8.

The Lennard-Jones constants used in this section are taken from Lee (2006) for the boron (B) and nitrogen (N) atoms and Girifalco (1992) for carbon (C); and the values for cisplatin are given in Table 8.1. Note that for the purposes of comparison, the same Lennard-Jones constants for carbon are used as in Section 8.3 so that a direct comparison can be made with this model. For the silicon nanotube, the Lennard-Jones constants are difficult to obtain from the available literature. Accordingly, the values for the interaction between silicon (Si) with sp^3 bonding are used. These values represent an approximation because in a hollow smooth-walled silicon nanotube, the bonding will in fact be sp^2 hybridised. A summary of all Lennard-Jones constants used in this section is given in Table 8.5. Note that the resulting constants ϵ and σ between the atoms on the nanotube and atoms on the drug molecule are determined from the standard empirical combining rules, given by Eq. (8.9).

8.4.1 ENCAPSULATION BEHAVIOUR

Following Section 8.3, three orientations of cisplatin are investigated. Table 8.6 provides the acceptance radii for each orientation and provides a comparison to the results for a carbon nanotube. As shown

Table 8.5 Lennard-Jones Constants

	ϵ (eV $\times 10^{-2}$)	σ (Å)
B-B	0.4116	3.453
N-N	0.6281	3.365
C-C	0.2864	3.469
Si-Si	1.74	4.295
Cl-Cl	2.724	4.217
H-H	0.5146	2.827
Pt-Pt	1.9833	3.92

Table 8.6 Minimum Nanotube Radius a_0 Which Provides Acceptance

Orientation	Radius a_0 (Å)			
	CNT	**BNNT**	**SiNT**	**BCNT**
(i)	4.795	4.7725	5.1325	4.794
(ii)	4.895	4.875	5.226	4.894
(iii)	4.785	4.75	5.127	4.784

Comparison between boron nitride (BNNT), silicon (SiNT), boron carbide (BCNT), and carbon (CNT) nanotubes for three orientations of cisplatin.

in Table 8.6, the minimum nanotube radius for acceptance for each orientation differs from that of carbon by at most 0.7% for boron nitride, 7.1% for silicon, and 0.02% for boron carbide. For all three orientations of cisplatin to be accepted into the boron nitride, silicon, and boron carbide nanotubes the tube radius must be at least 4.875 Å, 5.226 Å, and 4.894 Å, respectively. As shown in Fig. 8.12 and discussed in Section 8.3, for carbon all three orientations are accepted, provided the nanotube radius is at least 4.895 Å.

Alternatively, as an example, the differences between the boron nitride, silicon, boron carbide, and carbon nanotube encapsulation behaviour is illustrated in Fig. 8.18 for the acceptance and suction of cisplatin for orientation (i). It is clear from Fig. 8.18A that the profile of acceptance for the silicon nanotube is significantly different from the other nanotube materials. For all three orientations of cisplatin, the magnitude of the suction energy at its maximum in comparison to the carbon nanotube is lower for the boron nitride and boron carbide, but higher for the silicon nanotube, as shown in Fig. 8.18B. In particular, for orientation (i) the magnitude of the suction energy (Fig. 8.18B) at its maximum for the boron nitride, silicon, and boron carbide nanotubes is −0.41eV, 0.42eV and −0.55eV different from that of carbon. However, as shown in Table 8.7, the radius at which the maximum occurs is not significantly different. The nanotube radius providing maximum suction differs from that of carbon by, at most, 0.85% for boron nitride, 7.5% for silicon, and 0.09% for boron carbide.

Overall, the minimum radius of nanotube which will accept all three orientations is reduced in the case of a boron nitride and boron carbide tube. The boron nitride nanotube presents the smallest acceptance radius. In other words, a smaller radius is required to fill a boron nitride nanotube with the cisplatin molecule and therefore less material is required for delivery of the drug, which means a reduced toxicity in the system. Once the drug is ejected inside the cell, the remaining nanocapsule may either slowly clear from the body or may remain; thus it is vital to reduce the amount of material required for efficient encapsulation. In addition, reduced material may decrease manufacturing costs. On the other hand, for silicon the acceptance radius to accept all orientations increases by approximately 6.8% and therefore more material is required.

In addition, a cisplatin molecule entering both a boron nitride and boron carbide nanotube experiences a reduced suction energy, particularly at the maximum, regardless of orientation. This reduction will not significantly affect the uptake of the drug molecule into the nanotube interior. However, upon expulsion of cisplatin, the boron nitride or boron carbide nanotube will present a reduced energy barrier for cisplatin to overcome. In other words, it will be easier for the drug to be expelled from a boron nitride and boron carbide tube compared to carbon, and thus it is more easily

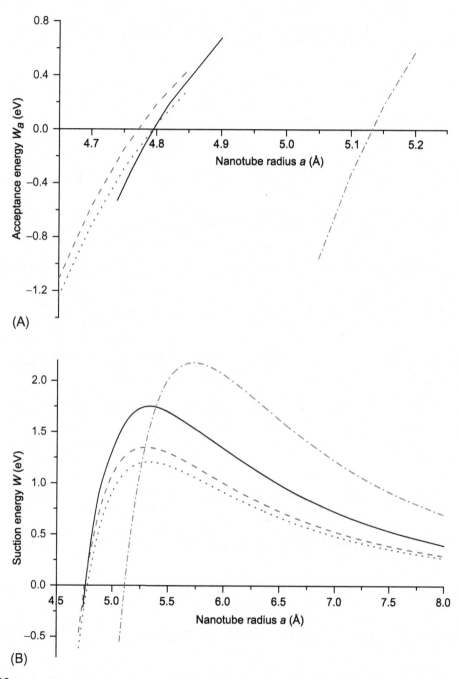

FIG. 8.18

Comparison of (A) acceptance, and (B) suction energy for orientation (i) of cisplatin entering a carbon (solid black line), boron nitride (red dash line), silicon (green dash-dot line), and boron carbide (blue dot line) nanotube.

Table 8.7 Nanotube Radius a_{max} Which Provides Maximum Suction Energy

Orientation	Radius a_{max} (Å)			
	CNT	BNNT	SiNT	BCNT
(i)	5.34	5.3	5.73	5.335
(ii)	5.34	5.3	5.73	5.335
(iii)	5.27	5.225	5.6675	5.27

Comparison between boron nitride (BNNT), silicon (SiNT), boron carbide (BCNT), and carbon (CNT) nanotubes for three orientations of cisplatin.

delivered to the target site. However, for the silicon nanotube, the suction energy is increased, which results in a more efficient uptake of the drug, though there is increased difficulty for expulsion.

Although the presented results are not significantly different from those for the carbon nanotube, there are significant implications, such as the reduced quantity of material required for efficient encapsulation in the case of boron nitride, which in turn reduces manufacturing costs and toxicity, and a reduced energy barrier for expulsion for both the boron nitride and boron carbide nanotubes. Overall the boron nitride nanotube presents the most ideal delivery capsule, as it requires the least amount of material to achieve encapsulation. Although silicon would provide the most efficient and fast uptake, it requires the most material and the most difficulty for expulsion; therefore it is the least favourable material. Some overall practical conclusions are as follows:

- Using the hybrid discrete-continuum approximation, it is possible to investigate the encapsulation of various drug molecules into nanotubes.
- It is possible to compare various nanotube materials for efficiency using simple mathematical models rather than conducting extensive experimental studies. Here, the boron nitride nanotube presented the most ideal drug delivery capsule when compared to carbon, boron carbide and silicon nanotubes.
- In principle, it is possible to load drug molecules into the interior of nanotubes and use mathematical models critical radii for acceptance and maximum uptake can be determined.
- Mathematical models, such as those proposed in this chapter, may be used to provide overall guidelines for future related experimental and molecular dynamics simulations.
- Nanotubes present exciting possibilities for future targeted drug delivery systems.

EXERCISES

8.1. List the advantages of using a carbon nanotube for drug delivery, as opposed to a nanoparticle.

8.2. What are the advantages of using the hybrid discrete-continuum approach to determine the interaction energy?

8.3. In this chapter, it is assumed that the encapsulation occurs in a vacuum environment. Typically, drug molecules are combined with solvents for delivery. How may a solvent be accounted for in this model?

8.4. Determine the Lennard-Jones parameters for cisplatin given in Table 8.2.

8.5. Determine the surface density of boron and carbon atoms in a boron carbide nanotube. Note that the bond lengths for a boron carbide tube are carbon–carbon of 1.42 Å and boron–carbon of 1.55 Å.

8.6. Plot the interaction force for a single water molecule entering a carbon nanotube of radii 3.39, 4.9, and 5.5 Å using the hybrid discrete-continuum model defined by Eq. (8.1), and detailed in Section 8.2. Use the constants $\epsilon = 5.133 \times 10^{-3}$ eV and $\sigma = 3.468$ Å for carbon–oxygen interactions, and use $\epsilon = 3.839 10^{-3}$ eV and $\sigma = 3.148$ Å for carbon–hydrogen interactions. Use a oxygen–hydrogen bond distance of 0.1 nm and an hydrogen–oxygen–hydrogen bond angle of 109.47°. Note that we assume that the water molecule enters with its hydrogen atoms first.

NEW FORMULAE FOR THE GEOMETRIC PARAMETERS OF CARBON NANOTUBES

9.1 INTRODUCTION

In the previous chapters, we have utilised the rolled-up cylindrical model of carbon nanotubes. We have assumed that the interaction of the carbon atoms situated at discrete locations can be approximated by replacement with an average atomic density (number of atoms per unit area of surface), which is distributed uniformly over the entire surface of the carbon nanotube. Accordingly in such an approach, the precise atomic locations have very little bearing on the mechanical interaction with other molecules and nanostructures. In this chapter, we are concerned with the precise atomic locations, and we have in mind situations where this is an important issue, namely small radii carbon nanotubes, which necessarily have large curvature. The new formulae obtained for nanotube radii are compared with other values obtained from ab initio calculations.

In this chapter an exact polyhedral model for the structure of carbon nanotubes is detailed, which is derived solely from the three geometric postulates:

(i) All bonds are equal in length.
(ii) All angles between adjacent bonds are equal.
(iii) All atoms lie equidistant from a central cylindrical axis.

In short, all bonds, bond angles, and atoms are equivalent, which differs from the conventional 'rolled-up' model in this respect, because certain bonds, bond angles, and atomic locations either differ from the others or play different roles for the others. These three postulates mean that all bonds and bond angles are unprivileged and give rise to new formulae which precisely determine bond angles and the locations of all atoms in a fully symmetric cylindrical structure. It is demonstrated that these formulae fully explain the effects of the curvature on small diameter carbon nanotubes. An asymptotic analysis of the equations is also given that leads to precisely the conventional formulae as the leading order terms. Lastly, a new expression for nanotube thickness is given, which emerges naturally from the analysis of the new model.

9.2 CONVENTIONAL 'ROLLED-UP' MODEL

Conventionally, carbon nanotubes are molecules which are considered to be formed from a sheet of graphene, which is rolled into a cylindrical configuration, as detailed in Section 1.2.2. To remind the reader, the orientation of the hexagons which make up the nanotube are characterised by two integers (n, m) which prescribe the chirality of the nanotube. This is done via the chiral vector $\mathbf{C} = n\mathbf{a} + m\mathbf{b}$, where $\mathbf{a}$ and $\mathbf{b}$ form a basis for the graphene sheet, as illustrated in Fig. 1.5. The direction of the chiral vector is termed the chiral angle θ_0, which is defined as the angle from the base vector for the graphene plane and is given by Eq. (1.1); again, to remind the reader

$$\cos\theta_0 = \frac{2n + m}{2\sqrt{n^2 + nm + m^2}}.$$

As mentioned in Section 1.2.2, there are two special cases of carbon nanotubes; these are when $m = 0$, which are termed zigzag nanotubes, and when $m = n$, which are called armchair nanotubes.

The conventional 'rolled-up' model implies that the magnitude of the chiral vector is equivalent to the circumference of the cylinder and therefore the radius r_0 is assumed to be given by Eq. (1.1). To remind the reader,

$$r_0 = \sigma\sqrt{3\left(n^2 + nm + m^2\right)}/2\pi, \tag{9.1}$$

where σ denotes the carbon–carbon covalent bond length. Additionally, the unit cell length L_0 is given by Eq. (1.2) and restated here

$$L_0 = 3\sigma\sqrt{n^2 + nm + m^2}/d_R, \tag{9.2}$$

where d_R is the greatest common divisor of $2n + m$ and $2m + n$, as outlined in Section 1.2.2.

However, when considering these formulae two problems are apparent. Firstly, the covalent bonds exist in three different directions and are therefore being applied over two or three different curvatures. This means that the straight-line distance in the three-dimensional space between atoms no longer matches the original bond length used to construct the graphene sheet; this discrepancy between bond lengths exists for differing bond angles in a single nanotube. The second issue is that the angles between the three bonds for a single atom (initially 120° for the flat sheet) also change by different amounts, which again, depending on the curvature involved leads to further asymmetry in the overall structure. These considerations lead to the conclusion that the conventional 'rolled-up' model is an approximation which ignores the effects of curvature. Specific in situ bond lengths and bond angles are tabulated by Jiang et al. (2003) for a range of particular carbon nanotubes.

The model proposed in this chapter is based on three postulates and addresses the shortcomings of the conventional model by first prescribing all bond lengths and all bond angles to be equal in three-dimensional space. That is, it is assumed that all the bond lengths and bond angles are equal for all bonds and atoms in the carbon nanotube in its cylindrical state. This assumption is equivalent to that of the conventional model for the flat graphene sheet and is justified physically because covalent bond energy is the overriding consideration for small diameter nanotubes, and the lowest energy state would naturally lead to an equalisation of all the bonds into a symmetric structure. Furthermore, it is assumed that all atoms lie on the surface of an ideal right circular cylinder as in the conventional model. However, due to the three-dimensional definition of distance, the radius and longitudinal extent for a carbon nanotube of a particular structure are different to those obtained from the 'rolled-up' model.

9.3 NEW 'POLYHEDRAL' MODEL
9.3.1 EXACT GEOMETRIC PARAMETERS

In order to fully describe the new model, two further angles need to be introduced, which are termed the subtend semiangle ψ, and the incline angle ω. Furthermore, three parameters λ, μ, and ν, are needed, which are defined in terms of the various angles by Eqs (9.34), (9.35), (9.36). The first step in determining the exact geometric structure is to determine the subtend angle 2ψ, subtended at the axis of the cylinder by the base of a single equilateral triangle, as shown in Section 9.4. This angle is determined as a root of the following transcendental equation

$$(n^2 - m^2)\sin^2([(n+m)\psi - \pi]/m) - n(n+2m)\sin^2((n\psi - \pi)/m)$$
$$+ m(2n+m)\sin^2\psi = 0. \tag{9.3}$$

This equation may have many roots, but specifically the one required here must also satisfy the inequalities

$$n\psi \leqslant \pi \leqslant (n+m)\psi. \tag{9.4}$$

The value of this root can be determined as accurately as required by a small number of iterations of Newton's method using the initial guess $\psi_0 = \pi(2n+m)/[2(n^2 + nm + m^2)]$. Once the correct root of Eq. (9.3) is found, all other expressions can be determined. One observes for armchair ($n = m$) and zigzag ($m = 0$) nanotubes that $\psi = \pi/2n$ and $\psi = \pi/n$, respectively, may be deduced from Eq. (9.3) in the appropriate limit.[1] It is worth commenting that the conventional formula ψ_0 is exact in these two special cases; and the inequalities given in Eq. (9.4) are satisfied in both cases, and in particular, the case of $m = 0$ Eq. (9.4) degenerates to the specific equality $\psi = \pi/n$.

The chiral angle θ is given by the expression

$$\cos^2\theta = \frac{n(n+2m)\sin^2\psi}{(n+m)^2\sin^2\psi - m^2\sin^2([(n+m)\psi - \pi]/m)}, \tag{9.5}$$

and the angle of incline ω of the pyramidal components of the surfaces is given by

$$\sin\omega = \left(\sqrt{\cot^2\psi + 4\cos^2\theta - 3} - \cot\psi\right)/\sqrt{3}\cos\theta. \tag{9.6}$$

With all the necessary angles now determined, the bond angle 2ϕ is determined from the formula

$$\sin\phi = \sqrt{3/(4+k^2)}, \tag{9.7}$$

where k is related to the perpendicular height of the constituent pyramids and is given by the positive root of $\lambda k^2 + \mu k + \nu = 0$, where the coefficients λ, μ, and ν are given by Eqs (9.34), (9.35), (9.36), respectively. In addition to the bond angle, the tube radius r is exactly given by the expression

$$r = \sigma\sin\phi\cos\theta/\sin\psi, \tag{9.8}$$

---

[1]The limit of m tending to zero must be done with care.

and a unit cell length L is given by

$$L = 4\sigma(n^2 + nm + m^2)\sin\phi\sin\theta/md_R, \tag{9.9}$$

recognising that the tube radius r is taken to be the common perpendicular distance of all atoms from the axis of the tube. Due to the facetted nature of the tube surface in this model, an inner radius may also be identified, and the conceptual or effective wall thickness δ of the carbon nanotube is proposed to be the difference between the tube radius and the inner radius[2] and is given by

$$\delta = \sigma\sin\phi\cos\theta\tan(\psi/2). \tag{9.10}$$

9.3.2 LEADING ORDER AND CORRECTION TERMS

In this section the leading two terms are given for the asymptotic expansions of the equations from the previous section. The details of the asymptotic expansions are contained in Worked Examples 9.1 and 9.2. Using the method of asymptotic expansions on Eq. (9.3) produces the following expression for the subtend semiangle

$$\psi = \frac{\pi(2n+m)}{2(n^2+nm+m^2)} + \frac{3\pi^3 nm^2(n^2-m^2)(2n+m)(n+2m)}{32(n^2+nm+m^2)^5} + O(1/n^5), \tag{9.11}$$

where the O term indicates the order of magnitude of the subsequent term; details for the evaluation of Eq. (9.11) are given in Worked Example 9.1. Note that the first term of this expansion is exactly what one would expect from the 'rolled-up' model and the second term is a correction term which takes into account the curvature of the cylinder in question. It is worth commenting that up to this order, Eq. (9.11) is entirely consistent with the armchair ($n = m$) and zigzag ($m = 0$) special cases noted previously, namely $\psi = \pi/2n$ and $\psi = \pi/n$, respectively.

WORKED EXAMPLE 9.1

Using series and asymptotic expansions, show that the subtend semiangle ψ is given by Eq. (9.11).

Solution

The expansion begins by determining ψ as a root of Eq. (9.3) using a series expansion in powers of $1/n$ and then using this as the basis for determining series expansions for all of the quantities derived in the previous section. In this example, we determine the expansion for the subtend semiangle ψ. Writing Eq. (9.3) in the form

$$(1-h^2)\sin^2\left(\frac{1+h}{h}\psi - \frac{\pi}{hn}\right) - (1+2h)\sin^2\left(\frac{\psi}{h} - \frac{\pi}{hn}\right) + h(2+h)\sin^2\psi = 0, \tag{9.12}$$

where $h = m/n$. Note that h is of order one and ψ becomes small as n increases; therefore Eq. (9.12) may be expanded in terms of ψ and $1/n$. The following series is defined:

$$\psi = \frac{\psi_0(h)}{n} + \frac{\psi_1(h)}{n^3} + \frac{\psi_2(h)}{n^5} + \cdots,$$

---

[2]From this definition the conceptual or effective wall thickness δ can also be considered as the degree of corrugation of the tube wall.

$$\cos^2 \theta = a_0(h) + \frac{a_1(h)}{n^2} + \frac{a_2(h)}{n^4} + \cdots,$$

then by the method of asymptotic expansions one may derive

$$\psi_0(h) = \frac{\pi(2+h)}{2(1+h+h^2)}, \tag{9.13}$$

$$\psi_1(h) = \frac{3\pi^3 h^2 (h^2 - 1)(1 + 2h)(2 + h)}{32(1 + h + h^2)^5}, \tag{9.14}$$

and substituting for h in Eqs (9.13), (9.14) produces the expansion for ψ given by Eq. (9.11).

WORKED EXAMPLE 9.2

Determine the asymptotic expansions for the following quantities: $\sin \omega$, λ, μ, ν, k, $\sin \phi$, r/σ, L/σ, and δ/σ.

The leading order of ψ is $1/n$ and that of $\cos \theta$ is of order one. Therefore by using the following expansion

$$\cot \psi = \frac{1}{\psi} - \frac{\psi}{3} - \frac{\psi^3}{45} - \cdots + \frac{(-1)^n 2^{2n} B_{2n}}{(2n)!} \psi^{2n-1} \cdots,$$

where B_{2n} is a Bernoulli number, one may derive the following asymptotic expansion for $\sin \omega$ from Eq. (9.6)

$$\sin \omega = \frac{\sqrt{3}(4C^2 - 3)}{6C} \psi - \frac{\sqrt{3}(4C^2 - 3)(12C^2 - 13)}{72C} \psi^3 + O(\psi^5), \tag{9.15}$$

where $C = \cos \theta$. Further substitution of the expansion (9.15) into the expressions for λ, μ, and ν given by Eqs (9.34), (9.35), (9.36) produces

$$\lambda = 1 - \frac{(4C^2 - 3)^2}{12} \psi^2 + \frac{(4C^2 - 3)^2(12C^2 - 13)}{72} \psi^4 + O(\psi^6),$$

$$\mu = 2\sqrt{3}C\psi^{-1} + \frac{\sqrt{3}(4C^2 - 9)}{12C} \psi + O(\psi^3),$$

$$\nu = -2 - \frac{(4C^2 - 3)^2}{6} \psi^2 + \frac{(4C^2 - 3)^2(12C^2 - 13)}{36} \psi^4 + O(\psi^6).$$

By substituting these expansions for the coefficients into the standard quadratic formula produces the following expansion for the pyramid height k, namely

$$k = \frac{1}{\sqrt{3}C} \psi + \frac{\sqrt{3}(32C^6 - 48C^4 + 14C^2 + 5)}{72C^3} \psi^3 + O(\psi^5). \tag{9.16}$$

The following expansion may be derived from Eqs (9.7), (9.16)

$$\sin \phi = \frac{\sqrt{3}}{2} - \frac{\sqrt{3}}{48C^2} \psi^2 + O(\psi^4). \tag{9.17}$$

Substitution and use of the usual series for $\sin \psi$ in Eq. (9.8) produces the following expansion for the exact radius:

$$\frac{r}{\sigma} = \frac{\sqrt{3}C}{2} \psi^{-1} + \frac{\sqrt{3}\left(4C^2 - 1\right)}{48C} \psi + O(\psi^3). \tag{9.18}$$

Making the appropriate substitutions in Eq. (9.9) yields the following expansion for the unit cell length

$$\frac{L}{\sigma} = 3\sqrt{1 + h + h^2}\, n$$

$$- \frac{\pi^2 \left[4(1 + h + h^2)^3 + (2 + h)^2(1 + 2h)^2(1 - h)^2 \right]}{32(1 + h + h^2)^{7/2}} \frac{1}{n} + O(1/n^3). \tag{9.19}$$

From Eq. (9.10) and the series for $\tan(\psi/2)$, one may obtain the following expansion for the thickness δ,

$$\frac{\delta}{\sigma} = \frac{\sqrt{3}C}{4}\psi + \frac{\sqrt{3}(2C^2 - 1)}{96C}\psi^3 + O(\psi^5). \tag{9.20}$$

From Eqs (9.18), (9.19), (9.20) and the series expansions (Eqs 9.11, 9.21) for ψ and $C = \cos\theta$, Eqs (9.24), (9.27), (9.29) may be deduced, respectively.

Substituting Eq. (9.11) into Eq. (9.5) and then further expansion in terms of $1/n$ produces the following expansion for $\cos^2\theta$

$$\cos^2\theta = \frac{(2n + m)^2}{4(n^2 + nm + m^2)} + \frac{\pi^2 m^2 (2n + m)^2 (n + 2m)^2 (n - m)^2}{64(n^2 + nm + m^2)^5} + O(1/n^4). \tag{9.21}$$

Again noting that taking the square root of the first term yields precisely the conventional formula given by Eq. (1.1) and the second term of Eq. (9.21) constitutes a correction term which accounts for the nanotube curvature. By expanding Eq. (9.6) in terms of the small parameter ψ, and then substituting into the relevant equations to determine an expression for $\sin\phi$ and finally using Eqs (9.11), (9.21), one may derive the following

$$\sin\phi = \frac{\sqrt{3}}{2} - \frac{\sqrt{3}\pi^2}{48(n^2 + nm + m^2)} + O(1/n^4). \tag{9.22}$$

Here, the first term of Eq. (9.22) is simply the value of $\sin\phi$ in flat graphene, and the correction term gives the order $1/n^2$ adjustment to the sine of the bond angle that takes into account the curvature of the nanotube. The effects of curvature can also be expressed in terms of the conventional nanotube radius r_0, given by

$$\sin\phi = \frac{\sqrt{3}}{2} - \frac{\sqrt{3}\sigma^2}{64r_0^2} + O(1/n^4), \tag{9.23}$$

where it can be seen that the first order correction to the bond angle is independent of the chirality. With expansions for all the angles now determined, Eq. (9.8) can be used to give

$$r = \frac{\sigma\sqrt{3(n^2 + nm + m^2)}}{2\pi}$$

$$+ \frac{\sqrt{3}\sigma\pi \left[4(n^2 + nm + m^2)^3 - 9n^2 m^2 (n + m)^2 \right]}{64(n^2 + nm + m^2)^{7/2}} + O(1/n^3), \tag{9.24}$$

where again the leading order term is exactly the conventional expression (9.1). The second term in Eq. (9.24) is an order $1/n$ correction to the radius. Recognising that some experimentalists (for example, Jorio et al., 2005) attempt to fit curvature data to an empirical function of the form

$$r = r_0 + \frac{A}{r_0}[1 + B\cos(3\theta_0)] + \frac{C}{r_0^2}[1 + D\cos(3\theta_0)], \tag{9.25}$$

where A, B, C, and D are fitting coefficients. An examination of the second term of Eq. (9.24) reveals that the above functional form is not the most appropriate one to capture the effects of curvature on the nanotube radius. In fact, it can be shown that the second term of Eq. (9.24), which is equal to the difference between the exact and conventional radius $r - r_0$ to the order $1/n^3$, can be expressed in the form

$$r = r_0 + \frac{\sigma^2}{64r_0}[5 + \cos(6\theta_0)] + O(1/n^3), \tag{9.26}$$

which with fewer terms provides a much better fit to the effect of curvature on the radius than that which is possible using the form (9.25). Due to Eq. (9.26) being a particularly accurate approximation, a second term is not included here, but note that the expansion for r given in Eq. (9.24) indicates that it would be proportional to r_0^{-3} and not r_0^{-2}. Appropriate substitution of the derived expansion into Eq. (9.9) yields

$$L = \frac{3\sigma\sqrt{n^2 + nm + m^2}}{d_R}$$
$$- \frac{\sigma\pi^2\left[4(n^2 + nm + m^2)^3 + (2n + m)^2(n + 2m)^2(n - m)^2\right]}{32d_R(n^2 + nm + m^2)^{7/2}} + O(1/n^3). \tag{9.27}$$

Worthy of comment is that the form of Eq. (9.25) is also not the most natural to characterise the effect of curvature on the unit cell length. However, the effect of curvature can be approximated by the relation

$$L \approx L_0 - \frac{\sigma^2\pi\sqrt{3}}{32d_R r_0}[3 + \cos(6\theta_0)], \tag{9.28}$$

where $\approx$ indicates that Eq. (9.28) is not asymptotically exact, though it constitutes an excellent approximation suggested by the second term in Eq. (9.27). Finally, the conceptual nanotube thickness proposed in Eq. (9.10) has the asymptotic expansion

$$\delta = \frac{\sqrt{3}\sigma\pi(2n + m)^2}{16(n^2 + nm + m^2)^{3/2}}$$
$$+ \frac{\sqrt{3}\sigma\pi^3(2n + m)^2[4n(n + m) - 5m^2][n^2(n + m)^2 + 8nm^2(n + m) - 2m^4]}{1536(n^2 + nm + m^2)^{11/2}}$$
$$+ O(1/n^5). \tag{9.29}$$

Because there is presently no theory on nanotube thickness, both terms in Eq. (9.29) are new. The leading term is of order $1/n$, and so the thickness tends to approach zero as the size of the nanotube increases. This is expected because the thickness is a measure of the curvature of the facetted surface model. Note that the thickness can be approximated by the relation

$$\delta = \frac{3\sigma^2 \cos^2 \theta_0}{8r_0} + O(1/n^3).$$ (9.30)

A major advantage of the present theoretical model is the discovery of precise equations, such as Eqs (9.23), (9.26), (9.28), (9.30), which are not directly accessible by experimental or molecular dynamics simulations.

9.4 DETAILS OF THE POLYHEDRAL MODEL
9.4.1 CARBON NANOTUBES AS FACETTED POLYHEDRA

With all atoms of the carbon nanotube lying on the surface of a right circular cylinder, the hexagons themselves cannot be faces of a polyhedron because not all vertices of each hexagon will necessarily be coplanar. The description of nanotubes as a polyhedron begins with the usual tessellation of regular hexagons, where the vertices of the tessellation represent the point masses of the atoms and lines of the hexagons represent covalent bonds. A second tessellation of equilateral triangles is overlayed on this, where the vertices of the triangles are the point masses of the atoms and every second triangle also has an atom located at its centre. The net effect of these two tessellations is a single tessellation of equilateral and isosceles triangles. By fixing the side lengths of the isosceles triangles, which represent covalent bonds, but by varying the side lengths of the equilateral triangles, it is possible to construct a truly facetted polyhedron where all vertices are equidistant from an axis of symmetry, and all the bond lengths and bond angles are equal for all atoms.

In the conventional model, to satisfy the constraints of curvature, the bond angles become smaller than the 120° found in flat graphene. The effect of this reduction is to reduce the equilateral triangle side length and the condition that the covalent bond length remains fixed means that every equilateral triangle that contains an atom at its centre 'buckles' forming a right regular triangular pyramid. The analysis presented here is to take the nanotube as defined by the chiral vector (n, m) and use this to calculate the orientation of equilateral triangles to form the polyhedron and then from this, we can determine the height of each pyramid and therefore the angle between bonds in the carbon nanotube. This leads to an expression for the equilateral triangle side length, which in turn can be used to determine the radius and unit cell length of the nanotube. To illustrate the idea, $(3, 3)$, $(4, 2)$, and $(5, 0)$ nanotubes are shown as such polyhedra in Fig. 9.1.

A possibly significant consequence of considering a carbon nanotube as a facetted polyhedron is that this polyhedron has an intuitively obvious effective wall thickness δ, being defined simply as the difference between the outer and inner radii of the polyhedron. For the rolled-up model, many mechanical models based on classical elasticity theory have been proposed to describe the deformation of carbon nanotubes, but all such models have lacked a clearly obvious expression for the effective wall thickness, which accordingly gives rise to a high degree of uncertainty in the various elastic constants for carbon nanotubes. While it remains to be seen whether this concept of effective wall thickness proves useful, it is an immediate consequence of the geometric model proposed here.

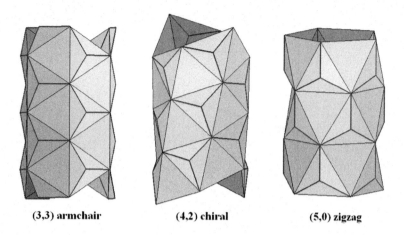

(3,3) armchair   (4,2) chiral   (5,0) zigzag

FIG. 9.1

Three carbon nanotubes represented as a facetted polyhedra with an atom at every vertex.

9.4.2 ORIENTATION OF THE EQUILATERAL TRIANGULAR FACES

To correctly arrange the equilateral triangular faces with all the vertices lying on the surface of a cylinder, the vertices are located on helices, which lie on the surface of the cylinder. The number of helices is determined by the chirality of the tube and the interhelix spacing is such that the distance in three-dimensional space between evenly spaced points on a helix, as well as between corresponding points on adjacent helices, are equal. Considering the first helix $\alpha(t)$ on the cylinder to have the parametric form in Cartesian coordinates

$$\alpha(t) = (r\cos(2\psi t/m), r\sin(2\psi t/m), bt/m),$$

where 2ψ is the angle subtended at the tube axis in the xy-plane of one edge of a triangle, b is the vertical spacing of the helix, and t is the parametric variable such that the vertices are spaced evenly at a distance m in this variable. The points $\mathbf{P} = \alpha(0) = (r,0,0)$ and $\mathbf{Q} = \alpha(m) = (r\cos 2\psi, r\sin 2\psi, b)$ are then taken, and therefore the distance ℓ in three-dimensional space between vertices is given by

$$\ell^2 = 4r^2\sin^2\psi + b^2.$$

This configuration of helices and points is shown in Fig. 9.2.

Now considering the adjacent helix $\beta(t)$, which due to the symmetry of the arrangement of helices around the tube, is given by the equation

$$\beta(t) = (r\cos(2(\psi t - \pi)/m), r\sin(2(\psi t - \pi)/m), bt/m),$$

and two points on this helix $\mathbf{R} = \beta(n)$ and $\mathbf{S} = \beta(n+m)$, which together with the point $\mathbf{P}$ define one equilateral triangle of the polyhedron, $\triangle\mathbf{PRS}$. Therefore the distances $|\mathbf{PR}| = |\mathbf{PS}| = \ell$, where these expressions are given by

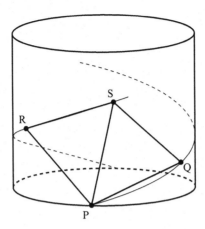

FIG. 9.2

Points lying on two helices forming equilateral triangles in three-dimensional space.

$$|PR|^2 = 4r^2 \sin^2 ((n\psi - \pi)/m) + n^2 b^2/m^2,$$

$$|PS|^2 = 4r^2 \sin^2 ([(n+m)\psi - \pi]/m) + (n+m)^2 b^2/m^2,$$

and by equating these expressions, the transcendental equation (9.3) is derived. The correct root of Eq. (9.3) provides the value of ψ which leads to the correct orientation of helices, and in turn leads to a polyhedron composed of equilateral triangles of the necessary chirality. Eq. (9.3) may have many roots, however the one specifically required is that which satisfies the inequalities $(n\psi - \pi)/m \leqslant 0 \leqslant [(n+m)\psi - \pi]/m$, which can be rewritten in the form given by the inequalities (9.4).

With the subtend semiangle ψ determined, the true chiral angle θ which is the angle the segment **PQ** makes with the xy-plane needs to be found. By projecting the point **Q** into the xy-plane and calculating the distance from **P**, which is called a here, and it follows that $a^2 = 4r^2 \sin^2 \psi$. Using this and equations for the triangle length, it can be shown that

$$b^2 = 4r^2 m^2 \left(\sin^2 \psi - \sin^2 ([(n+m)\psi - \pi]/m) \right) / [n(n+2m)] . \tag{9.31}$$

The hypotenuse of the right-angled triangle with perpendicular sides of length a and b, which is denoted by c, is given by

$$c^2 = 4r^2 \left[(n+m)^2 \sin^2 \psi - m^2 \sin^2 ([(n+m)\psi - \pi]/m) \right] / [n(n+2m)] . \tag{9.32}$$

It follows that the chiral angle θ is given by the relationship $\cos^2 \theta = a^2/c^2$, which after substitution from Eqs (9.31), (9.32) gives rise to Eq. (9.5), and therefore θ can easily be determined for a known value of ψ.

9.4.3 ARRANGEMENT OF THE PYRAMIDAL FACES

Now focussing on a single pyramid, where it is assumed that the lengths from the vertices that form the base to the apex are fixed at the bond length σ. If the bond angle is denoted by 2ϕ, then the perpendicular height of the pyramid can be determined from elementary triangular geometry and is given by $\sin\phi\sqrt{\csc^2\phi - 4/3}$. The approach is to position such a pyramid in space, rotate it according to the chiral angle θ and use the requirement that all atoms remain equidistant from a common axis to determine the actual value of the bond angle.

For algebraic convenience, all lengths are nondimensionalised by the distance $\sigma\sin\phi/\sqrt{3}$. Defining a pyramid by the points $\mathbf{A}, \mathbf{B}, \mathbf{C}$, and $\mathbf{D}$, according to a three-dimensional Cartesian coordinate system (x, y, z),

$$\mathbf{A} = \left(\sqrt{3}, 0, 0\right), \quad \mathbf{B} = \left(-\sqrt{3}, 0, 0\right), \quad \mathbf{C} = (0, 0, -3), \quad \mathbf{D} = (0, k, -1),$$

where $k = \sqrt{3\csc^2\phi - 4}$. These points are shown in Fig. 9.3, which is then rotated around the x-axis by an incline angle ω, the value of which is yet to be determined. The rotated points are given by

$$\mathbf{A}' = \left(\sqrt{3}, 0, 0\right), \quad \mathbf{B}' = \left(-\sqrt{3}, 0, 0\right), \quad \mathbf{C}' = (0, 3\sin\omega, -3\cos\omega),$$
$$\mathbf{D}' = (0, k\cos\omega + \sin\omega, k\sin\omega - \cos\omega).$$

Now the points are rotated around the y-axis by the value of the true chiral angle θ, as given by Eq. (9.5), yielding

$$\mathbf{A}'' = \left(\sqrt{3}\cos\theta, 0, \sqrt{3}\sin\theta\right), \quad \mathbf{B}'' = \left(-\sqrt{3}\cos\theta, 0, -\sqrt{3}\sin\theta\right),$$
$$\mathbf{C}'' = (3\sin\theta\cos\omega, 3\sin\omega, -3\cos\theta\cos\omega),$$
$$\mathbf{D}'' = (\sin\theta(\cos\omega - k\sin\omega), k\cos\omega + \sin\omega, \cos\theta(k\sin\omega - \cos\omega)).$$

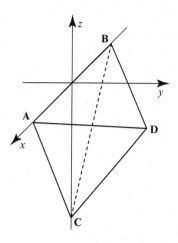

FIG. 9.3

Points forming a regular triangular pyramid.

The pyramid is now positioned as it would be in the cylindrical polyhedron with an axis of symmetry given by the line $\mathcal{L} = (0, -\gamma, t)$, where γ is a positive constant yet to be determined and t is the parametric variable for the line. Now, the condition that the perpendicular distances from each point to the line $\mathcal{L}$ are equal is imposed. As the line $\mathcal{L}$ is perpendicular to the xy-plane, this next step is facilitated by projecting all the points onto this plane and therefore reducing the number of dimensions of the problem by one. To this end the following points are defined in a two-dimensional Cartesian coordinate system (x, y):

$$\mathbf{a} = \left(\sqrt{3}\cos\theta, 0\right), \quad \mathbf{b} = \left(-\sqrt{3}\cos\theta, 0\right), \quad \mathbf{c} = (3\sin\theta\cos\omega, 3\sin\omega),$$

$$\mathbf{d} = (\sin\theta\,(\cos\omega - k\sin\omega), k\cos\omega + \sin\omega), \quad \mathbf{e} = (0, -\gamma).$$

The conditions imposed are that $|\mathbf{ae}| = |\mathbf{ce}| = |\mathbf{de}|$, and the unknowns γ, ϕ, and ω need to be determined. It would appear that one condition is lacking; however, this is provided by considering the angle subtended by one side of the base of the pyramid at the centre of the cylinder, where the subtend semiangle ψ is determined as a root of Eq. (9.3), and therefore $\gamma = \sqrt{3}\cos\theta\cot\psi$. The points $\mathbf{a}$, $\mathbf{b}$, $\mathbf{c}$, $\mathbf{d}$, and $\mathbf{e}$ and the angle ψ are shown in Fig. 9.4.

Firstly, considering the condition $|\mathbf{ae}| = |\mathbf{ce}|$ leads to the following quadratic equation in terms of $\sin\omega$

$$3\cos^2\theta\sin^2\omega + 2\sqrt{3}\cos\theta\cot\psi\sin\omega + 3 - 4\cos^2\theta = 0, \tag{9.33}$$

and as a positive value of $\sin\omega$ is required, the correct root of Eq. (9.33) is given by Eq. (9.6).

The next condition is $|\mathbf{ae}| = |\mathbf{de}|$, which leads to the quadratic equation $\lambda k^2 + \mu k + \nu = 0$, where λ, μ, and ν are defined by

$$\lambda = 1 - \sin^2\omega\cos^2\theta, \tag{9.34}$$

$$\mu = 2\cos\omega\cos\theta(\sin\omega\cos\theta + \sqrt{3}\cot\psi), \tag{9.35}$$

$$\nu = 1 - \cos^2\theta(4 - \sin^2\omega) + 2\sqrt{3}\sin\omega\cos\theta\cot\psi, \tag{9.36}$$

which in turn leads to $k = (-\mu \pm \sqrt{\mu^2 - 4\lambda\nu})/2\lambda$. As it is needed that $k > 0$ and for the angles under consideration $\mu > 0$, the positive square root must be taken in this case. Substituting for k

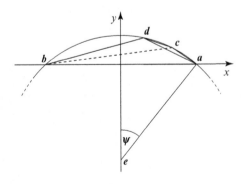

FIG. 9.4

Points forming rotated pyramid projected onto xy-plane.

yields Eq. (9.7), from which the angle ϕ can be determined. The task of determining all the angles for the problem is now complete, and therefore the true radius can be calculated from Eq. (9.8). A true unit cell length L is given by the longitudinal displacement of one hexagon along a single helix $b = 2\sigma \sin\phi \sin\theta$, multiplied by the number of hexagons in a unit cell $2(n^2 + nm + m^2)/d_R$, divided by the number of helices m, which yields Eq. (9.9). Additionally, the effective wall thickness δ of the carbon nanotube can now be determined. The inner radius is simply $r_{in} = r \cos\psi$, and therefore if it is assumed that the nanotube thickness is the radius of the nanotube less the inner radius, then Eq. (9.10) may be derived from $r - r_{in}$.

9.5 RESULTS

The magnitude of the discrepancies between the conventional 'rolled-up' model and the new model is a function of the curvature of the nanotube. When the effect of the curvature is examined, nanotubes with radii in the range from 2 to 5 Å are considered and that general idea is followed here. If the radius of a nanotube is much greater than 5 Å, then the effects of the curvature are quite small and generally in the range of less than 1%. Some typical numerical values of the various geometric parameters are shown in Table 9.1. We comment that such small carbon nanotubes such as (3,3), (4,2), and (5,0) have been shown to arise by formation in zeolite channels (see Wang et al., 2000). It may well be that the very smallest tubes are not realisable at the present time, but this is more a question of the limitations

Table 9.1 Results of the Polyhedral Model Using $\sigma = 1.44$ Å

Nanotube (n,m)	Subtend 2ψ (°)	Chiral θ (°)	Incline ω (°)	Bond 2ϕ (°)	Radius r (Å)	Length $L\,d_R$ (Å)	Thickness δ (Å)
(4, 0)	90.00	0.00	13.84	114.47	1.712	16.292	0.502
(3, 2)	76.32	23.33	5.01	115.59	1.811	18.339	0.387
(4, 1)	77.42	10.72	10.41	115.91	1.918	19.069	0.421
(5, 0)	72.00	0.00	10.81	116.57	2.084	20.840	0.398
(3, 3)	60.00	30.00	0.00	116.93	2.126	22.092	0.285
(4, 2)	64.60	19.01	6.05	117.01	2.172	22.396	0.336
(5, 1)	63.98	8.84	8.79	117.27	2.293	23.444	0.348
(6, 0)	60.00	0.00	8.90	117.65	2.464	25.300	0.330
(4, 3)	53.66	25.26	2.47	117.77	2.470	25.957	0.266
(5, 2)	55.56	16.02	6.04	117.86	2.544	26.547	0.293
(6, 1)	54.46	7.52	7.58	118.05	2.675	27.805	0.297
(4, 4)	45.00	30.00	0.00	118.29	2.797	29.668	0.213
(5, 3)	47.88	21.75	3.49	118.32	2.830	29.932	0.243
(7, 0)	51.42	0.00	7.57	118.29	2.849	29.715	0.282
(6, 2)	48.56	13.83	5.72	118.40	2.921	30.753	0.258
(7, 1)	47.40	6.54	6.65	118.54	3.060	32.157	0.258
(5, 4)	41.36	26.32	1.48	118.65	3.143	33.500	0.203
(6, 3)	42.94	19.07	3.86	118.69	3.198	33.991	0.222
(8, 0)	45.00	0.00	6.59	118.70	3.237	34.104	0.246

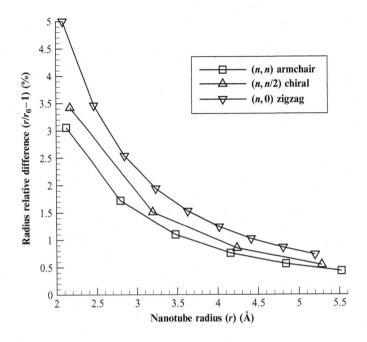

FIG. 9.5

Difference between new and conventional radius ($r - r_0$) for carbon nanotubes of type zigzag: $(5, 0)$–$(13, 0)$, chiral: $(4, 2)$–$(10, 5)$, and armchair: $(3, 3)$–$(8, 8)$.

of current technology. In the fullness of time the very smallest nanotubes such as (2,0) and (1,1) may well be achievable through new technology.

With respect to nanotube radius, the new model predicts an increase in radius of up to approximately 5% (for a (5, 0) zigzag nanotube). In general, zigzag nanotubes involve the greatest increase in the radius and armchair nanotubes the least, with chiral nanotubes increasing somewhere between these two extremes. The difference between radii calculated by the present new model and the conventional model for a number of nanotubes of chirality (n,0), (n,n/2), and (n,n) are shown in Fig. 9.5, and it is shown that the difference is proportional to r^{-1}.

The new model shows that there is up to approximately 3.5% of shortening in the longitudinal dimension and that this is also proportional to the reciprocal of the nanotube radius r^{-1}. The degree of longitudinal shortening is greatest for zigzag nanotubes and smallest for armchair nanotubes, with chiral nanotubes shortening somewhere between these extremes. The difference in the lengths of the unit cells calculated by the present new model and the conventional model for various carbon nanotubes is shown in Fig. 9.6, where the effect of the d_R term has been removed by multiplying the length by this term.

The angle between covalent bonds in the nanotube lattice is also a function of the curvature being proportional to r^{-2}. As the nanotube radius increases the bond angle approaches the expected 120° found in graphene. The bond angle shows little or no difference between nanotubes of the zigzag, armchair or chiral type. The bond angles for a number of nanotubes of chirality (n,0), (n,n/2), and (n,n) are shown in Fig. 9.7.

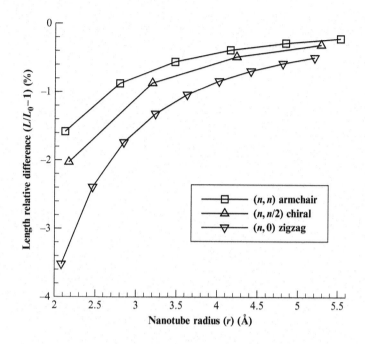

FIG. 9.6

Difference between new and conventional unit length ($L - L_0$) for carbon nanotubes of type zigzag: $(5, 0)$–$(13, 0)$, chiral: $(4, 2)$–$(10, 5)$, and armchair: $(3, 3)$–$(8, 8)$.

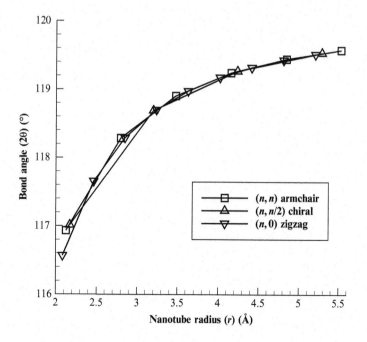

FIG. 9.7

Bond angle for carbon nanotubes of type zigzag: $(5, 0)$–$(13, 0)$, chiral: $(4, 2)$–$(10, 5)$, and armchair: $(3, 3)$–$(8, 8)$.

Table 9.2 Comparison of Radii From Conventional Model, New Model and Ab Initio Calculations of Cabria et al. (2003) Using $\sigma = 1.44\,\text{Å}$

Nanotube	r_0 (Å)	r (Å)	Cabria et al. (2003) (Å)
(4,0)	1.5878	1.7125	1.71
(3,2)	1.7303	1.8109	1.8
(4,1)	1.8191	1.9178	1.91
(5,0)	1.9848	2.084	2.06
(3,3)	2.0626	2.1258	2.12
(4,2)	2.1005	2.1724	2.17
(5,1)	2.2102	2.2934	2.28
(6,0)	2.3817	2.4641	2.45
(4,3)	2.4146	2.4703	2.46

Table 9.3 Comparison of Radii From Conventional Model, New Model and Ab Initio Calculations of Machón et al. (2002) Using $\sigma = 1.425\,\text{Å}$

Nanotube	r_0 (Å)	r (Å)	Machón et al. (2002) (Å)
(5,0)	1.9641	2.0623	2.035
(3,3)	2.0412	2.1037	2.1
(4,2)	2.0786	2.1498	2.14

Table 9.4 Comparison of Radii From Conventional Model, New Model and Ab Initio Calculations of Popov (2004) Using $\sigma = 1.42\,\text{Å}$

Nanotube	r_0 (Å)	r (Å)	Popov (2004) (Å)
(5,0)	1.9572	2.055	2.05
(3,3)	2.034	2.0963	2.12
(4,2)	2.0713	2.1423	2.14

The results of the new model are now compared with ab initio studies from the literature. Cabria et al. (2003) employ an initial bond length of $\sigma = 1.44\,\text{Å}$ and examine a number of nanotubes with radii less than 2.5 Å. Table 9.2 contains a comparison of the new model, the conventional model and the calculations of Cabria et al. (2003). Machón et al. (2002) use a value of bond length of $\sigma = 1.425\,\text{Å}$ and a similar comparison using this value of σ is presented in Table 9.3. Popov (2004) uses a value of $\sigma = 1.42\,\text{Å}$ and again the same comparison of results is performed using this value of the bond length in Table 9.4. The overall evaluation of these comparisons is that the new model agrees very well with ab initio calculations in the literature, reinforcing the view that the proposed new model reflects an underlying physical reality.

9.6 CONCLUSION

Assuming purely symmetric geometrical constraints, an analytical expression for the true radius of atoms (modelled as points) in a carbon nanotube is derived. This model takes into account the curvature of the tube and maintains the position of all atoms as equidistant from a common axis, all covalent bond lengths kept to a fixed constant and that the angle between all bonds is equal throughout the molecule. The analysis shows that significant differences may exist in the various dimensions of the carbon nanotubes as given by the conventional 'rolled-up' model and the new model proposed here. Furthermore, it is shown that as the radius of the nanotube decreases, the relative magnitude of these differences becomes more significant. In addition, as a consequence of the model proposed here, an expression for the effective wall thickness is provided, which is by no means a unique choice, but in some ways it is the most natural and may prove useful for modelling material properties, such as the elastic modulus of carbon nanotubes used in various continuum mechanical theories.

Asymptotic expansions of the analytical expressions lead to terms, which include the conventional formulae as their highest order term, but with corrections which are shown to contribute up to 5% for the tube radius and 3.5% for the unit length of nanotubes for radii in the 2–5 Å range. Finally, noting that the translation vector $\mathbf{T}$, which in the conventional model is always assumed to run parallel with the nanotube axis, will only be parallel in zigzag and armchair nanotubes in the new model. For chiral nanotubes the translation vector will generally trace out a helix on the surface of the nanotube, which means that to construct a chiral nanotube from unit cells, some rotational translation is necessary between cells to maintain the integrity of the nanotube structure.

The analysis of the results compared with existing ab initio calculations of three separate authors shows that the new formula for nanotube radius provides considerable agreement with the calculations. Indeed, this agreement with the ab initio calculations is so striking that it constitutes substantial evidence that the new model presented here provides an analytical description of the underlying physical geometry of carbon nanotubes, which is a significant improvement to that provided by the conventional 'rolled-up' model.

EXERCISES

9.1. What are the main differences between the new 'polyhedral' model proposed in this chapter and the 'rolled-up' model? In which instance is the difference between these two models the greatest?

9.2. Following Worked Example 9.1, show that the expansion for $\cos^2 \theta$ is given by Eq. (9.21).

TWO DISCRETE APPROACHES FOR JOINING CARBON NANOSTRUCTURES

10.1 INTRODUCTION

Since the discovery of carbon nanostructures, such as graphene sheets, C_{60} fullerenes, and carbon nanotubes, a number of researchers have investigated the topological properties for such structures by utilising Euler's theorem and molecular dynamics simulations. Kroto (1987) proposed that the stability of spherical carbon cages is primarily a result of pentagonal and hexagonal rings. Dunlap (1994) considered a pentagon–heptagon pair defect on a carbon nanotube; this defect can be introduced to connect two carbon nanotubes of different chirality. Moreover, the pentagon–heptagon pair is believed to strongly affect the electronic properties of carbon nanotubes, and therefore a connected structure might be useful as a building block for nanoelectrical devices.

Elbow structures are the basic components of certain nanotori and comprise the joining of two distinct carbon nanotubes, where they are considered in Section 10.2 as utilising the variation in the bond length method. We examine two different types of nanotori, which are formed from either two or three distinct types of carbon nanotube sections. In Section 10.3, the perpendicular joining between a carbon nanotube and a graphene sheet is examined using both the variation in bond length and bond angle methods. The resulting nanostructure might be considered as a basic unit necessary to transmit signals in nanoelectrical devices. Finally, the novel carbon nanostructure (see Nasibulin et al., 2007) formed from a C_{60} fullerene and a carbon nanotube, namely a nanobud, is also constructed by both the variation in bond length and the variation in bond angle approaches, which are presented in Section 10.4.

The discrete method described in this chapter exploits the idea that the basis of joining carbon nanostructures is an underlying requirement to take overall in a least squares sense for each interatomic distance be as close as possible to the pairwise carbon–carbon bond length and possess an ideal bond angle. The justification for this approach is that in the absence of geometrical constraints, each bond length would be as close as possible to the pairwise carbon–carbon bond length arising in graphene. This leads to the question, 'To what extent are carbon nanostructures dominated by geometric issues rather than energetic issues'? Although this approach appears to be purely geometric in nature, the requirement to minimise the energy is in a sense already taken into account by attempting to make each interatomic bond length as close as possible to the pairwise bond length. In this chapter, two variation approaches, namely the variation in bond length and the variation in bond angle, are undertaken to determine the joining of two carbon nanostructures, and the details of these two methods are given in the following sections. We note that to a certain extent the adoption of either variation approach is arbitrary. Our main purpose is the adaption of semianalytical geometrically inspired procedures which may be implemented with a minimum of computational effort.

10.1.1 VARIATION IN BOND LENGTH

We denote the *i*th terminal atoms at a joint location by the position vectors $\mathbf{a}_i = (a_{ix}, a_{iy}, a_{iz})$ and $\mathbf{b}_i = (b_{ix}, b_{iy}, b_{iz})$ for the two carbon nanostructures. We perform an appropriate translation and rotation for both nanostructures, then determine the Euclidean distance between corresponding atoms at the junction. Given these distances between matching atoms, the procedure determines optimal translational and rotational parameters by minimising the least squares variation of these distances from the pairwise carbon–carbon bond length σ. Therefore we seek to minimise the objective function given by

$$f(X, Y, \ell, \theta) = \sum_i (|\mathbf{a}_i - \mathbf{b}_i| - \sigma)^2, \tag{10.1}$$

where $|\mathbf{a}_i - \mathbf{b}_i|$ is a Euclidean or 'ordinary' distance between two points, which is given explicitly by $\sqrt{(a_{ix} - b_{ix})^2 + (a_{iy} - b_{iy})^2 + (a_{iz} - b_{iz})^2}$ in three-dimensional space and Cartesian coordinates, and $X, Y, \ell,$ and θ are certain parameters of the translation and rotation of the constituent structures, which we subsequently prescribe.

10.1.2 VARIATION IN BOND ANGLE

In this method, we assume the bond lengths are fixed to the ideal graphene bond length σ and allow the bond angles at connection sites to vary. We then seek to minimise the least square deviations from the ideal bond angle. We assume the bond length is $\sigma = 1.42\,\text{Å}$ and the ideal bond angle for a hexagonal rings are assumed to be 120°. The bond angles on the carbon nanotubes are taken from the new model of carbon nanotubes which properly incorporates curvature, as described in Chapter 9. As the atomic networks for carbon nanostructures are formed from hexagonal rings, a general procedure is proposed to determine the position vectors of all atoms at the junction with reference to Fig. 10.1 through the following steps:

  (i) Find the point $\mathbf{M}$ which is the midpoint of $\mathbf{A}_1$ and $\mathbf{A}_2$.
  (ii) Find the vector $\mathbf{U} = \overrightarrow{\mathbf{MA}_3}$.
  (iii) Find the unit vector $\widehat{\mathbf{V}} = \overrightarrow{\mathbf{A}_1\mathbf{A}_2}/|\overrightarrow{\mathbf{A}_1\mathbf{A}_2}|$, which is perpendicular to $\mathbf{U}$.
  (iv) Determine the vector $\mathbf{W}$, which is perpendicular to both $\mathbf{U}$ and $\widehat{\mathbf{V}}$ and has the same magnitude as $\mathbf{U}$; namely $\mathbf{W} = \mathbf{U} \times \widehat{\mathbf{V}}$.
  (v) The atom position is then given by $\mathbf{M} + \mathbf{U}\cos\phi + \mathbf{W}\sin\phi$.

Here, $\mathbf{A}_1, \mathbf{A}_2,$ and $\mathbf{A}_3$ are the atomic positions as shown in Fig. 10.1, and we adopt the boldface vector notation as a dual notation to designate either the location or the Cartesian vector representation of that location. The atomic position $\mathbf{A}_3$ lies on a circular path and its precise position is determined by the angle ϕ. We note that each bond that connects between the joining atoms of the two structures is assumed to be of fixed length σ. We also comment that in this approach, atoms may move out of plane (i.e. a_{zi} may be nonzero). The number of parameters in each system depends on both size and symmetry of the defect.

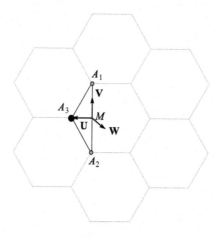

FIG. 10.1

Position vectors for the variation in bond angle approach.

WORKED EXAMPLE 10.1

Find the circular path for A_3, where the atomic positions for A_1, A_2 and A_3 are as defined in Fig. 10.2.

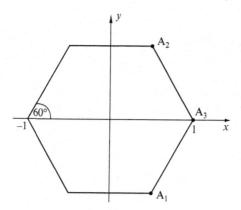

FIG. 10.2

Atomic positions for Worked Example 10.1.

We assume that a hexagonal ring of unit radius is centred at the origin.

Solution

From Fig. 10.2 the coordinates for A_1, A_2, and A_3 can be obtained as

$$\mathbf{A}_1 = \left(\frac{1}{2}, -\frac{\sqrt{3}}{2}, 0\right),$$

$$\mathbf{A}_2 = \left(\frac{1}{2}, \frac{\sqrt{3}}{2}, 0\right),$$

$$\mathbf{A}_3 = (1, 0, 0).$$

Following the five steps, we may deduce

$$\mathbf{M} = \text{mid}(\mathbf{A}_1, \mathbf{A}_2) = \left(\frac{1}{2}, 0, 0\right),$$

$$\mathbf{U} = \overrightarrow{\mathbf{MA}_3} = \left(\frac{1}{2}, 0, 0\right),$$

$$\hat{\mathbf{V}} = \frac{\overrightarrow{\mathbf{A}_1\mathbf{A}_2}}{|\overrightarrow{\mathbf{A}_1\mathbf{A}_2}|} = (0, 1, 0),$$

$$\mathbf{W} = \mathbf{U} \times \hat{\mathbf{V}} = \left(0, 0, \frac{1}{2}\right),$$

where $|\overrightarrow{\mathbf{A}_1\mathbf{A}_2}| = \sqrt{3}$. Therefore a circular path for $\mathbf{A}_3$ is given by

$$\mathbf{A}_3 = \left(\frac{1}{2} + \frac{1}{2}\cos\phi, 0, \frac{1}{2}\sin\phi\right).$$

The two approaches described here can be related to a minimum energy principle adopted by a number of researchers. According to this approach, we may assume that the bonded potential energy for small deformations is given by

$$E_{bonded} = \frac{1}{2}\sum_i \left\{ k_r(r - r_0)^2 + k_\phi(\phi - \phi_0)^2 + k_\tau[1 - \cos(n\tau - \tau_0)] \right\}, \tag{10.2}$$

where k_r, k_ϕ, and k_τ are certain bond stretching, bending angle, and torsional constants, respectively; r_0, ϕ_0, and τ_0 are equilibrium values of the bond length, bond angle, and ideal phase angle for this bond type, respectively; and n is an integer relating to the periodicity of the bonding and for sp^2 bonding $n = 3$. The parameters r, ϕ, and τ are shown in Fig. 10.3. The variation in bond length approach corresponds to including only the bond stretching energy. Similarly, the variation in bond angle approach corresponds to fixing all bond lengths and including only the angle bending energy from Eq. (10.2). In terms of the relative magnitudes of the three force constants, the torsional term is the smallest and therefore plays only a minor effect on the system. Thus the purely geometrical

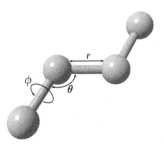

FIG. 10.3

Definitions of variables r, ϕ, and τ for bonded potential energy given by Eq. (10.2).

approaches adopted here are similar to existing energy minimisation schemes in situations which are dominated by either bond stretching or angle bending.

10.2 NANOTORI

Dunlap (1992) first proposed the nanotorus as a stable form of graphitic carbon constructed by joining two carbon nanotubes of different chirality, but with matching radii or by an introduction of the pentagon–heptagon pair. He concluded that the toroidal molecule comprises 12 connecting sections occurring for the 360° turn, and therefore the tubule bend angle is 30° for each section. Although toroidal carbon structures have been observed experimentally, there is as yet no consensus on their structure. However, they are believed to give rise to fascinating electrical, magnetic and elastic properties arising from the pattern of the hexagonal and pentagonal rings.

In this section, we examine the geometry of the basic repeatable units, made up of two and three distinct carbon nanotubes which comprise nanotori. These repeating units are assumed to be joined according to the least-squares minimisation of the deviations of the interatomic spacing from the ideal spacing $\sigma = 1.42$ Å. We note that for simplicity in this example, we only adopt the variation in the bond stretching approach, but we could also assemble toroidal structures based on the variation in the bond angle approach as well. Here, all the carbon nanotube sections are assumed to be either zigzag or armchair because from previous studies only these two types of nanotubes are thought to form nanotori. At present, there is no experimental evidence to indicate that chiral tubes can form these toroidal structures.

10.2.1 NANOTORI FORMED FROM TWO DISTINCT CARBON NANOTUBES

The elbow structure is modelled by positioning atoms on ideal cylinders representing the two species of nanotube and then identifying the terminal atoms on each structure, which bond with the corresponding terminal atoms on the other nanotube. The tubes are originally defined with coincided bases centred on the origin and aligned with their axes on the z-axis. We then perform a translation of Tube A in the positive z-direction by a length ℓ_1 and a corresponding translation of Tube B in the negative z-direction by a length ℓ_2. The lengths ℓ_1 and ℓ_2 are the half-lengths of the tubes comprising the basic units used to assemble the toroidal structure. With the tubes so situated, we then perform a rotation of Tube A by an angle ϕ around the y-axis, as shown in Fig. 10.4. After the translations and rotation, we derive the following expression for the Euclidean distance between two connecting atoms:

$$|\mathbf{a}_i - \mathbf{b}_i| = \{[a_{ix} \cos \phi + (a_{iz} + \ell_1) \sin \phi - b_{ix}]^2 + (a_{iy} - b_{iy})^2 + [(a_{iz} + \ell_1) \cos \phi - a_{ix} \sin \phi - (b_{iz} - \ell_2)]^2\}^{1/2},$$

where the ith terminal atom on Tube A is denoted by the position vector $\mathbf{a}_i = (a_{ix}, a_{iy}, a_{iz})$, and the corresponding terminal atom on Tube B is denoted by $\mathbf{b}_i = (b_{ix}, b_{iy}, b_{iz})$.

With this distance between bonded atoms so defined, the next step is to minimise the variation of this distance from the bond length σ. Therefore, in a least-squares sense, the goal is to minimise the objective function (10.1). Once the parameters ℓ_1, ℓ_2, and ϕ are determined, the 'natural' bend angle ϕ for the elbow configuration under consideration is prescribed. This angle will be of interest if the

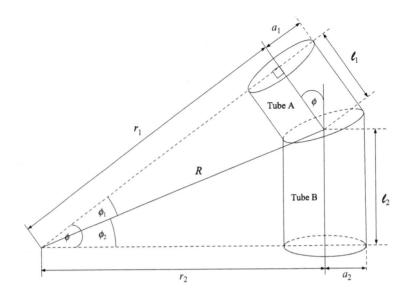

FIG. 10.4

Geometry of elbow formed from two nanotube sections.

elbow is not constrained; for example, if it is situated as a nanotube with a bend or a spiral. However, in the case of nanotori, and assuming that the torus remains symmetric in the plane with no buckling, the bend angle ϕ would necessarily be constrained to a value $\phi = 180°/n$, where $n \in \{2, 3, 4, \ldots\}$. A separate analysis of Eq. (10.1) with this constraint on ϕ leads to slightly different values for ℓ_1 and ℓ_2, which apply when the elbow is fixed into a toroidal configuration. Following the variation in bond length procedure, three elbows formed from two nanotube sections are shown in Fig. 10.5. We comment that Yao et al. (1999) show examples of atomic force microscope images of nanotube junction devices, where each nanotube consists of two straight segments of elbows connected by a sharp kink of about $\phi = 40°$.

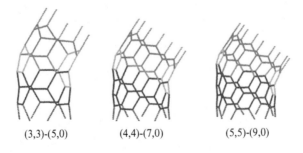

(3,3)-(5,0)   (4,4)-(7,0)   (5,5)-(9,0)

FIG. 10.5

Three elbows each formed from two nanotube sections.

When modelling an entire nanotorus, it is useful to make the continuum approximation that the discrete atoms, from which the nanotorus is comprised, are replaceable by an average atomic density uniformly distributed over some surface and also to assume that the polygonal torus can be considered a perfect toroidal surface. In doing so, it is necessary to be able to assign representative values for the generating torus radius c and the tube radius a.

The first step in this process is to determine the distance from the centre of the torus to the centre of each type of nanotube, which is denoted by r_1 and r_2, noting that an elbow can be considered as two right triangles with a common hypotenuse R, as shown in Fig. 10.4. The angle subtended at the centre of the elbow is ϕ. Denoting the angle subtended by the first nanotube by ϕ_1, and likewise the angle subtended by the second nanotube by ϕ_2, as well as noting that $\phi = \phi_1 + \phi_2$ from the compound angle formula and elementary trigonometry, it can be shown that

$$\sin\phi = \sin\phi_1 \cos\phi_2 + \sin\phi_2 \cos\phi_1 = (\ell_1 r_2 + \ell_2 r_1)/R^2,$$

and therefore

$$r_2 = \left(R^2 \sin\phi - \ell_2 r_1\right)/\ell_1. \tag{10.3}$$

Similarly, the compound angle formula for cosines gives

$$\cos\phi = \cos\phi_1 \cos\phi_2 - \sin\phi_1 \sin\phi_2 = (r_1 r_2 - \ell_1 \ell_2)/R^2.$$

Rearranging this equation for $\cos\phi$ and substituting for r_2 from Eq. (10.3) gives

$$\ell_1 R^2 \cos\phi = r_1 R^2 \sin\phi - \ell_2 \left(r_1^2 + \ell_1^2\right),$$

where $R^2 = r_1^2 + \ell_1^2$ is strictly positive. Subsequently dividing by R^2 and rearranging produces

$$r_1 = \ell_1 \cot\phi + \ell_2 \csc\phi, \tag{10.4}$$

and likewise

$$r_2 = \ell_2 \cot\phi + \ell_1 \csc\phi. \tag{10.5}$$

A noteworthy comment is that both Eqs (10.4), (10.5) are exact and strikingly simple, though they are not immediately obvious from the geometry of the problem.

With the value for the perpendicular distances r_1 and r_2 determined, the next step is to assign a representative value for the toroid generating radius c. One way to do this is to consider the following formula for a mean radius $\bar{r}$ for a circle:

$$\int_0^{\theta_0} r(\theta)\, d\theta = \bar{r}\theta_0.$$

In the case of a right triangle with sides r_1 and ℓ_1, $\theta_0 = \tan^{-1}(\ell_1/r_1)$, and $r(\theta) = r_1 \sec\theta$, and therefore

$$\bar{r_1}\phi_1 = r_1 \int_0^{\tan^{-1}(\ell_1/r_1)} \sec\theta\, d\theta = r_1 \sinh^{-1}(\ell_1/r_1),$$

where $\sinh^{-1}(x) = \ln(x + \sqrt{x^2 + 1})$. This can be repeated for the second right triangle and combining the two equations, then applying the same result to the combined section yields

$$c = \left[r_1 \sinh^{-1}(\ell_1/r_1) + r_2 \sinh^{-1}(\ell_2/r_2) \right]/\phi. \tag{10.6}$$

This process is then extended to determine an expression for the representative tube radius a. Here, a surface integral for a torus is undertaken to determine such a radius. The surface element for the tube is obtained by transforming the toroidal coordinate system (a, ϕ, ψ) into a Cartesian coordinate system via

$$x = (c + a \cos \psi) \cos \phi, \quad y = (c + a \cos \psi) \sin \phi, \quad z = a \sin \psi,$$

where c and a denote the mean radii for the torus and the tube, and ϕ and ψ are the torus and the tube angles, respectively (see Fig. 10.6). Considering the following expression for the mean tube radius $\bar{b}$

$$\int_0^{\theta_0} \int_0^{2\pi} b(\theta, \psi) \left[r(\theta) + b(\theta, \psi) \cos \psi \right] d\psi \, d\theta = 2\pi \bar{b} \theta_0 c, \tag{10.7}$$

where $b(\theta, \psi)$ is the radius of the tube and as before, $r(\theta)$ is the torus generating radius. For Tube A, $\theta_0 = \tan^{-1}(\ell_1/r_1)$, $r(\theta) = r_1 \sec \theta$, and $b(\theta, \psi) = a_1 \sqrt{\sec^2 \theta \cos^2 \psi + \sin^2 \psi}$. Using these definitions with equivalent expressions for Tube B and then combining them produces the following formula for the representative tube radius

$$a = \frac{1}{2c\pi\phi} \int_0^{2\pi} \left(a_1 r_1 \int_0^{\phi_1} \frac{\sqrt{1 - \sin^2 \theta \sin^2 \psi}}{\cos^2 \theta} d\theta \right.$$
$$\left. + a_2 r_2 \int_0^{\phi_2} \frac{\sqrt{1 - \sin^2 \theta \sin^2 \psi}}{\cos^2 \theta} d\theta \right) d\psi,$$

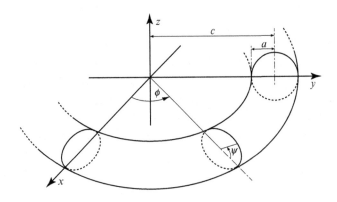

FIG. 10.6

Toroidal coordinate system (a, ϕ, ψ).

where the ψ integration is an elliptic integral, which upon the substitution of $k = \sin\theta$ gives

$$a = \frac{2}{c\pi\phi}\left[a_1r_1\int_0^{\ell_1/R}\frac{E(k)}{k'^3}\,dk + a_2r_2\int_0^{\ell_2/R}\frac{E(k)}{k'^3}\,dk\right], \tag{10.8}$$

where $E(k)$ is the complete elliptic integral of the second kind with modulus k, and $k' = \sqrt{1 - k^2}$ is the complementary modulus, as outlined in Section 2.10. The integrals in Eq. (10.8) may be expanded as an infinite series in which each term can be evaluated exactly. However, doing so increases the algebraic workload without elucidating the physical situation, therefore for these results, this integral is evaluated numerically using the software program MAPLE.

WORKED EXAMPLE 10.2

Derive an infinite series expansion for the formula (10.8).

Solution

In Eq. (10.8), one is required to integrate an elliptic function of the second kind in the form

$$I = \int_0^{\ell/R}\frac{E(k)}{(1 - k^2)^{3/2}}\,dk,$$

where ℓ is either ℓ_1 or ℓ_2, noting that the upper limit $0 < \ell/R < 1$. Using the series expansion of the elliptic function given as equation (900.07) in Byrd and Friedman gives

$$I = \frac{\pi}{2}\sum_{m=0}^{\infty}\frac{1}{1 - 2m}\binom{-\frac{1}{2}}{m}^2\int_0^{\ell/R}\frac{k^{2m}}{(1 - k^2)^{3/2}}\,dk.$$

Making the substitution $k = \sin x$ and $\alpha = \sin^{-1}(\ell/R)$ produces

$$I = \frac{\pi}{2}\sum_{m=0}^{\infty}\frac{1}{1 - 2m}\binom{-\frac{1}{2}}{m}^2\int_0^{\alpha}\frac{\sin^{2m}x}{\cos^2 x}\,dx.$$

To avoid the case $m = 0$ from the summation, this term is extracted from the series and in doing so yields

$$I = \frac{\pi\ell}{2r} - \frac{\pi}{2}\sum_{m=1}^{\infty}\frac{1}{2m - 1}\binom{-\frac{1}{2}}{m}^2\int_0^{\alpha}\frac{\sin^{2m}x}{\cos^2 x}\,dx,$$

where $r = (R^2 - \ell^2)^{1/2}$. Now adjusting the series by substituting $m + 1$ for m gives

$$I = \frac{\pi\ell}{2r} - \frac{\pi}{2}\sum_{m=0}^{\infty}\frac{1}{2m + 1}\binom{-\frac{1}{2}}{m + 1}^2\int_0^{\alpha}\frac{\sin^{2m+2}x}{\cos^2 x}\,dx.$$

Employing the equation 2.518(1) given in Gradshteyn and Ryzhik (2000) gives

$$I = \frac{\pi\ell}{2r} - \frac{\pi}{2}\sum_{m=0}^{\infty}\binom{-\frac{1}{2}}{m + 1}^2\left[\frac{\ell^{2m+1}}{(2m + 1)rR^{2m}} - \int_0^{\alpha}\sin^{2m}x\,dx\right].$$

Again, the $m = 0$ term must be avoided, so the first term is extracted from the series which gives

$$I = \frac{3\pi\ell}{8r} + \frac{\pi}{8}\sin^{-1}\left(\frac{\ell}{R}\right) - \frac{\pi}{2}\sum_{m=1}^{\infty}\binom{-\frac{1}{2}}{m + 1}^2\left[\frac{\ell^{2m+1}}{(2m + 1)rR^{2m}} - \int_0^{\alpha}\sin^{2m}x\,dx\right].$$

Employing equation 2.511(2) from Gradshteyn and Ryzhik (2000) finally yields

$$
I = \frac{3\pi\ell}{8r} + \frac{\pi}{8}\sin^{-1}\left(\frac{\ell}{R}\right) + \frac{\pi}{2}\sum_{m=1}^{\infty}\binom{-\frac{1}{2}}{m+1}^2\left\{\frac{(2m-1)!!}{2^m m!}\sin^{-1}\left(\frac{\ell}{R}\right)\right.
$$
$$
\left. - \frac{\ell^{2m+1}}{(2m+1)rR^{2m}}\left[1+\left(\frac{r}{\ell}\right)^2\sum_{n=0}^{m-1}\frac{(m-n-1)!(2m+1)!!}{2^{n+1}m!(2m-2n-1)!!}\left(\frac{R}{\ell}\right)^{2n}\right]\right\},
$$

where the double factorial $(2n-1)!!$ denotes $(2n-1)(2n-3)\cdots 5\cdot 3$; this equation can be used to numerically evaluate the integrals in Eq. (10.8). Note that the nanotori considered here exhibit the ratio $\ell/R \sim 1/3$. For these parameters, a result for I correct to eight significant digits can be determined by adding only the first term of the series up to $m = 5$, and the leading two terms alone, outside the summation, are sufficient for an answer accurate to three significant digits.

10.2.2 RESULTS AND DISCUSSION

Firstly, elbows made from the smallest possible nanotube sections are evaluated. In the case of an armchair section, this means that if two elbows are as close as possible, the heptagon defects on the inner side of the bend must share a common side. In the case of a zigzag section, the carbon atom forming the top of a heptagon ring must bond with the carbon atom forming the top of the next heptagon. In the armchair tube, two rings consisting of a total of $4n$ atoms can be added to extend the tube by the longitudinal dimension of two rings, where $n \in \{2, 3, 4, \ldots\}$. With zigzag tubes, a complete ring of hexagons consisting of $4n$ atoms can be added to the tube to change its length without changing the orientation of the atoms at the end of the elbow. Table 10.1 contains the basic parameters for elbows including the number of atoms in the basic section, as well as the increase in the number of atoms and the half length, which applies when each incremental unit is added.

We now examine some numerical results for this procedure when applied to various different species of nanotube elbows. Here, three species of nanotube elbow are analysed: one constructed from $(3, 3)$-$(5, 0)$ tubes, another constructed from $(4, 4)$-$(7, 0)$ nanotubes, and lastly one constructed from $(5, 5)$-$(9, 0)$ nanotubes. The tube radii are calculated using the geometrically precise method described in

Table 10.1 Fundamental Parameters for Nanotube Elbows Formed From Two Distinct Carbon Nanotube Sections

Nanotube	Radius (Å)	Base Unit	Incremental Unit	
		Number Atoms	Number Atoms	Half Length (Å)
$(3, 3)$	2.0963	18	+12	+1.2103
$(5, 0)$	2.0550	30	+20	+2.0550
$(4, 4)$	2.7586	32	+16	+1.2190
$(7, 0)$	2.8096	42	+28	+2.0930
$(5, 5)$	3.4273	50	+20	+1.2229
$(9, 0)$	3.5769	54	+36	+2.1079

Chapter 9 and using a value for the bond length of $\sigma = 1.42\,\text{Å}$. Once the positions of atoms are determined, the physical parameters ℓ_1, ℓ_2, and ϕ are determined by a minimisation process both for an unstrained ϕ and again with the constraint $\phi = 180°/n$, where $n \in \{2, 3, 4, \ldots\}$. The results are presented in Table 10.2. In the case of the $(5, 5)$-$(9, 0)$ elbow, the unconstrained bend angle ϕ is very close to $36°$. The $(4, 4)$-$(7, 0)$ elbow is also shown to have an unconstrained bend angle close to $36°$. The unconstrained bend angle of the $(3, 3)$-$(5, 0)$ elbow is shown to differ the most from a valid constrained angle. In this case, when the bend angle is constrained, the least squares minimisation process produces a preferred bend angle of $30°$. However, this is not an order of magnitude improvement over using a bend angle of $36°$, and so for this elbow type, nanotori constructed from both 10 and 12 elbows are considered.

The purpose of the remainder of this section is to introduce a nomenclature for the new carbon molecules. First, using the notation $(n, m)_p$ to refer to a section of an (n, m) nanotube that is constructed from p atoms, which it is assumed to consist of one base unit plus an integer multiple by incremental units arising from additional rings as specified. For example, from Table 10.1, one can see that $(3, 3)_{30}$ signifies a base unit (comprising 18 atoms) plus one increment unit (comprising 12 atoms) of a $(3, 3)$ armchair nanotube, whereas $(5, 0)_{30}$ signifies one base unit only (comprising 30 atoms) of the zigzag $(5, 0)$ nanotube. When constructing nanotori or other shapes from the elbow sections the nanotube chiral numbers are prefixed with the number of units making up the shape. Therefore $5(3, 3)_{18}5(5, 0)_{30}$ is a molecule consisting of five base units of $(3, 3)$ armchair and five base units of $(5, 0)$ zigzag nanotube. This molecule therefore comprises 240 atoms (that is, C_{240}), and this is the smallest nanotorus possible with the elbows considered here. Fig. 10.7 illustrates a C_{240} molecule with structure $5(3, 3)_{18}5(5, 0)_{30}$, and Fig. 10.8 illustrates a C_{360} molecule with structure $6(3, 3)_{30}6(5, 0)_{30}$.

Using the parameters for the constrained elbows, the toroidal parameters r_1 and r_2 are calculated from Eqs (10.4), (10.5), and finally values for the mean torus generating radius c and mean tube radius a are derived from Eqs (10.6), (10.8). These results are summarised in Table 10.11 for nanotori constructed from basic units and where $\ell_1 \sim \ell_2$ and $\ell_1, \ell_2 < 10\,\text{Å}$. The construction scheme presented here is limitless, and many more such caged molecules can be obtained using this scheme, both when ℓ_1 is not similar to ℓ_2 and when ℓ_1 or $\ell_2 > 10\,\text{Å}$. For reasons of space and divergence from the basic toroidal shape, which occurs when ℓ_1 and ℓ_2 differ significantly or when they become large, no attempt is made here to fully delineate all possible molecules. However, worthy of comment is that Table 10.11 describes four families of molecules:

Table 10.2 Bend Angles and Base Unit Section Half Lengths

Elbow Type	θ Unconstrained			θ Constrained		
	θ (°)	ℓ_1 (Å)	ℓ_2 (Å)	θ (°)	ℓ_1 (Å)	ℓ_2 (Å)
$(3, 3)$-$(5, 0)$	32.80	1.7755	3.1497	30	1.8242	3.0946
				36	1.7100	3.2196
$(4, 4)$-$(7, 0)$	35.39	2.4434	3.1700	36	2.4390	3.1758
$(5, 5)$-$(9, 0)$	35.89	2.9737	3.2599	36	2.9701	3.2636

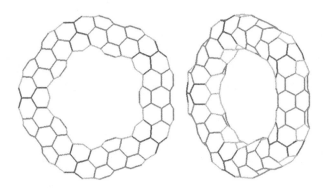

FIG. 10.7

C_{240} molecule with structure $5(3,3)_{18}5(5,0)_{30}$.

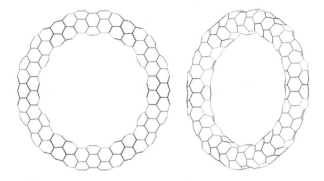

FIG. 10.8

C_{360} molecule with structure $6(3,3)_{30}6(5,0)_{30}$.

- $5(3,3)_p5(5,0)_q$, which contain molecules with $240 + 60j + 100k$ atoms,
- $6(3,3)_p6(5,0)_q$, which contain molecules with $288 + 72j + 120k$ atoms,
- $5(4,4)_p5(7,0)_q$, which contain molecules with $370 + 80j + 140k$ atoms,
- $5(5,5)_p5(9,0)_q$, which contain molecules with $520 + 100j + 180k$ atoms,

for any $j, k \in \{0, 1, 2 \ldots\}$.

For the toroidal results in Table 10.11, it is noted that the mean generating radius c lies between the values of $\min(r_1, r_2)$ and R. Furthermore, the values of the mean tube radii of the nanotori are also generally between a_1 and a_2. This is because the constituent tubes are sliced further from the perpendicular point, the tube cross-section becomes more elliptical which has a tendency to increase the mean value of a.

In addition to circular nanotori, the fivefold symmetry of the $5(i, i)_p5(j, 0)_q$ shape allows pentagonal shaped cage molecules to be generated by including only base units of one type of tube and many incremental units of other. By carefully matching the lengths of the armchair and zigzag nanotubes, other shapes such as rhombi and trapezia can also be achieved. The sixfold symmetry of the $6(3,3)_p6(5,0)_q$ shape has even greater versatility for generating regular shapes. Using three longer sections of one tube species alternately in the structure would lead to an equilateral triangular shape; for example, $6(3,3)_{18}3(5,0)_{30}3(5,0)_{90}$. Other shapes such as rectangles, rhombi, trapezia, hexagons, and ovals, are all achieved by simply varying the tube lengths appropriately. Some illustrative structures are listed in Table 10.3 and commenting that, for example, the first row of Table 10.3 can be readily generalised as follows:

$$C_{288+60k} = 6(3,3)_{18}3(5,0)_{30}3(5,0)_{30+20k},$$

where k is the number of incremental units used to form the triangular sides. The first row of Table 10.3 arises from the value $k = 3$, and there are similar generalisations for the other rows of the table.

10.2.3 NANOTORI FORMED FROM THREE DISTINCT CARBON NANOTUBES

An elbow structure for toroidal molecules is investigated by joining three distinct carbon nanotubes of lengths $2\ell_1, 2\ell_2$, and $2\ell_3$ and utilising the least squares in bond length approach. Again, only zigzag and armchair carbon nanotubes are examined. The proposed model assumes that the basic repeating unit comprises tubes A and C as half-unit lengths and Tube B as one unit length. Furthermore, it is assumed that the origin O of a rectangular Cartesian coordinate system (x, y, z) is located at the central point of Tube B, such that the axis of Tube B is aligned along the z-axis, as illustrated in Fig. 10.9.

The ith terminal atom at a joint location is defined by position vectors $\mathbf{a}_i = (a_{ix}, a_{iy}, a_{iz})$, $\mathbf{b}_i = (b_{ix}, b_{iy}, b_{iz})$, and $\mathbf{c}_i = (c_{ix}, c_{iy}, c_{iz})$ for tubes A, B and C, respectively. At the junction of tubes A and B with x'-axis as shown in Fig. 10.10A, a translation of the Tube B in the negative z-direction is performed by a length ℓ_{2A}, where $2\ell_2 = \ell_{2A} + \ell_{2B}$ and ℓ_{2B} is as defined later in the text. Tube A is also translated in the positive z-direction by a length ℓ_1 and rotated by an angle ϕ_1 about the y'-axis. Therefore the Euclidean distance between the atoms at the junction is given by

$$|\mathbf{a}_i - \mathbf{b}_i| = \{[a_{ix}\cos\phi_1 + (a_{iz} + \ell_1)\sin\phi_1 - b_{ix}]^2 + (a_{iy} - b_{iy})^2$$
$$[(a_{iz} + \ell_1)\cos\phi_1 - a_{ix}\sin\phi_1 - (b_{iz} - \ell_{2A})]^2\}^{1/2}.$$

Table 10.3 Examples of Other Ring Structures for the $(3,3)$-$(5,0)$ Elbow

a	b	c	d	Structure	Shape	Formula
$(3,3)_{18}$	$(5,0)_{30}$	$(5,0)_{90}$		abacabacabac	Triangle	C_{468}
$(3,3)_{18}$	$(3,3)_{90}$	$(5,0)_{30}$	$(5,0)_{90}$	acbcadacbcad	Rectangle	C_{552}
$(3,3)_{18}$	$(5,0)_{30}$	$(5,0)_{90}$		abacacabacac	Rhombus	C_{528}
$(3,3)_{18}$	$(5,0)_{30}$	$(5,0)_{90}$	$(5,0)_{150}$	abacacacabad	Trapezium	C_{588}
$(3,3)_{18}$	$(5,0)_{90}$			ababababab	Pentagon	C_{540}
$(3,3)_{18}$	$(5,0)_{90}$			ababababababab	Hexagon	C_{648}
$(3,3)_{30}$	$(5,0)_{30}$	$(5,0)_{90}$		ababacababac	Oval	C_{480}

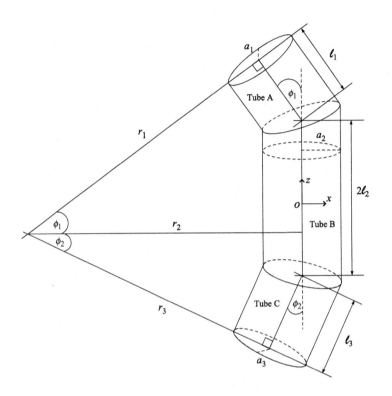

FIG. 10.9

Basic elbow unit formed from three nanotube sections.

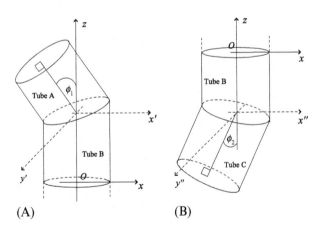

(A)   (B)

FIG. 10.10

Cartesian coordinate system for two single nanotube elbows.

Similarly, at the junction of tubes B and C with x''-axis as shown in Fig. 10.10B, Tube B is translated in the positive z-direction by a length ℓ_{2B}, while tube C is translated in the negative z-direction by a length ℓ_3 and rotated by an angle ϕ_2 about the y''-axis. The distance between the atoms at the joint location is then given by

$$|\mathbf{c}_i - \mathbf{b}_i| = \{[c_{ix}\cos\phi_2 + (c_{iz} - \ell_3)\sin\phi_2 - b_{ix}]^2 + (c_{iy} - b_{iy})^2$$
$$[(c_{iz} - \ell_3)\cos\phi_2 - c_{ix}\sin\phi_2 - (b_{iz} + \ell_{2B})]^2\}^{1/2}.$$

Given these distances between matching atoms, the procedure is to determine $\ell_1, \ell_2, \ell_3, \phi_1$, and ϕ_2 by minimising the least squares variation of these distances from the pairwise carbon–carbon bond length, which is taken to be $\sigma = 1.42\,\text{Å}$. Consequently, this approach is seeking to minimise the following objective functions,

$$f(\ell_1, \ell_{2A}, \phi_1) = \sum_i (|\mathbf{a}_i - \mathbf{b}_i| - \sigma)^2,$$
$$g(\ell_{2B}, \ell_3, \phi_2) = \sum_j (|\mathbf{c}_j - \mathbf{b}_j| - \sigma)^2.$$

Once the parameters $\ell_1, \ell_2, \ell_3, \phi_1$, and ϕ_2 are determined, the basic repeating unconstrained elbow unit can be obtained and is illustrated in Fig. 10.11. However, in the case of a nanotorus, an even number of elbow sections are required to form a symmetrical torus, so that the angles ϕ_1 and ϕ_2 must be

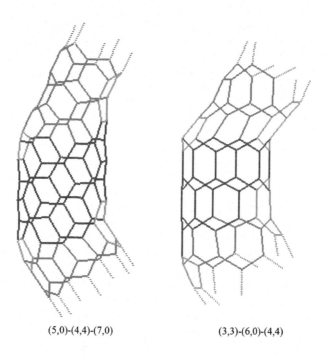

(5,0)-(4,4)-(7,0)   (3,3)-(6,0)-(4,4)

FIG. 10.11

Elbows formed from three distinct nanotube sections.

constrained to the value $\phi_1 + \phi_2 = 180°/n$ where $n \in \{2, 3, 4, \ldots\}$. In this case the objective function becomes

$$F(\ell_1, \ell_{2A}, \ell_{2B}, \ell_3, \phi_1) = f(\ell_1, \ell_{2A}, \phi_1) + g(\ell_{2B}, \ell_3, 180°/n - \phi_1).$$

Consequently with this additional constraint, slightly different values for ℓ_1, ℓ_2, and ℓ_3 might be obtained. The resulting nanotorus structure is achieved by translating the elbow in the x-direction by a distance r_2, which is obtained by the procedure given below.

A representative radius of the toroidal shapes is determined by connecting the basic elbow units with ϕ_1 and ϕ_2 constrained for the 360° turn. Firstly, the upper quadrilateral shown in Fig. 10.12 is considered. It comprises four sides, namely r_1, ℓ_1, ℓ_{2A}, and r_2, and the configuration also depends on the angle ϕ_1. Utilising the same technique as described in Section 10.2.1, we have

$$r_1 = \ell_1 \cot \phi_1 + \ell_{2A} \csc \phi_1, \tag{10.9}$$
$$r_2 = \ell_{2A} \cot \phi_1 + \ell_1 \csc \phi_1. \tag{10.10}$$

By precisely the same process for the quadrilateral comprising the sides r_2, ℓ_{2B}, ℓ_3, and r_3, then r_2 and r_3 can be deduced

$$r_2 = \ell_{2B} \cot \phi_2 + \ell_3 \csc \phi_2, \tag{10.11}$$
$$r_3 = \ell_3 \cot \phi_2 + \ell_{2B} \csc \phi_2.$$

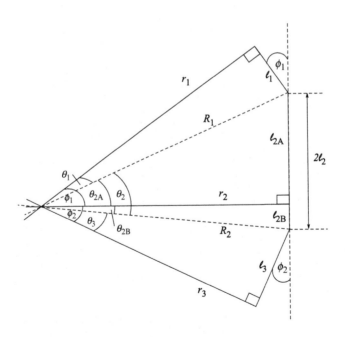

FIG. 10.12

Elbow skeleton formed from three distinct nanotube sections.

The parameters ℓ_{2A} and ℓ_{2B} can be rearranged from Eqs (10.10), (10.11), respectively,

$$\ell_{2A} = \frac{r_2 - \ell_1 \csc\phi_1}{\cot\phi_1} = r_2 \tan\phi_1 - \ell_1 \sec\phi_1,$$

$$\ell_{2B} = \frac{r_2 - \ell_3 \csc\phi_2}{\cot\phi_2} = r_2 \tan\phi_2 - \ell_3 \sec\phi_2.$$

As $2\ell_2 = \ell_{2A} + \ell_{2B}$, we obtain

$$r_2 = \frac{\ell_1 \sec\phi_1 + 2\ell_2 + \ell_3 \sec\phi_2}{\tan\phi_1 + \tan\phi_2}. \tag{10.12}$$

By substituting Eq. (10.12) into the above equations, r_1 and r_3 are obtained:

$$r_1 = \ell_1 \cot(\phi_1 + \phi_2) + 2\ell_2 \csc(\phi_1 + \phi_2)\cos\phi_2 + \ell_3 \csc(\phi_1 + \phi_2), \tag{10.13}$$

$$r_3 = \ell_1 \csc(\phi_1 + \phi_2) + 2\ell_2 \csc(\phi_1 + \phi_2)\cos\phi_1 + \ell_3 \cot(\phi_1 + \phi_2). \tag{10.14}$$

These two formulae provide the appropriate generalisation of those given in Section 10.2.1 for the case of two distinct tubes. The corresponding equations given in Section 10.2.1 can be obtained from Eqs (10.13), (10.14) with the formal identification $\ell_2 \equiv \phi_2 \equiv 0$.

WORKED EXAMPLE 10.3

By considering a quadrilateral which comprises the four sides r_1, ℓ_1, ℓ_{2A}, and r_2 as shown in Fig. 10.12, show that Eqs (10.9), (10.10) can be obtained.

Solution

On using the compound angle formula for sine, we can deduce

$$\sin\phi_1 = \sin\theta_1 \cos\theta_{2A} + \sin\theta_{2A} \cos\theta_1 = (\ell_1 r_2 + \ell_{2A} r_1)/R_1^2,$$

and therefore

$$r_1 = (R_1^2 \sin\phi_1 - \ell_1 r_2)/\ell_{2A}. \tag{10.15}$$

Similarly, from the compound angle formula for cosine

$$\cos\phi_1 = \cos\theta_1 \cos\theta_{2A} - \sin\theta_1 \sin\theta_{2A} = (r_1 r_2 - \ell_1 \ell_{2A})/R_1^2,$$

and therefore r_1 simplifies to become

$$r_1 = (R_1^2 \cos\phi_1 + \ell_1 \ell_{2A})/r_2. \tag{10.16}$$

By equating Eqs (10.15), (10.16), r_2 can be rearranged and is given by

$$r_2 R_1^2 \sin\phi_1 = \ell_{2A} R_1^2 \cos\phi_1 + \ell_1(r_2^2 + \ell_{2A}^2),$$

where $R_1^2 = r_2^2 + \ell_{2A}^2$ is strictly positive. By dividing by R_1^2 and rearranging, r_2 simplifies to obtain

$$r_2 = \ell_{2A} \cot\phi_1 + \ell_1 \csc\phi_1,$$

and likewise

$$r_1 = \ell_1 \cot\phi_1 + \ell_{2A} \csc\phi_1.$$

We also use the same procedure as described in Section 10.2.1 to determine a representative radius a and a representative generating radius c in terms of the perpendicular distances r_1, r_2, and r_3. For a right-angled triangle which consists of r_1, ℓ_1, and R_1 sides, θ_0 and r are obtained as $\theta_0 = \tan^{-1}(\ell_1/r_1)$ and $r(\theta) = r_1 \sec \theta$. It can be deduced that

$$\overline{r_1}\theta_1 = r_1 \sinh^{-1}(\ell_1/r_1).$$

The same procedure is repeated to obtain the mean radii for r_2 and r_3 and finally by averaging, the representative toroidal generating radius c is obtained to be given by

$$c = \frac{r_1 \sinh^{-1}\left(\frac{\ell_1}{r_1}\right) + r_2\left[\sinh^{-1}\left(\frac{\ell_{2A}}{r_2}\right) + \sinh^{-1}\left(\frac{\ell_{2B}}{r_2}\right)\right] + r_3 \sinh^{-1}\left(\frac{\ell_3}{r_3}\right)}{\phi_1 + \phi_2}. \tag{10.17}$$

Then, we extend this process to determine a representative expression for the representative tube radius a, which is described in Section 10.2.1. For the section of Tube A, $\theta_1 = \tan^{-1}(\ell_1/r_1)$, $r(\phi) = r_1 \sec \phi$, and $b(\phi, \psi) = a_1\sqrt{\sec^2 \phi \cos^2 \psi + \sin^2 \psi}$ and by Eq. (10.7), it can be deduced

$$\overline{a_1} = \frac{1}{2\pi c\theta_1} \int_0^{\theta_1} \int_0^{2\pi} a_1 r_1 \sec \phi \sqrt{\sec^2 \phi \cos^2 \psi + \sin^2 \psi}\, d\psi\, d\phi,$$

$$= \frac{a_1 r_1}{2\pi c\theta_1} \int_0^{\theta_1} \int_0^{2\pi} \frac{\sqrt{1 - \sin^2 \phi \sin^2 \psi}}{\cos^2 \phi}\, d\psi\, d\phi.$$

Upon the substitution of $k = \sin \phi$, the above integral can be written as

$$\overline{a_1} = \frac{2a_1 r_1}{\pi c\theta_1} \int_0^{\pi/2} \int_0^{\ell_1/R_1} \frac{\sqrt{1 - k^2 \sin^2 \psi}}{k'^3}\, dk\, d\psi = \frac{2a_1 r_1}{\pi c\theta_1} \int_0^{\ell_1/R_1} \frac{E(k)}{k'^3}\, dk,$$

where $E(k)$ is the complete elliptic integral of the second kind with modulus k, and $k' = \sqrt{1 - k^2}$ is the complementary modulus, as outlined in Section 2.10. Using these definitions with equivalent expressions for Tubes B and C, the following formula may be derived for the representative tube radius

$$a = \frac{2}{\pi c(\phi_1 + \phi_2)}\{a_1 r_1 h(\ell_1/R_1) + a_2 r_2[h(\ell_{2A}/R_1) + h(\ell_{2B}/R_2)] + a_3 r_3 h(\ell_3/R_2)\}, \tag{10.18}$$

where

$$h(x) = \int_0^x \frac{E(k)}{k'^3}\, dk. \tag{10.19}$$

The analytical expression in terms of an infinite series for Eq. (10.19) is given by

$$h(x) = \frac{3\pi x}{8r} + \frac{\pi}{8}\sin^{-1}x + \frac{\pi}{2}\sum_{m=1}^{\infty}\binom{-\frac{1}{2}}{m+1}^2\left\{\frac{(2m-1)!!}{2^m m!}\sin^{-1}x\right.$$

$$\left. - \frac{x^{2m+1}}{(2m+1)r}\left[1 + \left(\frac{r^2}{x^2}\right)\sum_{k=0}^{m-1}\frac{(m-k-1)!(2m+1)!!}{2^{k+1}m!(2m-2k-1)!!}\left(\frac{1}{x}\right)^{2k}\right]\right\},$$

where $r = (1 - x^2)^{1/2}$, and the double factorial $(2n - 1)!!$ denotes $(2n - 1)(2n - 3) \cdots 5 \cdot 3$. We note that the derivation of this equation can be found in Worked Example 10.2. The above procedures for the determination of the representative parameters a and c are by no means unique, but appear as the most natural and simplest for the determination of these quantities.

10.2.4 RESULTS AND DISCUSSION

Here, elbows made from the smallest possible nanotube sections are determined. By precisely the same procedure as that given in Section 10.2.2, the basic parameters for elbows are given in Table 10.4. The smallest possible nanotube sections which can be formed from the elbows are referred to as the base unit, and other possible structures can be obtained by adding further incremental units. The same nomenclature formulated in Section 10.2.2 for toroidal shaped molecules is employed by utilising the notation $N(n, m)_p$, where (n, m) refers to a section of nanotube which is constructed from p atoms and N is a number of base units.

Numerical results from the least squares procedure when applied to various distinct nanotube elbows are presented here. Two different elbow structures, which are $(5, 0)$-$(4, 4)$-$(7, 0)$ and $(3, 3)$-$(6, 0)$-$(4, 4)$, are considered. Using the new polyhedral facetted model for carbon nanotubes presented in Chapter 9 and the value of the bond length $\sigma = 1.42$ Å, the tube radii are obtained. Once the atom positions are determined, the physical parameters $\ell_1, \ell_2, \ell_3, \phi_1$, and ϕ_2 are determined by the minimisation process, both for no constraints and again with the constraint $\phi_1 + \phi_2 = 180°/n$, where $n \in \{2, 3, 4, \ldots\}$. In Table 10.5, the results for the unconstrained case and the constrained case, when $\phi_1 + \phi_2 = 60°$ (i.e. $n = 3$) for two different nanotori are presented. Note that the sum of the angles ϕ_1 and ϕ_2 needs to be exactly or close to a common factor of $360°$ for toroidal structures, and there is only one case arising for these particular two structures. Moreover, there is no straightforward procedure to choose the elbow structures for which $\phi_1 + \phi_2 \simeq 180°/n$, so that only $(5, 0)$-$(4, 4)$-$(7, 0)$ and $(3, 3)$-$(6, 0)$-$(4, 4)$ are presented here.

Using the parameters for the constrained elbows, the toroidal parameters r_1, r_2, and r_3 from Eqs (10.13), (10.12), (10.14), respectively, are calculated, and values for the mean torus generating radius c and the mean tube radius a are derived from the expressions (10.17) and (10.18). These results are presented in Table 10.6. In Fig. 10.13, the toroidal structure of $3(5, 0)_{176}6(4, 4)_{48}3(7, 0)_{19}$

Table 10.4 Fundamental Parameters for Nanotube Elbows Formed From Three Distinct Carbon Nanotube Sections				
		Base Unit	**Incremental Unit**	
Nanotube	**Radius (Å)**	**Number Atoms**	**Number Atoms**	**Length (Å)**
$(5, 0)$	2.0551	17	+20	+4.7986
$(4, 4)$	2.7582	48	+16	+2.4380
$(7, 0)$	2.8094	19	+28	+4.2230
$(3, 3)$	2.0965	12	+12	+2.4206
$(6, 0)$	2.4298	32	+24	+4.1580
$(4, 4)$	2.7582	24	+16	+2.4380

Table 10.5 Bend Angles and Base Unit Section for Nanotube Elbows

	Elbow Type	
	$(5, 0)$-$(4, 4)$-$(7, 0)$	$(3, 3)$-$(6, 0)$-$(4, 4)$
$\phi_1 + \phi_2$ unconstrained		
ϕ_1 (°)	25.59	31.38
ϕ_2 (°)	36.00	33.80
ℓ_1 (Å)	3.7089	3.2202
ℓ_2 (Å)	3.6571	2.4085
ℓ_3 (Å)	3.1727	2.3430
$\phi_1 + \phi_2 = 60°$		
ϕ_1 (°)	24.00	26.20
ϕ_2 (°)	36.00	33.80
ℓ_1 (Å)	3.3396	3.2716
ℓ_2 (Å)	3.6571	2.4085
ℓ_3 (Å)	3.1727	2.3430

Table 10.6 Physical Parameters of Two Specific Toroidal Structures

Toroidal Structures	r_1 (Å)	r_2 (Å)	r_3 (Å)	c (Å)	a (Å)
$3(5, 0)_{17}6(4, 4)_{48}3(7, 0)_{19}$	12.4239	12.7083	13.4038	12.9935	2.6209
$3(3, 3)_{12}6(6, 0)_{32}3(4, 4)_{24}$	9.2163	9.7138	10.1213	9.7980	2.4780

is depicted, which can be referred to as a C_{396} molecule, and also in Fig. 10.14, the toroidal structure of $3(3, 3)_{12}6(6, 0)_{32}3(4, 4)_{24}$ is shown, which can be referred to as a C_{300} molecule.

10.3 JOINING CARBON NANOTUBES AND FLAT GRAPHENE SHEETS

In order to transmit signals from future nanoelectromechanical graphene sheets to other materials, connections with carbon nanotubes provide one means of achieving this. Here, three particular perpendicular connections of carbon nanotubes are examined employing two simple least squares approaches.

The possibility of connecting zigzag $(8, 0)$ and armchair $(4, 4)$ nanotubes with a flat graphene sheet is investigated by fixing the atomic positions on the tube end and the graphene sheet. The variation in the distance between an atom on the open end of the tube and an atom on the sheet from the bond length σ between two carbon atoms is minimised. In both cases, there are eight atoms which are connected to the tube by two carbon bonds so that they require one other bond to complete the sp^2 structure. Consequently, the defect on the graphene sheet must have eight atoms and each requires one further

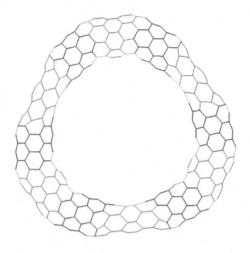

FIG. 10.13

Nanotorus formed from $3(5,0)_{17}6(4,4)_{48}3(7,0)_{19}$ where $\phi_1 + \phi_2 = 60°$.

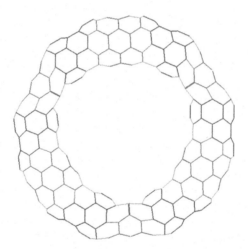

FIG. 10.14

Nanotorus formed from $3(3,3)_{12}6(6,0)_{32}3(4,4)_{24}$ where $\phi_1 + \phi_2 = 60°$.

bond to complete the structure. In Fig. 10.15, 16 possible defects are depicted, to which the $(8,0)$ and the $(4,4)$ tubes might be joined. The first atom on the sheet is denoted by a black square, while atoms indicated by grey circles are numbered sequentially and in the anticlockwise direction from the first atom. As a result of the symmetric locations of all atoms on the open end of the zigzag tube, there are 16 possible configurations denoted by an integer #n from 1 to 16, corresponding to the 16 possible

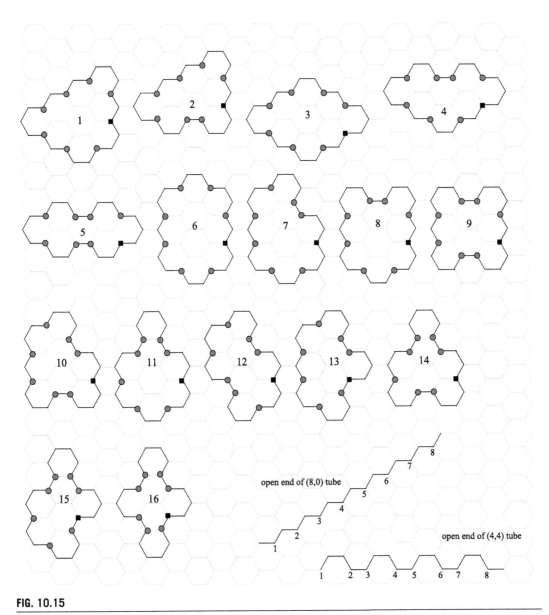

FIG. 10.15

Sixteen possible defects which require another eight bonds for the sp^2 network.

defects. For the armchair tube, there are two connected forms for the adjacent atoms on the tube end, which are either connected by three bonds, as in the first atom to the second atom, or by a single bond, as in the second atom to the third atom. This pattern alternates around the tube end as shown in Fig. 10.15. Therefore the joining of the first atom on the tube with the first atom on the sheet is denoted by #n-a, and the joining of the second atom on the tube with the first atom on the sheet is denoted by #n-b, where #n is again an integer from 1 to 16 corresponding to the 16 possible defects.

10.3.1 VARIATION IN BOND LENGTH

As in Section 10.1.1, the variation in bond length for joining a flat graphene sheet and a carbon nanotube can be determined. In a Cartesian coordinate system, the graphene sheet is assumed to be located in the (x, y) plane and the ith atom on the sheet is assumed to have the position vector $\mathbf{a}_i = (a_{ix}+X, a_{iy}+Y, 0)$, where $i = 1, 2, 3, \ldots, 8$. The sheet is allowed to move in both x- and y-directions by distances X and Y, respectively, which can be either positive or negative. To keep the approach as simple as is possible, the atoms in the graphene sheet are assumed to remain in the $z = 0$ plane. A refinement of the present approach would be to assume a displacement Z out of the $z = 0$ plane, but this additional complexity makes no substantial qualitative changes to the final geometric structure obtained, and in this section $z = 0$ is adopted because the bond angle is considered fixed.

The position vector of the ith atom on the open end of the tube is assumed to be given by $\mathbf{b}_i = (b_{ix}, b_{iy}, \ell)$, where ℓ is the spacing between the tube and the sheet in the positive z-direction. In addition, the tube can be rotated by an angle θ. The distance between the atom on the tube end and the atom on the sheet is then given by

$$|\mathbf{a}_i - \mathbf{b}_i| = \{[(a_{ix} + X) - (b_{ix} \cos\theta - b_{iy} \sin\theta)]^2$$
$$+ [(a_{iy} + Y) - (b_{ix} \sin\theta + b_{iy} \cos\theta)]^2 + \ell^2\}^{1/2}.$$

Using the least squares method, we aim to determine particular values of X, Y, ℓ, and θ which minimise the function

$$f(X, Y, \ell, \theta) = \sum_{i=1}^{8}(|\mathbf{a}_i - \mathbf{b}_i| - \sigma)^2. \tag{10.20}$$

Using the algebraic computer package MAPLE, the numerical values for the least squares function f defined by Eq. (10.20) and the distance ℓ are presented in Tables 10.7 and 10.8 for the zigzag and the armchair tubes, respectively, where the radii of the tubes are taken from Chapter 9. The global minimum values for the least squares are first graphed to ensure a genuine global minimum and the optimisation package in MAPLE is then utilised to find the values for each parameter which gives the minimum.

10.3.2 VARIATION IN BOND ANGLE

Next, we examine the joining of a carbon nanotube to a graphene sheet using the variation in the bond angle method. For the graphene sheet, the bond angle is assumed to be 120°, while the bond angle for the carbon nanotube is again taken from the geometric model proposed in Chapter 9. Firstly, the case of a $(6, 0)$ carbon nanotube connecting with a sixfold type symmetric defect shown in Fig. 10.16 is considered. Due to the sixfold type symmetry of the configuration, there is only one joining site that needs to be considered, for which atom A_3 on the sheet connects with atom C on the tube. Moreover, atoms A_1 and A_2 are assumed to be fixed, but atom A_3 can move around a circular path, and its position

Table 10.7 Values of Least Square Function f (Å^2) and Distance ℓ (Å) for 16 Configurations of an $(8, 0)$ Tube and Corresponding Polygons P_n Where n Is Number of Sides

Configurations #	f	ℓ	P_5	P_6	P_7	P_8	P_9
1	0.0239	1.3243	–	4	2	2	-
2	0.2092	1.1663	1	3	1	3	–
3	0.0018	1.3027	–	4	2	2	-
4	0.0676	0.9471	1	2	3	2	–
5	1.2631	0.0010	2	–	4	2	-
6	0.0817	1.3218	–	2	6	–	-
7	0.0729	1.1823	1	1	5	1	–
8	0.0020	1.3500	1	2	3	2	–
9	0.0592	1.3349	2	2	–	4	-
10	0.0793	1.2210	2	1	2	3	–
11	0.5513	0.9560	2	–	5	–	1
12	0.2498	0.9191	2	–	4	2	-
13	0.0404	1.1951	2	1	2	3	–
14	0.6005	0.9841	3	–	2	2	1
15	0.5262	0.9482	3	–	2	2	1
16	0.5284	0.9227	4	–	–	2	2

is determined by an angle θ. The five steps for determining atomic positions at the connection site can be found in Section 10.1.2. For the graphene sheet and by assuming that the bond length σ is 1.42 Å, the bond angle is 120° and from the diagram in Fig. 10.16, the coordinates for the atoms A_1, A_2, A_3, and A_4 are given by

$$\mathbf{A_1} = \sigma \left(\frac{5}{2}, \frac{\sqrt{3}}{2}, 0 \right), \quad \mathbf{A_2} = \sigma \left(\frac{5}{2}, -\frac{\sqrt{3}}{2}, 0 \right),$$

$$\mathbf{A_3} = \sigma \left(\frac{5}{2} - \frac{1}{2} \cos \theta, 0, \frac{1}{2} \sin \theta \right), \quad \mathbf{A_4} = \sigma \left(\frac{7}{2}, \frac{\sqrt{3}}{2}, 0 \right).$$

By precisely the same procedure, the coordinates for the atoms A, B, C, and D on the tube can be expressed as

$$\mathbf{A} = (2.104, -1.215, L + 1.42), \quad \mathbf{B} = (2.104, -1.215, L),$$
$$\mathbf{C} = (2.104 + 0.326 \cos \phi + 0.735 \sin \phi, 0, L - 0.735 \cos \phi + 0.326 \sin \phi),$$
$$\mathbf{D} = (2.104, 1.215, L),$$

where the atom C moves around a circular path given by the angle ϕ, and L is the spacing between the tube and the sheet, as shown in Fig. 10.16D. Noting that $L = \ell + 0.659$ Å, where ℓ is defined for the variation in bond length method. Again, adopting the geometric model described in Chapter 9, the

Table 10.8 Values of Least Square Function f (Å^2) and Distance ℓ (Å) for 32 Configurations of a (4, 4) Tube and Corresponding Polygons P_n Where n Is the Number of Sides (by Symmetry #1-a and #1-b are Equivalent)

Configurations #	f	ℓ	P_4	P_5	P_6	P_7	P_8	P_9	P_{10}
1-a	0.0526	1.1548	–	2	1	3	1	1	–
1-b	0.0526	1.1548	–	2	1	3	1	1	–
2-a	0.0638	1.1501	–	1	2	4	–	1	–
2-b	0.0528	1.1798	1	2	–	2	1	2	–
3-a	0.4494	1.2208	–	4	–	–	2	2	-
3-b	0.0643	1.1157	–	–	2	6	–	–	-
4-a	0.3751	1.0544	–	2	3	–	1	2	–
4-b	0.0847	0.9834	1	–	1	4	2	–	–
5-a	0.0435	0.6723	–	–	6	–	–	2	-
5-b	0.2458	0.6746	2	–	–	2	4	–	-
6-a	0.5177	1.1151	–	–	4	2	2	–	-
6-b	0.3802	1.1098	–	2	2	–	4	–	-
7-a	0.3132	1.0767	1	–	3	1	2	1	–
7-b	0.2498	1.1093	–	1	3	1	3	–	–
8-a	0.0273	1.1968	–	–	3	4	1	–	-
8-b	0.1637	1.2151	1	2	1	–	2	2	–
9-a	0.0127	1.2968	–	–	2	6	–	–	-
9-b	0.0161	1.3612	2	2	–	–	–	4	-
10-a	0.0687	1.1979	1	–	2	3	1	1	–
10-b	0.0393	1.1624	1	1	2	1	1	2	–
11-a	0.0272	1.1240	2	–	2	–	3	–	1
11-b	0.5537	1.0920	–	–	5	–	3	–	-
12-a	0.0430	0.6309	2	–	2	–	2	2	–
12-b	0.1764	1.0026	–	–	4	2	2	–	-
13-a	0.2451	0.8439	2	–	2	1	–	3	–
13-b	0.0425	1.1583	–	1	2	3	2	–	–
14-a	0.0176	0.7303	2	–	1	2	2	–	1
14-b	0.2985	1.1284	1	–	4	–	1	2	–
15-a	0.0494	1.0936	3	–	1	–	1	2	1
15-b	0.4425	1.1057	–	–	4	2	2	–	-
16-a	0.0561	1.0353	4	–	–	–	–	2	2
16-b	0.3448	1.1007	–	–	4	2	2	–	-

radius of the (6, 0) tube is 2.430 Å and the bond angle is given by 117.65°. We note that this technique can be used to study a larger nanotube, where there will be many choices of a defect on a graphene sheet which may be chosen.

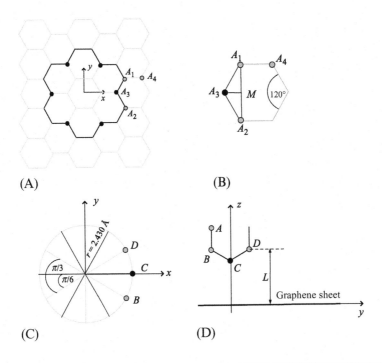

FIG. 10.16

Model formation for joining a (6, 0) tube with most symmetric defect where (A) and (B) are for a graphene sheet and (C) and (D) are for a (6, 0) tube.

WORKED EXAMPLE 10.4

Find the coordinates for atoms A, B, C, and D for a $(6, 0)$ carbon nanotube as shown in Fig. 10.16C and D.

Solution

In this case, we need to find the atomic positions for A, B, C, and D to determine a circular path of C. From Fig. 10.16, we may deduce

$$A = (r\cos(-\pi/6), r\sin(-\pi/6), L + 1.42),$$
$$B = (r\cos(-\pi/6), r\sin(-\pi/6), L),$$
$$C = (r\cos(0), r\sin(0), L - \sigma\cos(\phi_0/2)),$$
$$D = (r\cos(\pi/6), r\sin(\pi/6), L),$$

where $r = 2.43$ Å, bond length $\sigma = 1.42$ Å, and the bond angle $\phi_0 = 117.65°$.

In this example, we use MAPLE to numerically determine the five steps described in Section 10.1.2. The LinearAlgebra package is called to calculate the cross product for two vectors, and the geom3d package is called to work in three-dimensional Euclidean geometry.

```
> restart;
> with(LinearAlgebra):
> with(geom3d):
```

We then define the parameters.

```
> r:=2.43: sigma:=1.42: phi0:=117.65*Pi/180:
```

From Fig. 10.16C and D, we now input the atomic positions for the four atoms. coordinates() will return the coordinates of the given point X and point() is called to define the x-, y-, and z-coordinate of the given point X.

```
> coordinates(point(A,evalf(r*cos(-Pi/6)),evalf(r*sin(-Pi/6)),L+sigma));
```

$$[2.104441732, -1.215000000, L + 1.42]$$

```
> coordinates(point(B,evalf(r*cos(-Pi/6)),evalf(r*sin(-Pi/6)),L));
```

$$[2.104441732, -1.215000000, L]$$

```
> coordinates(point(C,evalf(r*cos(0)),evalf(r*sin(0)),
  evalf(L-sigma*cos(phi0/2))));
```

$$[2.43, 0., L - 0.7350683060]$$

```
> coordinates(point(D,evalf(r*cos(Pi/6)),evalf(r*sin(Pi/6)),L));
```

$$[2.104441732, 1.215000000, L]$$

We call the `midpoint()` command to determine the midpoint between the given two points, which is utilised to find **M**.

```
> coordinates(midpoint(M,B,D));
```

$$[2.104441732, 0., L]$$

Now we determine vector **U**.

```
> coordinates(point(U,xcoord(C)-xcoord(M),ycoord(C)-ycoord(M),
  zcoord(C)-zcoord(M)));
```

$$[0.325558268, 0., -0.7350683060]$$

We then determine the unit vector $\widehat{\mathbf{V}}$. The norm of a vector can be determined by using the `distance()` command.

```
> coordinates(point(V,(xcoord(D)-xcoord(B))/distance(B,D),
  (ycoord(D)-ycoord(B))/distance(B,D),(zcoord(D)-zcoord(B))
  /distance(B,D)));
```

$$[0., 1.000000000, 0.]$$

We determine the cross product between **U** and $\widehat{\mathbf{V}}$ using the `Matrix()` and `Determinant()` commands.

```
> W:=Matrix([[i,j,k],[xcoord(U),ycoord(U),zcoord(U)],[xcoord(V)
  ,ycoord(V),zcoord(V)]]);
```

$$W := \begin{bmatrix} i & j & k \\ 0.325558268 & 0. & -0.7350683060 \\ 0. & 1.000000000 & 0. \end{bmatrix}$$

```
> Determinant(W);
```

$$0.7350683060\,i + 0.3255582680\,k$$

Finally, a circular path for the atom C is obtained as

$$(2.014 + 0.326\cos\phi + 0.735\sin\phi, 0, L - 0.735\cos\phi + 0.326\sin\phi).$$

The variations of θ and ϕ from the normal physical bond angles, which are $120°$ and $117.65°$, are minimised, where each bond length which joins between an atom of the tube to one on the sheet is restricted to $1.42\,\text{Å}$. Therefore the constraints are the maximum distances of the hexagonal network, which are $|\overrightarrow{AC}| = |\overrightarrow{BA_3}| = 2.474\,\text{Å}$ and $|\overrightarrow{CA_1}| = |\overrightarrow{A_3A_4}| = 2.460\,\text{Å}$. Using the optimisation package

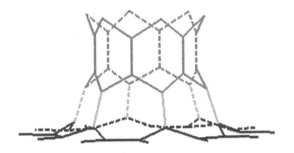

FIG. 10.17

Three-dimensional illustration for a $(6, 0)$ tube perpendicularly connected to a graphene sheet.

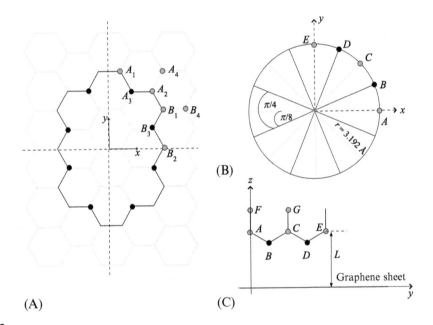

FIG. 10.18

Model formation for joining an $(8, 0)$ tube with defect #6, where (A) is for a graphene sheet and (B) and (C) are for an $(8, 0)$ tube.

in MAPLE, the parameters are obtained as $\theta = 30.73°$, $\phi = 9.39°$, and $L = 2.315\,\text{Å}$, and the three-dimensional figure is depicted in Fig. 10.17, where the dashed and solid lines represent bonds in the background and foreground, respectively.

The same method is exploited to examine the joining of the $(8, 0)$ tube with defect #6 identified as the most likely configuration. As the defect #6 has fourfold symmetry, only the atoms in the first quadrant are examined, and the positions of every atom are shown in Fig. 10.18. Note that the spacing L, illustrated in Fig. 10.18C, is equivalent to $L = \ell + 0.681\,\text{Å}$, where ℓ is defined for the variation in the bond length method. The calculation of the circular paths are examined and the coordinates for each atom are then given by:

$$\mathbf{A}_1 = \sigma \left(\frac{1}{2}, 2\sqrt{3}, 0 \right), \quad \mathbf{A}_2 = \sigma \left(2, \frac{3\sqrt{3}}{2}, 0 \right),$$

$$\mathbf{A}_3 = \sigma \left(\frac{5}{4} - \frac{1}{4} \cos \theta_1, \frac{7\sqrt{3}}{4} - \frac{\sqrt{3}}{4} \cos \theta_1, \frac{1}{2} \sin \theta_1 \right), \quad \mathbf{A}_4 = \sigma \left(\frac{5}{2}, 2\sqrt{3}, 0 \right),$$

$$\mathbf{B}_1 = \sigma \left(\frac{5}{2}, \sqrt{3}, 0 \right), \quad \mathbf{B}_2 = \sigma \left(\frac{5}{2}, 0, 0 \right),$$

$$\mathbf{B}_3 = \sigma \left(\frac{5}{2} - \frac{1}{2} \cos \theta_2, \frac{\sqrt{3}}{2}, \frac{1}{2} \sin \theta_2 \right), \quad \mathbf{B}_4 = \sigma \left(\frac{7}{2}, \sqrt{3}, 0 \right),$$

$$\mathbf{A} = (3.192, 0, L),$$

$$\mathbf{B} = (2.725 + 0.224 \cos \phi_1 + 0.629 \sin \phi_1, 1.129 + 0.093 \cos \phi_1 + 0.261 \sin \phi_1,$$
$$L - 0.681 \cos \phi_1 + 0.243 \sin \phi_1),$$

$$\mathbf{C} = (2.257, 2.257, L),$$

$$\mathbf{D} = (1.129 + 0.093 \cos \phi_2 + 0.261 \sin \phi_2, 2.725 + 0.224 \cos \phi_2 + 0.629 \sin \phi_2,$$
$$L - 0.681 \cos \phi_2 + 0.243 \sin \phi_2),$$

$$\mathbf{E} = (0, 3.192, L), \quad \mathbf{F} = (3.192, 0, L + 1.42), \quad \mathbf{G} = (2.257, 2.257, L + 1.42),$$

where the radius of the $(8, 0)$ tube is $3.192\,\text{Å}$, and the bond angle is given by $118.70°$. The computer package MAPLE is employed to minimise the bond angles of the system with the set of the constraints $|\overrightarrow{\mathbf{FB}}| = |\overrightarrow{\mathbf{AB}_3}| = |\overrightarrow{\mathbf{GD}}| = |\overrightarrow{\mathbf{CA}_3}| = 2.468\,\text{Å}$ and $|\overrightarrow{\mathbf{B}_3\mathbf{B}_4}| = |\overrightarrow{\mathbf{BB}_1}| = |\overrightarrow{\mathbf{A}_3\mathbf{A}_4}| = |\overrightarrow{\mathbf{DA}_2}| = 2.460\,\text{Å}$, where $|\overrightarrow{\mathbf{BB}_3}| = |\overrightarrow{\mathbf{DA}_3}| = 1.42\,\text{Å}$. For this configuration, the parameters are obtained as $L = 2.222\,\text{Å}$, $\theta_1 = 38.85°$, $\theta_2 = 17.16°$, $\phi_1 = -42.29°$ and $\phi_2 = 2.53°$, and the three-dimensional figure is presented in Fig. 10.19A. The corresponding structure previously obtained by minimisation of the bond length is shown in Fig. 10.19B, and it is again clear that the two approaches give closely related structures in terms of the atomic locations. In order to give a more mathematical measure for the difference between the atomic locations of these two structures, the mean absolute error is determined, which is defined by

$$Error = \frac{1}{n} \sum_{i=1}^{n} |\mathbf{r}_{1i} - \mathbf{r}_{2i}|, \tag{10.21}$$

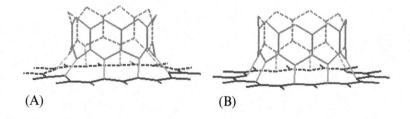

(A)　　　　　　　　　　(B)

FIG. 10.19

Three-dimensional illustrations for an $(8, 0)$ connection with a graphene sheet by (A) variation in bond angle and (B) variation in bond length.

where in this case $n = 16$ and $Error = 0.372\,\text{Å}$ (26.19% of bond length). Moreover, the difference in the spacing between the $(8, 0)$ tube and the graphene sheet in terms of L is obtained as $0.22\,\text{Å}$. Therefore in this case the two approaches give very similar outcomes.

WORKED EXAMPLE 10.5

From Fig. 10.18, show that $|\overrightarrow{FB}| = |\overrightarrow{AB_3}| = |\overrightarrow{GD}| = |\overrightarrow{CA_3}| = 2.468\,\text{Å}$ and $|\overrightarrow{B_3B_4}| = |\overrightarrow{BB_1}| = |\overrightarrow{A_3A_4}| = |\overrightarrow{DA_2}| = 2.460\,\text{Å}$, where we assume that $|\overrightarrow{BB_3}| = |\overrightarrow{DA_3}| = 1.42\,\text{Å}$.

Solution

The maximum distance between two atoms can be obtained by using Pythagoras' theorem.

All bond lengths are assumed to be $\sigma = 1.42\,\text{Å}$, bond angle for a $(8, 0)$ tube is $118.70°$, and the bond angle for a graphene sheet is assumed to be $120°$. Due to the symmetry of the tube and the sheet, we consider the joint between atoms B and B_3, and from Fig. 10.20, we have

$$|\overrightarrow{FB}| = \sqrt{[\sigma \sin(118.7°/2)]^2 + [\sigma + \sigma \cos(118.7°/2)]^2} = 2.467527231,$$

and

$$|\overrightarrow{B_3B_4}| = \sqrt{[\sigma \sin(120°/2)]^2 + [\sigma + \sigma \cos(120°/2)]^2} = 2.459512147,$$

where $|\overrightarrow{BB_3}| = 1.42\,\text{Å}$.

10.3.3 RESULTS AND DISCUSSION

Euler's polyhedra theorem is utilised to examine the joining of a graphene sheet and a carbon nanotube. The details for Euler's theorem are presented in Section 1.2.5, and here the zigzag $(8, 0)$ and the armchair $(4, 4)$ tubes, which topologically can be considered to be capped at one end with a hemispherical C_{60} fullerene comprising six pentagons, are considered. In order to maintain the Euler characteristic for any shape which is joined to a nanotube, the connection must necessarily consist of a certain number of other polygons, and the polygons which occur at the junction of the tube and the graphene sheet must satisfy

$$- 2P_4 - P_5 + P_7 + 2P_8 + 3P_9 + 4P_{10} = 6, \tag{10.22}$$

and this equation may be confirmed by the numerical values given in Tables 10.7 and 10.8. However, the square ring is very unstable for the carbon network, and heptagons and octagons can also be introduced into the system. Therefore, there are 11 and 12 possible configurations for joining the $(8, 0)$ and the $(4, 4)$ tubes with the graphene sheet, respectively. Nevertheless, defects in a nanotube are experimentally observed with only pentagonal and heptagonal rings, and imposing this physical requirement leaves only configuration #6 for the zigzag tube. Fig. 10.21A illustrates the three-dimensional structure, where the solid lines are for foreground bonds and the dashed lines indicate hidden or background bonds. In a least squares sense, the configuration #3 gives the minimum value of f, which is $0.0018\,\text{Å}^2$, and therefore this configuration might also be accepted despite the existence of octagonal rings as shown in Fig. 10.21B. Similarly by considering only pentagons and heptagons, the #3-b and #9-a configurations for the armchair $(4, 4)$ tube might be the most likely occurring structures. The three-dimensional

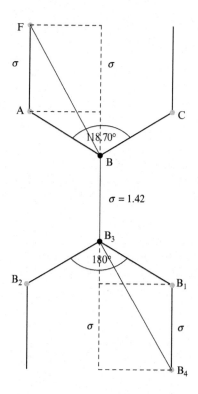

FIG. 10.20

Diagram for Worked Example 10.5.

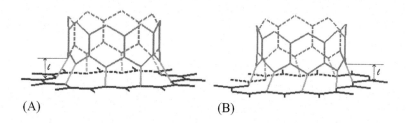

(A)   (B)

FIG. 10.21

Connecting an (8, 0) tube with a graphene sheet for the defect configurations (A) #6 and (B) #3.

illustrations for these two configurations are presented in Fig. 10.22. Noting that the #9-a defect configuration (see Fig. 10.22B) is more likely to occur from a comparison of the values of the least squares function given in Eq. (10.20).

From these results, the joining of (8, 0) tube to the defect #6 and the joining of (4, 4) tube to the defect #9-a are the most likely to occur. In Tables 10.9 and 10.10, numerical values of the atomic

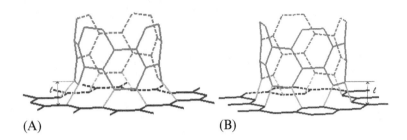

FIG. 10.22

Connecting a (4, 4) tube with a graphene sheet for (A) #3-b and (B) #9-a configurations. The second configuration has the smallest variation in bond length in a least squares sense.

Table 10.9 Eight Coordinate Positions for Joining an (8, 0) Tube With Defect #6 Using the Variation in Bond Length Method

Positions	1	2	3	4	5	6	7	8
(8, 0) tube								
x	2.949	2.949	1.222	−1.222	−2.949	−2.949	−1.222	1.222
y	−1.222	1.222	2.949	2.949	1.222	−1.222	−2.949	−2.949
z	1.322	1.322	1.322	1.322	1.322	1.322	1.322	1.322
#6 sheet								
x	2.840	2.840	1.420	−1.420	−2.840	−2.840	−1.420	1.420
y	−1.230	1.230	3.689	3.689	1.230	−1.230	−3.689	−3.689
z	0	0	0	0	0	0	0	0

Table 10.10 Eight Coordinate Positions for Joining a (4, 4) Tube With Defect #9-a Using the Variation in Bond Length Method

Positions	1	2	3	4	5	6	7	8
(4, 4) tube								
x	2.389	1.379	−1.379	−2.389	−2.389	−1.379	1.379	2.389
y	1.379	2.389	2.389	1.379	−1.379	−2.389	−2.389	−1.379
z	1.297	1.297	1.297	1.297	1.297	1.297	1.297	1.297
#9-a sheet								
x	2.840	0.710	−0.710	−2.840	−2.840	−0.710	0.710	2.840
y	1.230	2.460	2.460	1.230	−1.230	−2.460	−2.460	−1.230
z	0	0	0	0	0	0	0	0

Table 10.11 Geometric Parameters for New Nanotori Carbon Molecules Constructed From Elbows

Molecular Structure	Formula	ℓ_1 (Å)	ℓ_2 (Å)	r_1 (Å)	r_2 (Å)	R (Å)	a (Å)	c (Å)
$5(3,3)_{18}5(5,0)_{30}$	C_{240}	1.7100	3.2196	7.8311	7.3406	8.0156	2.0929	7.6730
$5(3,3)_{30}5(5,0)_{30}$	C_{300}	2.9203	3.2196	9.4970	9.3997	9.9358	2.0926	9.6064
$5(3,3)_{54}5(5,0)_{50}$	C_{520}	5.3409	5.2746	16.3248	16.3463	17.1763	2.0936	16.6111
$5(3,3)_{78}5(5,0)_{70}$	C_{740}	7.7615	7.3296	23.1526	23.2930	24.4190	2.0941	23.6134
$5(3,3)_{90}5(5,0)_{90}$	C_{900}	8.9718	9.3846	28.3147	28.1805	29.7021	2.0931	28.7232
$6(3,3)_{18}6(5,0)_{30}$	C_{288}	1.8242	3.0946	9.3488	9.0084	9.5251	2.0852	9.2620
$6(3,3)_{30}6(5,0)_{30}$	C_{360}	3.0345	3.0946	11.4451	11.4290	11.8406	2.0877	11.5699
$6(3,3)_{54}6(5,0)_{50}$	C_{624}	5.4551	5.1496	19.7477	19.8296	20.4873	2.0885	20.0180
$6(3,3)_{66}6(5,0)_{70}$	C_{816}	6.6654	7.2046	25.9540	25.8095	26.7962	2.0872	26.1811
$6(3,3)_{90}6(5,0)_{90}$	C_{1080}	9.0860	9.2596	34.2566	34.2101	35.4411	2.0877	34.6311
$5(4,4)_{32}5(7,0)_{42}$	C_{370}	2.4390	3.1758	8.7600	8.5206	9.0932	2.8127	8.7775
$5(4,4)_{48}5(7,0)_{42}$	C_{450}	3.6580	3.1758	10.4378	10.5945	11.0602	2.8066	10.6905
$5(4,4)_{64}5(7,0)_{70}$	C_{670}	4.8770	5.2688	15.6764	15.5491	16.4175	2.8091	15.8748
$5(4,4)_{96}5(7,0)_{98}$	C_{970}	7.3150	7.3618	22.5929	22.5777	23.7476	2.8081	22.9662
$5(4,4)_{128}5(7,0)_{126}$	C_{1270}	9.7530	9.4548	29.5093	29.6062	31.0793	2.8076	30.0559
$5(5,5)_{50}5(9,0)_{54}$	C_{520}[a]	2.9701	3.2636	9.6404	9.5450	10.0875	3.5358	9.7533
$5(5,5)_{70}5(9,0)_{54}$	C_{620}	4.1930	3.2636	11.3235	11.6255	12.0749	3.5245	11.6578
$5(5,5)_{90}5(9,0)_{90}$	C_{900}[a]	5.4159	5.3715	16.5929	16.6073	17.4544	3.5318	16.8801
$5(5,5)_{130}5(9,0)_{126}$	C_{1280}	7.8617	7.4794	23.5454	23.6696	24.8232	3.5303	24.0049
$5(5,5)_{150}5(9,0)_{162}$	C_{1560}	9.0846	9.5873	28.8148	28.6514	30.2129	3.5342	29.2166

[a] Denotes the two molecules studied by Fonseca et al. (1995).

positions in the Cartesian coordinate system are presented for the purpose of future comparisons. The configurations given are the 16 atom positions at the junctions after the optimisation process for the $(8,0)$ and the $(4,4)$ tubes with the #6 and the #9-a defects, respectively. For these two particular configurations, X, Y, and θ can be taken to be zero by an appropriate choice of the coordinate axes, as might be expected from the symmetries of both configurations.

10.4 NANOBUDS

Following the methodology presented in Section 10.1 for the two least squares approaches, we determine the connecting structure between a C_{60} fullerene and a carbon nanotube to form a hybrid carbon nanostructure, which is referred to as a nanobud. Two distinct defects on the C_{60} fullerene are assumed, as shown in Fig. 10.23, where the defect illustrated in Fig. 10.23A is denoted as DA and the defect illustrated in Fig. 10.23B is denoted as DB. We assume that the pairwise bond length between carbon atoms is taken to be $\sigma = 1.42$ Å, and that all atoms are connected in a sp^2 structure. Throughout this section, all the atomic positions and bond angles for the carbon nanotubes are calculated using the

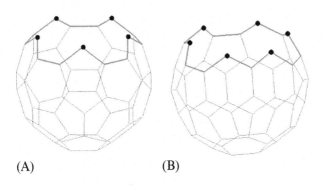

(A)   (B)

FIG. 10.23

Two defect patterns, DA and DB, taken out of the C_{60} fullerene.

new geometric model of carbon nanotubes presented in Chapter 9, whereas all 60 atom coordinates for a C_{60} fullerene are determined utilising `TruncatedIcosahedron()` command in MAPLE with a radius of 3.55 Å.

10.4.1 VARIATION IN BOND LENGTH

The centre of the C_{60} fullerene is assumed to be located at the origin, and the tube axis is assumed to be parallel to the y-axis. The ith terminal atom at a joint location is defined by the position vectors $\mathbf{a}_i = (a_{ix}, a_{iy}, a_{iz})$ and $\mathbf{b}_i = (b_{ix}, b_{iy}, b_{iz})$ for the open end of the tube and for the C_{60} fullerene, respectively. The spacing between the tube and the centre of the fullerene in the y-direction is assumed to be ℓ, and we assume that the tube can be rotated about the y-axis through an angle θ. Moreover, the tube is allowed to move in both the x- and z-directions by distances X and Z, respectively, which can be either positive or negative. Consequently, the position vector for the atom locations at the open end of the tube can be written as $\mathbf{a}_i = (X + a_{ix}\cos\theta - a_{iz}\sin\theta, \ell, Z + a_{ix}\sin\theta + a_{iz}\cos\theta)$. The Euclidean distance between the atoms at the junction is then given by

$$|\mathbf{a}_i - \mathbf{b}_i| = \{[(X + a_{ix}\cos\theta - a_{iz}\sin\theta) - b_{ix}]^2$$
$$+ (\ell - b_{iy})^2 + [(Z + a_{ix}\sin\theta + a_{iz}\cos\theta) - b_{iz}]^2\}^{1/2}.$$

Given these distances between matching atoms, our procedure is to determine X, Z, ℓ, and θ by minimising the least squares variation of these distances from the pairwise carbon–carbon bond length $\sigma = 1.42$ Å. In other words, we wish to minimise the objective function (10.1).

10.4.2 VARIATION IN BOND ANGLE

In this section, we assume that the bond lengths are fixed to be σ but we vary the bond angles at connection sites so as to minimise the least squares deviations from the physical bond angles. The carbon–carbon bond length is taken to be $\sigma = 1.42$ Å, the ideal bond angles of the pentagon and the hexagon for the fullerene are assumed to be 108° and 120°, respectively, and the ideal bond angles for the carbon nanotubes are taken from the geometric model of carbon nanotubes presented in Chapter 9 which properly incorporates curvature. We employ the five steps in Section 10.1.2 to determine the atomic positions for the atoms at the open end of the tube and the atoms at the defect on the C_{60} fullerene; the numerical results are presented in the following section.

10.4.3 RESULTS AND DISCUSSION

There are five and six terminal atoms for the defects DA and DB, respectively, which require one other bond to complete the sp^2 structure. Therefore the connected carbon nanotubes must have five and six atoms at the open end which also need one other bond to complete such sp^2 structure. In this study, a $(5,0)$ and a $(6,0)$ carbon nanotubes are assumed to join with the defects DA and DB, respectively.

All numerical calculations are carried out using the algebraic computer package MAPLE. The objective function (10.1) is graphed to ensure a genuine global minimum, and the optimisation package is then utilised to find the values for each parameter which gives the minimum. The radii of the $(5,0)$ and the $(6,0)$ tubes are taken to be $a = 2.055\,\text{Å}$ and $a = 2.430\,\text{Å}$, respectively, and the radius of the C_{60} fullerene is assumed to be $b = 3.55\,\text{Å}$. Three-dimensional figures for a $(5,0)$ and a $(6,0)$ carbon nanotubes joining with the defects DA and DB are shown in Fig. 10.24 for the two variation models. For both methods, the objective function is approximately zero. This means that the bond lengths and the bond angles for the connected structures are very close to the ideal physical values, and the new hybrid nanobuds might be considered as the ideal structures in the sense that the objective function is as close to zero as possible.

In terms of polygonal rings which occur at the connection site, Euler's theorem is used to verify that the proposed structures are geometrically sound. A carbon nanotube can be considered capped at one end with a hemispherical C_{60} fullerene comprising six pentagons. In order to maintain the Euler characteristic, the connection must necessarily involve six pentagons with six heptagons or an equivalent number of other polygons. As a result, the polygons which occur at the junction must satisfy $2P_4 + P_5 - P_7 - 2P_8 = 6$. In this study the defect DA represents one pentagon and the defect DB represents three pentagons; therefore, there are 11 and 9 pentagons remaining on the C_{60} fullerene in each case. We observe that there are five and three heptagons occurring for the connected structures

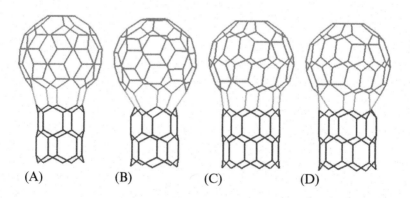

(A)   (B)   (C)   (D)

FIG. 10.24

Three-dimensional illustrations for a $(5,0)$ tube joining with defects DA using (A) variation in bond length and (B) variation in bond angle, and for a $(6,0)$ tube joining with defects DB using (C) variation in bond length and (D) variation in bond angle.

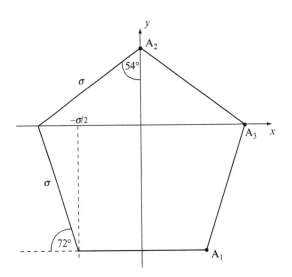

FIG. 10.25

Atomic positions of A_1, A_2, and A_3 on a pentagon.

of the $(5,0)$ tube and the defect DA and the $(6,0)$ tube and the defect DB, respectively, which satisfy Euler's theorem.

EXERCISES

10.1. Explain the two least squares approaches. What are their main differences?

10.2. Is the least squares approach a good approximation to the minimisation energy approach? Why?

10.3. What is meant by the terminology $N(5,5)_p$? How many atoms would a $5(5,5)_{18}5(9,0)_{30}$ elbow have?

10.4. Show that the representative toroidal generating radius c and the representative tube radius a for a toroidal molecule formed from three distinct carbon nanotube sections can be given by Eqs (10.17), (10.18), respectively.

10.5. Find a circular path for atom A_3 as shown in Fig. 10.25. We assume that a pentagon is centred at the origin with a bond length $\sigma = 1$ Å.

10.6. Find 10 possible defects on a graphene sheet which require another 15 bonds for the sp^2 network to connect with a $(15,0)$ carbon nanotube.

10.7. Boron nitride is a binary chemical compound consisting of equal proportions of boron and nitrogen. The hexagonal structure of boron nitride is geometrically similar to that of graphite in carbon structures. Given that boron and nitrogen atoms are alternately bonded to form a hexagonal lattice, two triangular sublattices may be thought of as the basis units superimposing to form such a hexagonal network. Find five possible defect on a boron nitride sheet which require another 15 bonds for the sp^2 network to connect with a $(15,0)$ boron nitride nanotube. We note that a boron atom is most likely to bond with a nitrogen atom.

CONTINUOUS APPROACH FOR JOINING CARBON NANOSTRUCTURES

11.1 INTRODUCTION

In Chapter 10, a numerical method is proposed to determine the geometric parameters for connecting two carbon nanostructures. That method is based on a discrete model for the carbon atoms and the covalent bonds comprising the joint. The particular geometries that are generated are obtained by the principle of minimising the difference of the square of the distance between atoms which are separated by two covalent bonds and their maximum possible distance. The discrete geometric method presented in Chapter 10 is related to certain bonded potential energy methods. The question arises as, 'To what extent can such intrinsically discrete problems be approximated by a continuous model'? Any continuous model which accounts for the dominant qualitative features of the discrete problems would, from a theoretical perspective, be highly desirable.

In this chapter, we employ variational calculus to formulate a continuous model to examine the connection between two carbon nanostructures. The calculus of variations, detailed in Section 11.2, is utilised to minimise the curvature subject to a length constraint and to obtain an Euler–Lagrange equation, which determines the connection between two carbon nanostructures. Throughout this chapter the defects on the nanostructures are assumed to be axially symmetric, so that the problem can be considered in the two dimensional plane. As the curvature can be both positive and negative depending on the gap between the two nanostructures, two models are examined. In Section 11.3, we determine the continuous joining between a carbon nanotube and a flat graphene sheet, where the final structure is comparable with the one obtained by the discrete method presented in Section 10.3 of Chapter 10. The nanobud structures are also determined utilising the calculus of variations, and this is presented in Section 11.4. Finally, in Section 11.5, we construct a novel carbon nanostructure formed from two fullerenes, namely a nanopeanut.

11.2 CALCULUS OF VARIATIONS

This section gives the basic equations of the variational calculus, which we use to determine the curve adopted by a line connecting two carbon nanostructures. In the terminology of the calculus of variations, we seek to determine the function $y(x)$, which has an element of arc length ds such that the function

$$J[y] = \int_0^\ell \kappa^2 ds + \lambda \int_0^\ell ds,$$

where κ is the curvature, and λ is a Lagrange multiplier corresponding to the fixed length constraint. The first term arises from the general belief that carbon nanotubes are perfectly elastic and deform rather like the elastica, while the fixed length constraint arises from the fact that the joint involves an integer number bonds. For a two-dimensional curve $y = y(x)$, we have $\kappa = y''/(1+y'^2)^{3/2}$, $ds = (1+y'^2)^{1/2}dx$, and this equation becomes

$$J[y] = \int_a^{x_0} \frac{y''^2}{(1+y'^2)^{5/2}}dx + \lambda \int_a^{x_0} (1+y'^2)^{1/2}dx,$$

where x_0 and x_1 are two endpoints, and primes denote differentiation with respect to x. Applying the delta variational operator, we may derive the standard equation

$$\delta J[y] = \left[\left(F_{y'} - \frac{d}{dx}F_{y''} \right) \delta y + F_{y''}\delta y' \right]_a^{x_0}$$
$$+ \int_a^{x_0} \left(F_y - \frac{d}{dx}F_{y'} + \frac{d^2}{dx^2}F_{y''} \right) \delta y dx, \tag{11.1}$$

where subscripts denote partial derivatives and here F is given by

$$F(y',y'') = \frac{y''^2}{(1+y'^2)^{5/2}} + \lambda(1+y'^2)^{1/2}. \tag{11.2}$$

For the present problems, we require the natural or alternative boundary condition, which applies when the value of $y(a)$ is not prescribed but is derived from the first term in Eq. (11.1) and is given by

$$\left(F_{y'} - \frac{d}{dx}F_{y''} \right) |_{x=a} = 0 . \tag{11.3}$$

WORKED EXAMPLE 11.1

On applying the delta variational operator, show that Eq. (11.1) can be obtained.

Solution

Firstly, we assume that $F = F(x, y, y', y'')$ and

$$y = y_0 + \epsilon\delta y, \quad y' = y_0' + \epsilon\delta y', \quad y'' = y_0'' + \epsilon\delta y''.$$

We have

$$J[y] = \int_{x_0}^{x_1} F(x, y, y', y'')dx,$$

and we apply the delta variational operator to deduce

$$\delta J[y] = \frac{dJ[y]}{d\epsilon}|_{\epsilon=0}$$
$$= \int_{x_0}^{x_1} \left(\frac{\partial F}{\partial x}\frac{\partial x}{\partial \epsilon} + \frac{\partial F}{\partial y}\frac{\partial y}{\partial \epsilon} + \frac{\partial F}{\partial y'}\frac{\partial y'}{\partial \epsilon} + \frac{\partial F}{\partial y''}\frac{\partial y''}{\partial \epsilon} \right) dx$$
$$= \int_{x_0}^{x_1} (F_y\delta y + F_{y'}\delta y' + F_{y''}\delta y'')dx.$$

Using integration by parts for the second and third terms of the above equation, as well as the fact that the variation and the derivative operations commute, we may deduce (e.g. Elsgolc, 1961, pp. 42–45):

$$\delta J[y] = \left[\left(F_{y'} - \frac{d}{dx} F_{y''} \right) \delta y + F_{y''} \delta y' \right]_{x_0}^{x_1} + \int_{x_0}^{x_1} \left(F_y - \frac{d}{dx} F_{y'} + \frac{d^2}{dx^2} F_{y''} \right) \delta y \, dx.$$

The usual Euler–Lagrange equation is obtained from the second term in Eq. (11.1), which can be written as

$$F_y - \frac{d}{dx} F_{y'} + \frac{d^2}{dx^2} F_{y''} = 0.$$

As F given in Eq. (11.2) does not depend on y, we may immediately perform an integration to deduce

$$F_{y'} - \frac{d}{dx} F_{y''} = C, \tag{11.4}$$

where C is an arbitrary constant. By considering

$$\frac{dF}{dx} = F_x + y' F_y + y'' F_{y'} + y''' F_{y''},$$

and taking C in Eq. (11.4) equal zero to accommodate the alternative boundary condition (11.3), we may deduce

$$\frac{dF}{dx} = y'' \frac{d}{dx} F_{y''} + y''' F_{y''} = \frac{d}{dx} \left(y'' F_{y''} \right),$$

because in this case F does not depend on both x and y. By integrating the above equation, we may deduce

$$F - y'' F_{y''} = -\alpha, \tag{11.5}$$

where α is an arbitrary constant of integration. We now substitute Eq. (11.2) into Eq. (11.5) to obtain

$$\frac{y''^2}{(1 + y'^2)^3} = \lambda + \frac{\alpha}{(1 + y'^2)^{1/2}} = \kappa^2,$$

so that the curvature κ is given by

$$\kappa = \pm \left(\lambda + \frac{\alpha}{(1 + y'^2)^{1/2}} \right)^{1/2}. \tag{11.6}$$

11.3 JOINING CARBON NANOTUBES AND FLAT GRAPHENE SHEETS

In this section, we use the continuous approach outlined in Section 11.2 to model the perpendicular joining of carbon nanotubes and a flat graphene sheet. In particular, we use variational calculus to determine the curve adopted by the line connecting a horizontal plane and a vertical carbon nanotube, such that the arc length of the curve and the size of the defect in the graphene sheet are specified. It is important to note that the distance of the carbon nanotube from the graphene sheet is not prescribed but is determined as part of the solution.

We position the graphene sheet in the (x, z)-plane, assuming a circular defect of radius x_0 centred on the origin. We also assume that a nanotube of radius a is located with its axis colinear with the y-axis, starting from an unknown positive distance above the (x, z)-plane, which we will denote by y_0. As the defect and the nanotube are assumed to be rotationally symmetric about the y-axis, we can consider this as a problem in the two dimensional (x, y)-plane. The connecting covalent bonds are assumed to join the points on the graphene defect $(x_0, 0)$, and the nanotube (a, y_0) and have a total prescribed arc length ℓ. Two likely configurations are illustrated in Fig. 11.1.

We consider two distinct models, as shown in Fig. 11.1; one for which the joint curvature remains positive (Model I), and another for which the joint comprises two regions, one of positive curvature and one of negative curvature (Model II). We subsequently show that the two models are natural extensions of each other in the sense that the parameter spaces overlap as indicated in Figs 11.2 and 11.3. For both models, we impose the continuity boundary conditions at the graphene sheet:

$$y(x_0) = 0, \quad y'(x_0) = 0. \tag{11.7}$$

In Model I, the value of y' ranges from 0 at $x = x_0$ to $-\infty$ at $x = a$. Therefore for Model I, the boundary condition is $y'(a) = -\infty$. In Model II, y' ranges from 0 to $-\infty$, where it changes sign and then ranges from ∞ down to some finite positive value before returning to ∞. Therefore in Model II

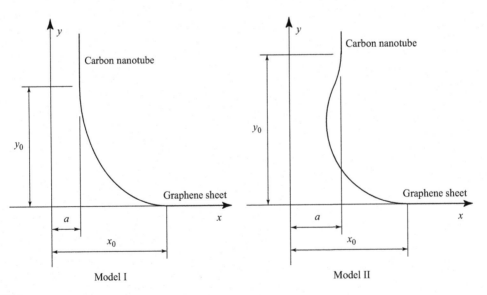

FIG. 11.1

Problem geometries for Model I, in which the joint contains only a positive curvature, and Model II, which has both positive and negative curvatures in the joint region.

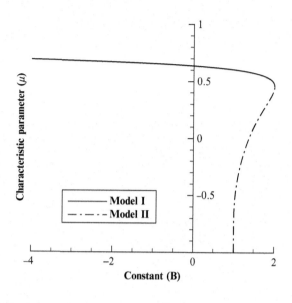

FIG. 11.2

Relation between characteristic parameter $\mu = (x_0 - a)/\ell$ and constant $B = 1/k^2$ for both models, obtained from Eqs (11.20), (11.27).

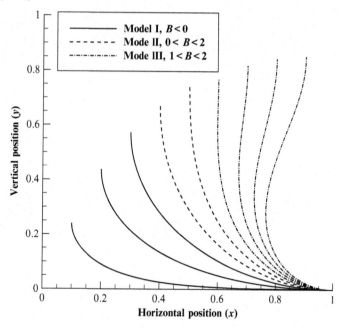

FIG. 11.3

Nondimensional plots of joints $y = y(x)$ for various values of μ.

the boundary condition is $y'(a) = \infty$. Thus for the two models at $x = a$, we have Eq. (11.3) along with the boundary conditions

$$y'(a) = -\infty, \quad \text{Model I} \tag{11.8}$$

$$y'(a) = \infty, \quad \text{Model II} \tag{11.9}$$

and in both cases y_0 is determined from the value $y_0 = y(a)$. As for both models the value of $y(x)$ at $x = a$ is not prescribed, so the alternative or natural boundary condition (11.3) applies. The major implication of Eq. (11.3) is that the constant C arising in Eq. (11.4) is zero, and therefore the solution (11.6) applies.

11.3.1 MODEL I: POSITIVE CURVATURE

By making the substitution $\tan \theta = y'$ and assuming positive curvature, Eq. (11.6) becomes

$$\kappa = (\lambda + \alpha \cos \theta)^{1/2}. \tag{11.10}$$

From the definition of curvature $\kappa = y''/(1 + y'^2)^{3/2}$ and making the same substitution for y', we obtain

$$\frac{dy}{d\theta} = \frac{\sin \theta}{(\lambda + \alpha \cos \theta)^{1/2}}.$$

In order to simplify the algebra, we now introduce the constant $k = [(\lambda + \alpha)/2\alpha]^{1/2}$ and a new parametric variable ϕ, which is defined by

$$\cos \theta = 1 - 2k^2 \sin^2 \phi. \tag{11.11}$$

It may be shown from these substitutions that

$$\frac{d\theta}{d\phi} = \frac{2k \cos \phi}{(1 - k^2 \sin^2 \phi)^{1/2}}, \tag{11.12}$$

and on introducing a second new constant $\beta = (2/\alpha)^{1/2}$, we may deduce

$$\frac{dy}{d\phi} = 2\beta k \sin \phi,$$

which on integration yields

$$y(\phi) = 2\beta k (1 - \cos \phi), \tag{11.13}$$

on noting that the constant of integration arises from the boundary conditions (11.7). Here, we consider Eq. (11.13) as a parametric equation for y in terms of the parameter ϕ, and we denote the value of ϕ at the point where $\theta = -\pi/2$ with $\phi_0 = \sin^{-1}(1/\sqrt{2}k)$. We also note that for Model I, $-\phi_0 < \phi \leqslant 0$.

WORKED EXAMPLE 11.2

Derive Eq. (11.12).

Solution

From the new parametric variable ϕ, which is given by Eq. (11.11) and restated here

$$\cos\theta = 1 - 2k^2 \sin^2\phi,$$

we may deduce

$$-\sin\theta d\theta = -4k^2 \sin\phi \cos\phi d\phi, \qquad (11.14)$$

and

$$\frac{d\theta}{d\phi} = \frac{4k^2 \sin\phi \cos\phi}{\sin\theta}.$$

We then square both sides of the new parametric variable equation (11.11) to obtain

$$\cos^2\theta = 1 - 4k^2 \sin^2\phi + 4k^4 \sin^4\phi,$$

which on using the trigonometric identity $\cos^2\theta + \sin^2\theta = 1$ becomes

$$\sin^2\theta = 4k^2 \sin^2\phi(1 - k^2 \sin^2\phi).$$

Substituting this value for $\sin\theta$ into Eq. (11.14) and rearranging gives

$$\frac{d\theta}{d\phi} = \frac{2k \cos\phi}{(1 - k^2 \sin^2\phi)^{1/2}}.$$

Now we determine the corresponding parametric equation for x by taking Eq. (11.10) and substituting $dy = \tan\theta \, dx$, which yields

$$\frac{dx}{d\theta} = \frac{\cos\theta}{(\lambda + \alpha \cos\theta)^{1/2}}. \qquad (11.15)$$

After changing to the new parameter ϕ, we obtain

$$\frac{dx}{d\phi} = \beta \frac{1 - 2k^2 \sin^2\phi}{\left(1 - k^2 \sin^2\phi\right)^{1/2}},$$

or alternatively,

$$\frac{dx}{d\phi} = \beta \left[2\left(1 - k^2 \sin^2\phi\right)^{1/2} - \frac{1}{\left(1 - k^2 \sin^2\phi\right)^{1/2}} \right],$$

which upon integration yields

$$x(\phi) = x_0 + \beta \left[2E(\phi, k) - F(\phi, k) \right], \qquad (11.16)$$

where the constant of integration arises from the boundary condition (11.7), and $F(\phi, k)$ and $E(\phi, k)$ denote the usual Legendre incomplete elliptic integrals of the first and second kinds, respectively. We use the form of these integrals as defined in Byrd and Friedman (1971), p. 8, and also outlined in Section 2.10.

By substituting the values from the boundary condition (11.8) for Model I at the point $x = a$ where the joint meets the tube into Eqs (11.13), (11.16) yields

$$y_0 = 2\beta k(1 - \cos \phi_0), \tag{11.17}$$

$$x_0 - a = \beta \left[2E(\phi_0, k) - F(\phi_0, k)\right]. \tag{11.18}$$

From the arc length constraint we have

$$\ell = \int_a^{x_0} (1 + y'^2)^{1/2} dx,$$

so that on substituting $y' = \tan \theta$ according to Eq. (11.15), we have

$$\ell = \int_{-\pi/2}^0 \frac{d\theta}{(\lambda + \alpha \cos \theta)^{1/2}},$$

and the substitution $\cos \theta = 1 - 2k^2 \sin^2 \phi$ yields

$$\ell = \beta \int_{-\phi_0}^0 \frac{d\phi}{(1 - k^2 \sin^2 \phi)^{1/2}},$$

from which we may deduce by changing ϕ to $-\phi$,

$$\ell = \beta F(\phi_0, k). \tag{11.19}$$

Thus for prescribed a, x_0, and ℓ, Eqs (11.18), (11.19) constitute two equations for the determination of the two unknowns β and k from which the attachment height y_0 may be obtained from Eq. (11.17).

By substituting Eq. (11.19) into Eq. (11.18) we may derive

$$\mu = 2 \left(\frac{E(\phi_0, k)}{F(\phi_0, k)} \right) - 1, \tag{11.20}$$

where $\mu = (x_0 - a)/\ell$ and $-1 < \mu < 1$. As $\phi_0 = \sin^{-1}(1/\sqrt{2}k)$, Eq. (11.20) must be solved numerically for a given μ to determine the value for k, hence ϕ_0. Then by substitution back into Eq. (11.19), the value of β is determined, and therefore y_0 can be determined from Eq. (11.17).

11.3.2 MODEL II: POSITIVE AND NEGATIVE CURVATURE

In this section, we proceed exactly as in Section 11.3.1 for positive curvature for the first region, from the point of attachment to the graphene sheet $(x_0, 0)$ up until the critical point (x_c, y_c) where the curvature changes sign. We then consider the second region from the critical point (x_c, y_c) to the point of attachment to the carbon nanotube (a, y_0), throughout which the curvature is negative. We denote the value of the parameter θ as defined in the previous section at the critical point to be θ_c, and from geometrical considerations we have that $-\pi < \theta_c < -\pi/2$.

The same considerations apply to the region of positive curvature, as used in Section 11.3.1 and therefore Eqs (11.13), (11.16) are valid in the first region of Model II. This region is bounded by the point where the curvature $\kappa = 0$, and from Eq. (11.10), we may derive

$$\theta_c = -\cos^{-1}(-\lambda/\alpha).$$

Now, employing the new parametric variable ϕ as defined by Eq. (11.11), we determine that $\phi = -\pi/2$ when $\theta = \theta_c$. By substituting $\phi = -\pi/2$ into Eqs (11.13), (11.16), we may derive

$$y_c = 2\beta k, \quad x_c = x_0 - \beta \left[2E(k) - K(k) \right], \tag{11.21}$$

where β and k are as defined in Section 11.3.1, and $K(k)$ and $E(k)$ are the usual complete elliptic integrals of the first and second kinds, respectively, as outlined in Section 2.10.

In the second region, we take the negative sign of Eq. (11.6) and following a similar procedure to that described in Section 11.3.1, we may obtain

$$y(\phi) = 2\beta k (1 + \cos \phi), \tag{11.22}$$

where the constant of integration arises from the condition that $y = y_c$ when $\phi = -\pi/2$, then we use the expression $(11.21)_1$ for y_c (see Exercise 2 at the end of this chapter). We note that Eq. (11.22) differs from Eq. (11.13) only by a change in the sign of one term. Similarly, by taking the negative sign of Eq. (11.6) and solving for the parametric form of x, we may derive

$$x(\phi) = 2x_c - x_0 - \beta \left[2E(\phi, k) - F(\phi, k) \right],$$

where the constant of integration is determined from the boundary condition at $\phi = -\pi/2$. Substituting for x_c using Eq. $(11.21)_2$ (see Exercise 2 at the end of this chapter) yields

$$x(\phi) = x_0 - \beta \left\{ 2 \left[2E(k) - E(-\phi, k) \right] - \left[2K(k) - F(-\phi, k) \right] \right\}. \tag{11.23}$$

We again note that we follow throughout the usual convention that $E(k)$ refers to the complete elliptic integral of the second kind, while $E(-\phi, k)$ denotes the corresponding incomplete elliptic integral.

From the boundary conditions at the point of attachment to the carbon nanotube (Eq. 11.9) we now know that $\phi = -\phi_0$ at the point (a, y_0). By substitution into Eq. (11.22), we may derive

$$y_0 = 2\beta k (1 + \cos \phi_0), \tag{11.24}$$

and similarly, substitution in Eq. (11.23) gives

$$x_0 - a = \beta \left\{ 2 \left[2E(k) - E(\phi_0, k) \right] - \left[2K(k) - F(\phi_0, k) \right] \right\}. \tag{11.25}$$

The arc length constraint is obtained from the two regions and we have

$$\ell = \int_{\theta_c}^{0} \frac{d\theta}{(\lambda + \alpha \cos \theta)^{1/2}} + \int_{\theta_c}^{-\pi/2} \frac{d\theta}{(\lambda + \alpha \cos \theta)^{1/2}},$$

which by essentially the same procedure to that employed in Section 11.3.1, we may derive

$$\ell = \beta \left[2K(k) - F(\phi_0, k) \right]. \tag{11.26}$$

Thus for a prescribed a, x_0, and ℓ, Eqs (11.25), (11.26) constitute two equations for the two unknowns β and k, remembering that $\phi_0 = \sin^{-1} \left(1/\sqrt{2}k \right)$. Once these two constants are determined, the attachment height y_0 may be obtained from Eq. (11.24).

By straightforward algebra, we may deduce

$$\mu = 2 \left(\frac{2E(k) - E(\phi_0, k)}{2K(k) - F(\phi_0, k)} \right) - 1, \tag{11.27}$$

whereas before, $\mu = (x_0 - a)/\ell$. As $\phi_0 = \sin^{-1}(1/\sqrt{2}k)$, Eq. (11.27) only involves the single unknown k. We may solve this equation numerically for a prescribed μ to determine the value for k which in turn gives the value for ϕ_0. Further substitution into Eq. (11.26) gives the value for β, which allows by substitution into Eq. (11.24) the attachment height y_0 to be deduced.

We observe that formally Eq. (11.20) coincides with Eq. (11.27) for the value $k = 1/\sqrt{2}$. We denote the value of μ at this point by μ_0 and we have

$$\mu_0 = 2\left(\frac{E(1/\sqrt{2})}{K(1/\sqrt{2})}\right) - 1 = 0.4569465810\ldots, \tag{11.28}$$

where $K(k)$ and $E(k)$ are the complete elliptic integrals of the first and second kinds, respectively.

11.3.3 RESULTS AND DISCUSSION

We begin by examining some general features of the solutions to Models I and II, and then we subsequently examine a particular carbon nanotube–graphene joint.

The first observation is that the solution is characterised by the nondimensional characteristic parameter $\mu = (x_0 - a)/\ell$. The different regions of the solution can be seen by examining Eqs (11.20), (11.27), as shown in Fig. 11.2 in which for convenience we plot μ against a new constant $B = 1/k^2$. In this plot, three distinct regions are evident. First, there is the region $2/\pi < \mu < 1$, which corresponds to join with arc length ℓ less than a quarter of a circle circumference of radius $x_0 - a$. In this case the constant B is negative, which corresponds to a negative value of α and an imaginary modulus k for the elliptic integrals. The solution asymptotes with the line $\mu = 1$ and crosses the vertical axis at the point $\mu = 2/\pi$, which corresponds to the solution degenerating to constant curvature (i.e. a circular joint). The second region exists for $\mu_0 \leqslant \mu < 2/\pi$, where μ_0 is given by Eq. (11.28), and corresponds to Model I, where the arc length ℓ of the joint is greater than the quarter circumference of a circle of radius $x_0 - a$. In this region, $0 < B \leqslant 2$, and therefore α is always positive and the modulus k is strictly real. The third and final region applies to the range $-1 < \mu < \mu_0$ and corresponds to Model II. This model is invoked when the arc length ℓ is much greater than the quarter circumference of a circle of radius $x_0 - a$, and therefore a change of curvature is necessary to accommodate the joint. For Model II, the constant B is restricted to the range $1 < B \leqslant 2$, and therefore again α is strictly positive. As can be seen from Fig. 11.2, the constant B never takes a value greater than two for any of the solution regions.

We now apply the solutions derived in Sections 11.3.1 and 11.3.2 to a nondimensionalised situation. We assume a fixed arc length $\ell = 1$ and a graphene attachment point $x_0 = 1$. We then allow the tube radius a to take values between 0.1 and 0.9 in increments of 0.1. In this configuration $\mu = 1 - a$ and the resulting joints are shown in Fig. 11.3. As can be seen from this figure, for the three cases when $2/\pi < \mu < 1$, Model I is used with a negative value of $1/k^2$. For the cases when $\mu_0 < \mu < 2/\pi$, Model I is again used, however in these cases, $0 < 1/k^2 < 2$. Finally, the cases where $-1 < \mu < \mu_0$, Model II provides the solution. For various values of the characteristic parameter μ, numerical values of the constant k and the attachment height y_0 for all the curves in the figure are listed in Table 11.1.

We now compare our results with those described in Section 10.3 for the symmetric joint for a $(6, 0)$ nanotube with a graphene sheet for which a least squares approach was used. As we are comparing

> **Table 11.1 Numerical Values for the Parameter μ, Corresponding Constant k and Attachment Height y_0 for Various Tube Radii a, Assuming $x_0 = \ell = 1$**
>
> | a | μ | Model | k | y_0 |
> |-----|-------|-------|------|-------|
> | 0.1 | 0.9 | I | 0.00473980i | 0.23985778 |
> | 0.2 | 0.8 | I | 0.09913584i | 0.43683709 |
> | 0.3 | 0.7 | I | 0.42121651i | 0.57207331 |
> | 0.4 | 0.6 | I | 0.95463759 | 0.66838632 |
> | 0.5 | 0.5 | I | 0.71416531 | 0.73872974 |
> | 0.6 | 0.4 | II | 0.71422872 | 0.78957689 |
> | 0.7 | 0.3 | II | 0.74414001 | 0.82443860 |
> | 0.8 | 0.2 | II | 0.78101861 | 0.84529525 |
> | 0.9 | 0.1 | II | 0.81876073 | 0.85321208 |

an essentially discrete asymmetric surface with a continuous axially symmetric one, there is no unique procedure. To compare with Section 10.3, we consider the comparison surface to be that which is defined by the two central atomic locations, and the two midpoints between the two atoms at the endpoints on the graphene sheet and the carbon nanotube. In Section 10.3, the radius of the fixed part of the graphene sheet is 3.55 Å, and the fixed part of the tube radius is taken to be 2.105 Å. It is a little more difficult to specify the arc length ℓ for this comparison as in the least squares study, the joint comprises three straight sections of lengths 0.71 Å, 1.42 Å, and 0.735 Å. These sections add up to a total length of 2.865 Å. However, we cannot simply use this as a value for ℓ because it does not account for the curvature inherent in the variational solution. Therefore we assume that the joint curve will be approximately a quarter circle, and we also assume that the longest straight section (1.42 Å) subtends an angle of $\pi/4$ radians at the centre of the circle. In this way, we may derive a more comparable arc length $\ell \approx 2.914$ Å. Using these values we derive a value $\mu = 0.4959$ and an attachment height $y_0 = 2.16$ Å which compares reasonably well with the figure for the least squares approach, which is $y_0 = 2.315$ Å, as it differs by only 0.155 Å or 6.7%.

WORKED EXAMPLE 11.3

Assuming that the length of the longest straight is 1.24 Å and subtends an angle of $\pi/4$ radians at the centre of the circle, show that the arc length is $\ell \approx 2.914$ Å.

Solution

From the cosine rule, $a^2 = b^2 + c^2 - 2bc \cos A$ and from Fig. 11.4, we may deduce

$$\sigma^2 = r^2 + r^2 - 2r^2 \cos(\pi/4) = 2r^2(1 - \cos(\pi/4)).$$

Now, we apply the trigonometric identity $\cos 2\theta = 1 - 2\sin^2 \theta$ to obtain

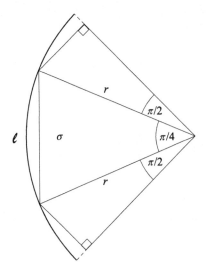

FIG. 11.4

Diagram illustrating the method for calculating arc length ℓ.

$$r = \frac{\sigma}{2\sin(\pi/8)}.$$

In this example, we assume that $\sigma = 1.42\,\text{Å}$ and the arc length ℓ is approximately a quarter circle. Therefore we may deduce

$$\ell = \frac{1.42\pi}{4\sin(\pi/8)} = 2.914\text{Å}.$$

In Fig. 11.5 we graph the joint shapes predicted by the variational and least squares methods. Also shown in the figure are the assumed atomic locations for the variational approach, which are denoted by squares that are joined by straight lines for comparison purposes. It can been seen from this figure that the y_0 height is similar in both cases, but the atomic locations which participate in the joint as compared to those predicted by the least squares method are not so similar (see Table 11.2). The third column of Table 11.2 is the absolute distance between the two predictions, while the fourth column gives the corresponding parameter value for ϕ. Despite this apparent discrepancy, the absolute error for the participating atomic positions is still within $0.18\,\text{Å}$ of that predicted by the least squares method. For larger radii nanotubes, we expect the agreement of the continuum and discrete approaches to be improved, therefore the continuum method gives a readily obtained reliable estimate.

11.4 NANOBUDS

Nanobuds arise by combining a single-walled carbon nanotube and a fullerene, and they are believed to possess many advantageous physical and electrical properties. Nanobuds were first experimentally

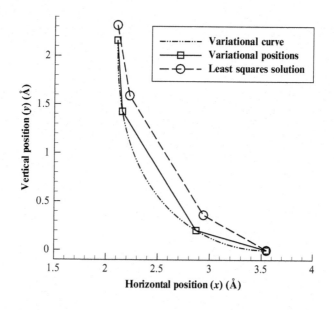

FIG. 11.5

Comparison of the joint shape for a (6, 0) tube joining graphene using the variational and least squares methods.

Table 11.2 Atomic Positions for the Variational Method Compared With the Least Squares Procedure for a (6, 0) to Graphene Joint

Least Squares		Variational		Error	Parameter
x	y	x	y	(Å)	Value ϕ
3.55	0	3.55	0	0	0
2.94	0.363	2.87	0.204	0.174	−0.40979250
2.225	1.59	2.155	1.427	0.177	−1.13494449
2.105	2.315	2.105	2.16	0.155	−1.44517978

observed in a ferrocene–carbon monoxide system by Nasibulin et al. (2007) using transmission electron microscopy. In their experiment the diameter of the fullerenes is in the range of 0.39–0.98 nm, where the most frequently occurring diameter is 0.70 nm which is the C_{60} fullerene. However, we note that a dynamic process involving a combination of these two carbon nanostructures was first shown by Zhao et al. (2003). They consider a fullerene encapsulated inside the nanotube, and a nanobud forming during this process. Nanobuds are believed to be promising field emission devices since the off-plane nature of the fullerenes on the carbon nanotubes increases the surface area.

In this section, we employ the continuous approach to connect a C_{60} fullerene and a carbon nanotube. Variational calculus is utilised to determine the curve adopted by a line smoothly connecting a circular arc defect plane with a vertical carbon nanotube, such that the arc length of the curve and the

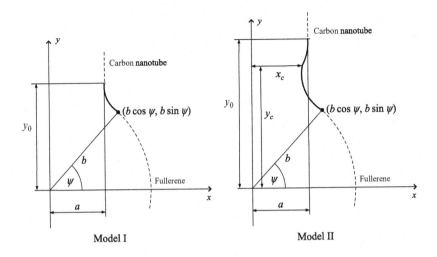

Model I   Model II

FIG. 11.6

Geometries for Model I, for which the joint contains only a positive curvature, and Model II, which has both positive and negative curvatures in the joint region.

size of the defect on the fullerene are specified. However, the distance in the y-direction, which is y_0 of the joint to the cylindrical tube, is not prescribed but rather determined as part of the solution.

Two distinct models are determined here, which are illustrated in Fig. 11.6. In terms of Model I, the joint curvature remains positive, whereas there are two joint regions for Model II, one of positive curvature and one of negative curvature. We position a C_{60} fullerene of radius b in the (x, z)-plane centred at the origin. A carbon nanotube of radius a is located with its axis colinear with the y-axis starting from an unknown positive distance above the (x, z)-plane which we denote by y_0. As the defect and the nanotube are rotationally symmetric about the y-axis, we can consider this to be a problem in the two dimensional (x, y)-plane. The total prescribed arc length ℓ is assumed to connect the defect at $(b \cos \psi, b \sin \psi)$ and the tube at (a, y_0).

By imposing the continuity of the function y and its derivative, boundary conditions at the joint to the C_{60} fullerene are obtained, namely

$$y(b \cos \psi) = b \sin \psi, \quad y'(b \cos \psi) = - \cot \psi.$$

For the carbon nanotube, y_0 is unknown, so we require the natural or alternative boundary condition (11.3). In Model I, the value of y' ranges from $- \cot \psi$ at $x = b \cos \psi$ to $-\infty$ at $x = a$, and therefore the boundary condition is $y'(a) = -\infty$. In Model II, y' ranges from $- \cot \psi$ to $-\infty$, where it changes sign and then ranges from ∞ down to some finite positive value before returning to ∞. Therefore in Model II, the boundary condition in $y'(a) = \infty$.

11.4.1 MODEL I: POSITIVE CURVATURE

From Fig. 11.6 the join curvature of Model I remains positive along an arc length ℓ, therefore only the positive case of Eq. (11.6) is considered. By employing the same technique as described in

Section 11.3.1 and using the boundary condition at the point $(b \cos \psi, b \sin \psi)$ of attachment to the fullerene to determine the constant of the integration, we obtain

$$y(\phi) = 2\beta k(\cos \phi_f - \cos \phi) + b \sin \psi, \tag{11.29}$$

$$x(\phi) = b \cos \psi - \beta \{2[E(\phi, k) - E(\phi_f, k)] - [F(\phi, k) - F(\phi_f, k)]\}, \tag{11.30}$$

where ψ is an initial angle of the attachment to the C_{60} fullerene and $\phi_f = \sin^{-1}(\sqrt{(1 - \sin \psi)/2k^2})$ corresponds to $\theta = \psi - \pi/2$ at the point $(b \cos \psi, b \sin \psi)$. By using the boundary condition at the point (a, y_0) of the open end of the tube with $\phi_t = \sin^{-1}(1/\sqrt{2}k)$, where $\theta = -\pi/2$ we may deduce

$$y_0 = 2\beta k(\cos \phi_f - \cos \phi_t) + b \sin \psi, \tag{11.31}$$

$$b \cos \psi - a = \beta \{2[E(\phi_t, k) - E(\phi_f, k)] - [F(\phi_t, k) - F(\phi_f, k)]\}. \tag{11.32}$$

The derivation of Eqs (11.31), (11.32) can be found in Exercise 4 at the end of this chapter.

WORKED EXAMPLE 11.4

Show that $\phi_f = \sin^{-1}(\sqrt{(1 - \sin \psi)/2k^2})$ and $\phi_t = \sin^{-1}(1/\sqrt{2}k)$.

Solution

We know that the boundary conditions are given by

$$y'(b \cos \psi) = -\cot \psi, \quad y'(y_0) = -\infty.$$

We then make the substitution $y' = \tan \theta$, which leads to

$$\theta(b \sin \psi) = \psi - \pi/2, \quad \theta(y_0) = -\pi/2.$$

From Eq. (11.11), the variable θ has been transformed to be the parametric variable ϕ, hence we may deduce

$$\phi(b \sin \psi) = \sin^{-1}\left(\sqrt{\frac{1 - \sin \psi}{2k^2}}\right), \quad \phi(y_0) = \sin^{-1}\left(\frac{1}{k\sqrt{2}}\right).$$

From the arc length constraint, we obtain

$$\ell = \int_a^{b \cos \psi} (1 + y'^2)^{1/2} dx.$$

By making the substitution $y' = \tan \theta$ and changing to the parameter ϕ, we may deduce

$$\ell = \beta[F(\phi_t, k) - F(\phi_f, k)]. \tag{11.33}$$

By substituting Eq. (11.33) into Eq. (11.32), we may derive

$$\mu = 2\left(\frac{E(\phi_t, k) - E(\phi_f, k)}{F(\phi_t, k) - F(\phi_f, k)}\right) - 1, \tag{11.34}$$

where $\mu = (b \cos \psi - a)/\ell$. Thus, for a prescribed a, b, ψ and ℓ, Eq. (11.34) can be numerically solved to determine the value for k. By substitution of k back into Eq. (11.33), the value of β is determined, and therefore y_0 can be obtained from Eq. (11.31).

11.4.2 MODEL II: POSITIVE AND NEGATIVE CURVATURES

In this section, we proceed exactly as in Section 11.4.1 for the first region of positive curvature from the point of attachment to the C_{60} fullerene $(b \cos \psi, b \sin \psi)$ up until the critical point (x_c, y_c) where the curvature changes sign. We then consider the second region from the critical point (x_c, y_c) to the point of attachment to the carbon nanotube (a, y_0), throughout which the curvature is negative. We note that the solution for the case when the curvature changes in sign from negative to positive is also included in this analysis and it will be noted briefly later in the text.

The same procedure is applied to the region of positive curvature, and this region is bounded by the point where the curvature $\kappa = 0$ and where $\theta_c = -\cos^{-1}(\lambda/\alpha)$. Employing the parameter variable ϕ as defined by Eq. (11.11), we have $\phi = -\pi/2$ when $\theta = \theta_c$, and from Eqs (11.29), (11.30), the parametric equations for y_c and x_c may be obtained as

$$y_c = 2\beta k \cos \phi_f + b \sin \psi, \tag{11.35}$$

$$x_c = b \cos \psi - \beta\{2[E(k) - E(\phi_f, k)] - [K(k) - F(\phi_f, k)]\}. \tag{11.36}$$

In the second region, we take the negative sign in Eq. (11.6) and from similar considerations described in Section 11.3.2, we may deduce

$$y(\phi) = 2\beta k(\cos \phi_f + \cos \phi) + b \sin \psi,$$

where the constant of integration arises from the condition that $y = y_c$ when $\phi = -\pi/2$, and then we use the expression (11.35) for y_c. From the boundary condition at the point of attachment to the carbon nanotube, we know that $\phi = \phi_t$ at the point (a, y_0), so that y_0 is given by

$$y_0 = 2\beta k(\cos \phi_f + \cos \phi_t) + b \sin \psi.$$

We note that the solution for y_0 in this model differs only by a change in the sign of one term. Similarly, by taking the negative sign of Eq. (11.6) and solving for the parametric form of x with the expression (11.36) for x_c, we may deduce

$$x(\phi) = b \cos \psi - \beta\{2[2E(k) - E(\phi, k) - E(\phi_f, k)] - [2K(k) - F(\phi, k) - F(\phi_f, k)]\}.$$

At the point (a, y_0), where $\phi = \phi_t$ we have

$$b \cos \psi - a = \beta\{2[2E(k) - E(\phi_t, k) - E(\phi_f, k)] - [2K(k) - F(\phi_t, k) - F(\phi_f, k)]\}.$$

The arc length constraint is obtained from the two regions, so we have

$$\ell = \int_{b \cos \psi}^{x_c} (1 + y'^2)^{1/2} dx + \int_{x_c}^{a} (1 + y'^2)^{1/2} dx.$$

Again, by making the substitution $y' = \tan \theta$ and changing to the parameter ϕ, we may deduce

$$\ell = \beta[2K(k) - F(\phi_t, k) - F(\phi_f, k)]. \tag{11.37}$$

We note that the derivation of Eq. (11.37) can be found in Exercise 5 at the end of this chapter. As before, the equation for $\mu = (b \cos \psi - a)/\ell$ is obtained by straightforward algebra, and it can be written as

$$\mu = 2 \left(\frac{2E(k) - E(\phi_t, k) - E(\phi_f, k)}{2K(k) - F(\phi_t, k) - F(\phi_f, k)} \right) - 1. \tag{11.38}$$

We observe that Eq. (11.38) only involves the single unknown k, and can be numerically solved for some prescribed μ to determine the value for k. By further substitution into Eq. (11.37), the value for β can be obtained, and therefore the attachment height y_0 can be calculated.

We observe that Eq. (11.34) of Model I coincides with Eq. (11.38) of Model II for the values $k = 1/\sqrt{2}$ and $k = [(1 - \sin \psi)/2]^{1/2}$. When $k = 1/\sqrt{2}$, the value of μ is denoted by μ_1, and it is given by

$$\mu_1 = 2 \left(\frac{E(1/\sqrt{2}) - E(\sin^{-1}(\sqrt{1 - \sin \psi}), 1/\sqrt{2})}{K(1/\sqrt{2}) - F(\sin^{-1}(\sqrt{1 - \sin \psi}), 1/\sqrt{2})} \right) - 1.$$

When $k = [(1 - \sin \psi)/2]^{1/2}$, the value of μ is denoted by μ_2, and it is given by

$$\mu_2 = 2 \left(\frac{E(\sqrt{(1 - \sin \psi)/2}) - E(\sin^{-1}(1/\sqrt{1 - \sin \psi}), \sqrt{(1 - \sin \psi)/2})}{K(\sqrt{(1 - \sin \psi)/2}) - F(\sin^{-1}(1/\sqrt{1 - \sin \psi}), \sqrt{(1 - \sin \psi)/2})} \right) - 1.$$

A detailed numerical analysis of these equations is given in the following section.

11.4.3 RESULTS AND DISCUSSION

Some general features of the solutions for Model I and Model II are examined in this section. Firstly, we examine the solution which is characterised by the nondimensional parameter $\mu = (b \cos \psi - a)/\ell$ subject to the constraint $-1 < \mu < 1$. In Fig. 11.7, we show the relationship between the parameter μ and a new parameter $B = 1/k^2$, and this graph can be divided into two main regions.

The first region is $\mu < \mu_0$, which corresponds to $B < 2$, and the second region is $\mu > \mu_0$, which corresponds to $B > 2/(1 - \sin \psi)$, where μ_0 is the asymptotic value for μ as k tends to zero. The value of μ_0 is analytically determined in Worked Example 11.5, where it is shown to be

$$\mu_0 = 1 + \frac{\sqrt{2}(1 - \sqrt{2} \cos \omega)}{\ln[(\sqrt{2} - 1)/\tan(\omega/2)]}, \tag{11.39}$$

where $\omega = \pi/4 - \psi/2$. We note that these two values of B arise when Eqs (11.34), (11.38) coincide, as mentioned earlier.

WORKED EXAMPLE 11.5

Derive the asymptotic value μ_0 as given by Eq. (11.39).

Solution

This formula is not immediately apparent from Eq. (11.34) and follows only after rearrangement of Eq. (11.34) to the particular form Eq. (11.42) given below. Firstly, we consider the usual Legendre incomplete elliptic integral of the first

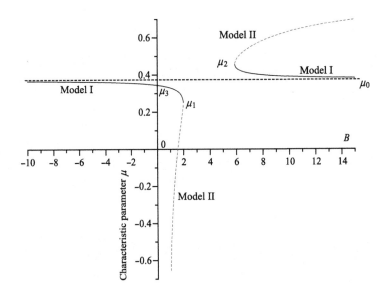

FIG. 11.7

Relation between the parameters $\mu = (b\cos\psi - a)/\ell$ and $B = 1/k^2$.

kind $F(\phi, k)$, as defined by Byrd and Friedman (1971), and for the value of ϕ when $\phi \in (\phi_t, \phi_f)$, we may deduce

$$F(\phi_t, k) - F(\phi_f, k) = \int_{\phi_f}^{\phi_t} \frac{d\phi}{\sqrt{1 - k^2 \sin^2\phi}},$$

where $\phi_t = \sin^{-1}(1/\sqrt{2}k)$ and $\phi_f = \sin^{-1}(\sqrt{(1-\sin\psi)/2k^2})$, and ψ is the initial angle of the attachment to the C_{60} fullerene. By making the substitution $k\sin\phi = \sin\lambda$, we may derive

$$F(\phi_t, k) - F(\phi_f, k) = \int_{\pi/4-\psi/2}^{\pi/4} \frac{d\lambda}{\sqrt{k^2 - \sin^2\lambda}}. \tag{11.40}$$

By precisely the same considerations for the incomplete elliptic integral of the second kind $E(\phi, k)$ for $\phi \in (\phi_t, \phi_f)$, we obtain

$$E(\phi_t, k) - E(\phi_f, k) = \int_{\pi/4-\psi/2}^{\pi/4} \frac{\cos^2\lambda}{\sqrt{k^2 - \sin^2\lambda}} d\lambda. \tag{11.41}$$

By substitution of Eqs (11.40), (11.41) into Eq. (11.34), we may derive

$$\mu = \left(2\int_{\pi/4-\psi/2}^{\pi/4} \frac{\cos^2\lambda \, d\lambda}{\sqrt{k^2 - \sin^2\lambda}} - \int_{\pi/4-\psi/2}^{\pi/4} \frac{d\lambda}{\sqrt{k^2 - \sin^2\lambda}}\right) \Big/ \int_{\pi/4-\psi/2}^{\pi/4} \frac{d\lambda}{\sqrt{k^2 - \sin^2\lambda}}. \tag{11.42}$$

For μ_0, which is the asymptotic value as k tends to zero, by using $\cos 2\theta = 2\cos^2\theta - 1 = 1 - 2\sin^2\theta$, Eq. (11.42) can be formally reduced, and μ_0 becomes

$$\mu_0 = 1 - 2\int_\omega^{\pi/4} \sin\lambda \, d\lambda \Big/ \int_\omega^{\pi/4} \frac{d\lambda}{\sin\lambda},$$

where $\omega = \pi/4 - \psi/2$. By evaluating the above equation, we obtain

$$\mu_0 = 1 + \frac{\sqrt{2}(1 - \sqrt{2}\cos\omega)}{\ln[(\sqrt{2} - 1)/\tan(\omega/2)]}.$$

The region of $\mu < \mu_0$ can be divided into three subregions. The first subregion is $\mu_3 < \mu < \mu_0$, where μ_3 is the asymptotic value of μ as k tends to infinity. We find that the solution asymptotes with the line $\mu = \mu_0$ and crosses the vertical axis at the point μ_3. The value of μ_3 is analytically determined in Worked Example 11.6 where it is found to be

$$\mu_3 = (1 - \cos\psi)/\psi. \tag{11.43}$$

In this case, the parameter B is negative, which corresponds to a negative value of α and an imaginary modulus k for the elliptic functions. The second subregion exists for $\mu_1 < \mu < \mu_3$, which corresponds to $0 < B \leq 2$, so that α is always positive and the modulus k is strictly real. The final subregion, which is obtained from Model II, applies for the range $-1 < \mu < \mu_1$ which corresponds to $1 < B \leq 2$, and again, α is always positive.

WORKED EXAMPLE 11.6

Derive the asymptotic value μ_3 as given by Eq. (11.43).

Solution

Following Worked Example 11.5, the equation for μ_3 is determined. μ_3 is the asymptotic value as k tends to infinity, we have from Eq. (11.42)

$$\mu_3 = \left(2\int_\omega^{\pi/4} \cos^2\lambda\, d\lambda - \int_\omega^{\pi/4} d\lambda\right) \Big/ \int_\omega^{\pi/4} d\lambda,$$

and from which we may deduce

$$\mu_3 = \frac{(1 - \cos\psi)}{\psi}.$$

The region of $\mu > \mu_0$ can be divided into two subregions, which are $\mu_0 < \mu < \mu_2$ from Model I and $\mu_2 < \mu < 1$ from Model II. The value of B is always positive in both of these subregions, and it corresponds to a negative value of α and a complex value of angle ϕ of the form $\phi = -\pi/2 + i\varphi$. We note that in this region the curvature changes in sign from negative to positive.

We also note that the values μ given in Eqs (11.34), (11.38) depend on the initial angle ψ of a C_{60} fullerene, as depicted in Fig. 11.8. For the case of $\psi = 0$, a hemispherical fullerene, the gradients at the attachment points of the fullerene and the carbon nanotube are ∞ ($-\infty$), therefore the curvature of the arc has to change sign, and consequently, only Model II is applicable. This is shown in Fig. 11.8A, where μ_0 becomes zero, and μ_1 and μ_2 meet at $B = 2$. In the case of $0 < \psi < \pi/2$, the solution for μ can occur in both of the two main regions of Model I and Model II as described previously. However, in the second region, $\mu > \mu_0$ moves and goes to infinity as ψ tends to $\pi/2$, as shown in Fig. 11.8B and C. Finally, for the extreme case where $\psi = \pi/2$, only the first region $\mu < \mu_0$ can be found with $\mu_0 = 1$. The point of attachment at the C_{60} fullerene has a gradient of ∞ and the solution is similar to the case

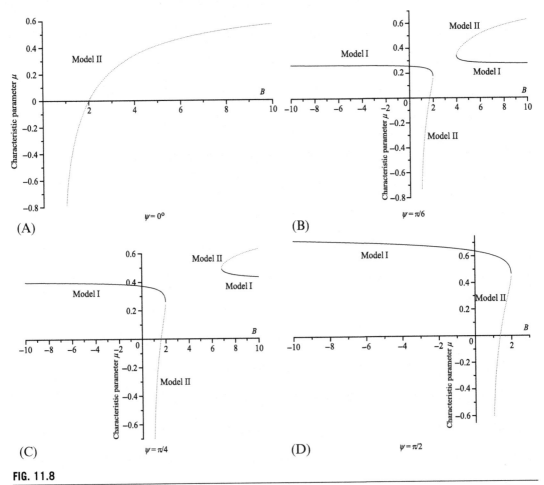

FIG. 11.8

Relation between the parameters μ and B when (A) $\psi = 0$, (B) $\psi = \pi/6$, (c) $\psi = \pi/4$, and (D) $\psi = \pi/2$.

of joining a carbon nanotube to a flat graphene sheet, as described in Section 11.3. We now apply the solution of the continuous approach to a nondimensionalised situation, which is shown in Fig. 11.9. In this case, the fixed arc length is assumed to be $\ell = 1$, an initial angle of the fullerene is assumed to be $\psi = \pi/6$ and the values of B are taken from the graph shown in Fig. 11.8B. We choose five values of B for the five possible cases of Model I and Model II.

In order to compare the continuous model with the discrete model as described in Section 10.4, we consider two cases of nanobuds, which are the joining of the defect DA on the fullerene with the $(5, 0)$ nanotube and the joining of the defect DB on the fullerene with the $(6, 0)$ nanotube. All parameter values for the C_{60} fullerene for the two distinct defects are presented in Table 11.3. The radii of the $(5, 0)$ and the $(6, 0)$ carbon nanotubes are taken from Chapter 9.

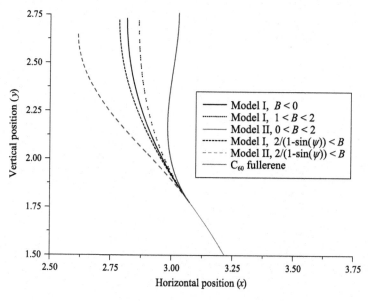

FIG. 11.9

Plots of joints $y = y(x)$ for different values of B belonging to the shown ranges, and ψ is assumed to be $\pi/6$.

Table 11.3 Parameter Values		
Constants	**(5, 0) Nanobud**	**(6, 0) Nanobud**
Radii of tubes, a (Å)	2.055	2.430
Radius of C_{60}, b (Å)	3.55	3.55
Initial angles, ψ (°)	30.99	29.61
μ_0	0.27945	0.26676
μ_1	0.17755	0.16987
μ_2	0.60418	0.58301
μ_3	0.26392	0.25272
ℓ (Å)	2.84867	2.84792
μ	0.34690	0.23045

There is no unique procedure for this comparison because we are comparing an essentially discrete and asymmetric model with a continuous axially symmetric one. In particular, it is difficult to specify the arc length ℓ for the comparison because in the discrete approximation, the joint comprises three straight sections. However, we cannot simply use the sum of these lengths as a value for ℓ because it does not account for the curvature inherent in the variational solution. Therefore we assume that the joint curve may be approximated by an arc of a circle, and the longest straight section σ is assumed to subtend an angle of $\psi/2$ at the centre of such a circle, as illustrated in Fig. 11.10. We may then derive

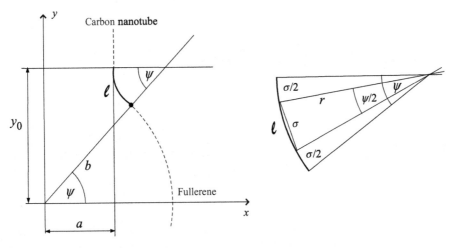

FIG. 11.10

Diagram illustrating the method for calculating arc length ℓ.

the arc length ℓ using the cosine rule, which is found to be given by

$$\ell = \frac{\sigma \psi}{2 \sin(\psi/4)},$$

where ψ is the initial angle of the C_{60} fullerene and σ is assumed to be 1.42 Å. The numerical values for ℓ and the corresponding values for μ are shown in Table 11.3. We find that in the case of the defect DA and the $(5,0)$ tube, Model I applies with $\mu > \mu_0$, and in the case of the defect DB and the $(6,0)$ tube, Model I applies but with $\mu_1 < \mu < \mu_3$.

In Fig. 11.11, we graph the joint shapes predicted by the two discrete approaches and the continuous approach.

Also shown in the figure for comparison purposes are the assumed atomic locations for the continuous approach, which are denoted by triangles that are joined by straight lines. It can be seen from this figure that the atom locations of the nanobud formed from the $(5,0)$ tube and the defect DA do not quite match for each method; this can be explained by the curvature effect of the tube. Furthermore, four atom locations at the connection site for the two nanobuds and for the three approximations are given in Table 11.4. The term 'maximum difference' refers to the difference between the greatest and the least values at each location. We find that the overall percentage differences are less than 10% for all cases.

11.5 NANOPEANUTS

The structure consisting of two connected fullerenes, namely a peanut-like nanostructure, are hollow-sphere structures which have the potential to encapsulate other molecules. The coalescence of two fullerenes was first reported by Ueno et al. (1998), who employed the Stone–Wales calculation and

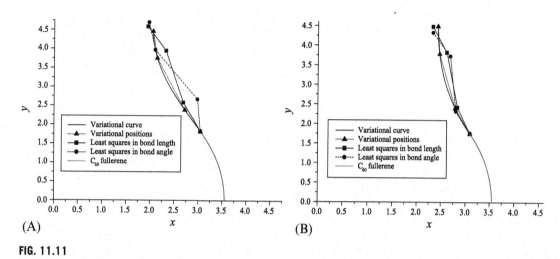

FIG. 11.11

Comparisons of the joint shape for (A) (5, 0) tube joining with defect DA and (B) (6, 0) tube joining with defect DB on the C_{60} fullerenes.

Table 11.4 Atomic Positions for Two Discrete Approaches Compared With the Continuous Approach for Two Nanobuds

	x_1	y_1	x_2	y_2	x_3	y_3	x_4	y_4
Defect DA—(5, 0) tube								
Least squares in bond length	3.043	1.828	2.685	2.581	2.322	3.949	1.948	4.584
Least squares in bond angle	3.043	1.828	2.980	2.668	2.100	3.971	1.971	4.700
Average discrete model	3.043	1.828	2.833	2.625	2.211	3.960	1.960	4.642
Continuous model	3.043	1.828	2.716	2.387	2.143	3.759	2.055	4.461
Maximum difference (Å)	0	0	0.295	0.281	0.222	0.212	0.107	0.239
Percentage difference (%)	0	0	4.11	9.05	3.08	5.15	4.87	3.90
Defect DB—(6, 0) tube								
Least squares in bond length	3.086	1.752	2.831	2.421	2.612	3.823	2.330	4.482
Least squares in bond angle	3.086	1.752	2.798	2.320	2.682	3.729	2.330	4.327
Average discrete model	3.056	1.752	2.818	2.371	2.647	3.776	2.330	4.405
Continuous model	3.086	1.754	2.785	2.397	2.465	3.789	2.430	4.493
Maximum difference (Å)	0	0.002	0.046	0.101	0.217	0.094	0.100	0.166
Percentage difference (%)	0	0.11	1.05	1.12	6.88	0.34	4.29	2.01

found that any two C_{60} fullerenes can rearrange themselves to form a C_{120} molecule. Other authors use molecular dynamics simulations to investigate the processes of dimerisation and fusion reactions induced by two C_{60} molecules, finding that dumbbell-shaped molecules can be formed at low collision energies, and when the energy is high enough, a large C_{120} can be obtained. Finding the joint between two buckyballs is a challenging task because computational calculations can hardly go beyond several hundred steps. Therefore in order to solve this difficult problem an applied mathematical approach is needed.

Here, we employ the continuous method to connect two fullerenes based on the variational calculus to determine the shape adopted by a surface smoothly connecting two arbitrary circular arc defect planes, such that the size of the two defects on the two fullerenes are specified. However, the distance between the two fullerene centres in the y-direction, y_0 is not prescribed and it is determined as part of the solution. The fullerenes are modelled as spheres and the defects on both fullerenes are assumed to be rotationally symmetric and axially positioned, so that the problem can be reduced to two dimensions.

Again, two distinct models are determined here, as illustrated in Fig. 11.12. In Model I, the joint curvature remains positive throughout, whereas in Model II there are two regions, one of positive curvature and one of negative curvature. Again, we comment that the two models are subsequently shown to be part of the same overlapping parameter space. To represent the first fullerene, we position a circle of radius a centred at the origin. The other fullerene is represented by a circle of radius b located on the y-axis at a positive distance y_0 from the origin. The defects are assumed to be symmetric about the y-axis and defined by an angle ψ_1 anticlockwise from the x-axis for the first fullerene, and at an angle ψ_2 anticlockwise from the x-axis for the second fullerene. A total prescribed arc length ℓ is assumed to connect the defect at $(a \cos \psi_1, a \sin \psi_1)$ to the defect at $(b \cos \psi_2, y_0 + b \sin \psi_2)$.

By imposing the continuity of the function y and its derivative, boundary conditions at the joint to the fullerene located at the origin are obtained and given by

$$y(a \cos \psi_1) = a \sin \psi_1, \quad y'(a \cos \psi_1) = - \cot \psi_1.$$

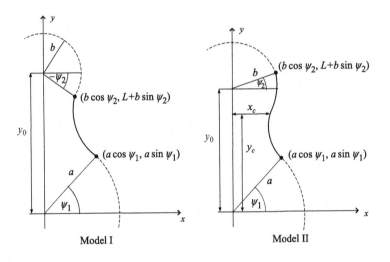

FIG. 11.12

Geometries for Model I, for which the joint contains only a positive curvature, and Model II, which has both positive and negative curvatures in the joint region.

For the other fullerene, y_0 is unknown, and therefore we require the natural or alternative boundary condition, which is derived from the first term in Eq. (11.1). In Model I, y' ranges in value from $y' = -\cot\psi_1$ at $x = a\cos\psi_1$ to $y' = -\cot\psi_2$ at $x = b\cos\psi_2$, and therefore the boundary condition is $y'(b\cos\psi_2) = -\cot\psi_2$. In Model II, y' ranges in value from $y' = -\cot\psi_1$ to $y' = -\infty$, where it changes sign and then ranges from $y' = \infty$ down to some finite positive value before returning to $y' = -\cot\psi_2$ at $x = b\cos\psi_2$. Therefore in Model II, the boundary condition in $y'(b\cos\psi_2) = -\cot\psi_2$.

11.5.1 MODEL I: POSITIVE CURVATURE

As can be seen in Fig. 11.12, the joint curvature of Model I remains positive throughout the arc length ℓ, and therefore only the positive case of Eq. (11.6) need be considered. By employing the same technique as described in Section 11.3.1 and using the boundary condition at the point $(a\cos\psi_1, a\sin\psi_1)$ to determine the constant of the integration, we obtain

$$y(\phi) = 2\beta k(\cos\phi_1 - \cos\phi) + a\sin\psi_1, \tag{11.44}$$

$$x(\phi) = a\cos\psi_1 - \beta\{2[E(\phi,k) - E(\phi_1,k)] - [F(\phi,k) - F(\phi_1,k)]\}, \tag{11.45}$$

where ψ_1 is an initial angle of the attachment to the fullerene, which is located at the origin, and $\phi_1 = \sin^{-1}(\sqrt{(1-\sin\psi_1)/2k^2})$. Using the boundary condition at the point $(b\cos\psi_2, y_0 + b\sin\psi_2)$ with $\phi_2 = \sin^{-1}(\sqrt{(1-\sin\psi_2)/2k^2})$, the point attached to the second fullerene can be obtained as

$$y_0 = 2\beta k(\cos\phi_1 - \cos\phi_2) + a\sin\psi_1 - b\sin\psi_2, \tag{11.46}$$

$$a\cos\psi_1 - b\cos\psi_2 = \beta\{2[E(\phi_2,k) - E(\phi_1,k)] - [F(\phi_2,k) - F(\phi_1,k)]\}, \tag{11.47}$$

where $F(\phi,k)$ and $E(\phi,k)$ denote the usual Legendre incomplete elliptic integrals of the first and second kinds, respectively, as defined by Byrd and Friedman and in Section 2.10.

From the arc length constraint, we obtain

$$\ell = \int_{b\cos\psi_2}^{a\cos\psi_1} (1 + y'^2)^{1/2}dx,$$

and making the substitution $y' = \tan\theta$ and changing to the new parameter ϕ, we may deduce

$$\ell = \beta[F(\phi_2,k) - F(\phi_1,k)]. \tag{11.48}$$

By substituting Eq. (11.48) into Eq. (11.47), we may derive

$$\mu = 2\left(\frac{E(\phi_2,k) - E(\phi_1,k)}{F(\phi_2,k) - F(\phi_1,k)}\right) - 1, \tag{11.49}$$

where $\mu = (a\cos\psi_1 - b\cos\psi_2)/\ell$. Thus for a prescribed a, b, ψ_1, ψ_2, and ℓ, Eq. (11.49) can be numerically solved to determine the value for k. By substituting k back into Eq. (11.48), the value of β may be determined and therefore y_0 can be obtained from Eq. (11.46).

11.5.2 MODEL II: POSITIVE AND NEGATIVE CURVATURES

In this section, we proceed exactly as in Section 11.5.1 for the first region of positive curvature, from the point of attachment to the fullerene located at the origin $(a\cos\psi_1, a\sin\psi_2)$ up until the critical point (x_c, y_c), where the curvature changes sign. We then consider the second region, from the critical point (x_c, y_c) to the point of attachment to the other fullerene $(b \cos \psi_2, y_0 + b \sin \psi_2)$, throughout which the curvature is negative. The same procedure as used in the previous section is applied to the region of positive curvature, and this region is bounded by the point where the curvature $\kappa = 0$ and $\theta_c = \cos^{-1}(-\lambda/\alpha)$. Employing the parameter variable ϕ as defined by Eq. (11.11), we have $\phi = -\pi/2$ when $\theta = \theta_c$, and from Eqs (11.44), (11.45), the parametric equations for y_c and x_c may be obtained and are given by

$$y_c = 2\beta k \cos \phi_1 + a \sin \psi_1, \tag{11.50}$$

$$x_c = a \cos \psi_1 - \beta\{2[E(k) - E(\phi_1, k)] - [K(k) - F(\phi_1, k)]\}, \tag{11.51}$$

where β, k are defined in Section 11.3.1 and ϕ_1 is defined in Section 11.5.1, $F(\phi_f, k)$ and $E(\phi_f, k)$ are the usual incomplete elliptic integrals of the first and second kinds, respectively, and $K(k)$ and $E(k)$ are complete elliptic integrals of the first and second kinds, respectively.

In the second region, we take the negative sign in Eq. (11.6), and from similar considerations described in Section 11.5.1, we may deduce

$$y(\phi) = 2\beta k(\cos \phi_1 + \cos \phi) + a \sin \psi_1, \tag{11.52}$$

where the constant of integration arises from the condition that $y = y_c$ when $\phi = -\pi/2$, we use the expression (11.50) for y_c. From the boundary condition at $(b \cos \psi_2, y_0 + b \sin \psi_2)$ on the second fullerene, we know that $\phi = \phi_2$ at this boundary, and therefore y_0 is given by

$$y_0 = 2\beta k(\cos \phi_1 + \cos \phi_2) + a \sin \psi_1 - b \sin \psi_2.$$

We note that this solution for y_0 differs from that in Eq. (11.46) only by a change in the sign of one term. Similarly, by taking the negative sign of Eq. (11.6) and solving for the parametric form of x with the expression (11.51) for x_c, we may deduce

$$x(\phi) = a \cos \psi_1 - \beta \{2[2E(k) - E(\phi, k) - E(\phi_1, k)] \tag{11.53}$$
$$-[2K(k) - F(\phi, k) - F(\phi_1, k)]\}.$$

At the point $(b \cos \psi_2, y_0 + b \sin \psi_2)$ where $\phi = \phi_2$, we have

$$a \cos \psi_1 - b \cos \psi_2 = \beta \{2[2E(k) - E(\phi_2, k) - E(\phi_1, k)]$$
$$-[2K(k) - F(\phi_2, k) - F(\phi_1, k)]\}.$$

The arc length constraint is obtained from the two regions and we have

$$\ell = \int_{a \cos \psi_1}^{x_c} (1 + y'^2)^{1/2} dx + \int_{x_c}^{b \cos \psi_2} (1 + y'^2)^{1/2} dx,$$

and again, making the substitution $y' = \tan \theta$ and changing to the parameter ϕ, we may deduce

$$\ell = \beta[2K(k) - F(\phi_2, k) - F(\phi_1, k)]. \tag{11.54}$$

As before, the equation for $\mu = (a\cos\psi_1 - b\cos\psi_2)/\ell$ is obtained by straightforward algebra and is given by

$$\mu = 2\left(\frac{2E(k) - E(\phi_2, k) - E(\phi_1, k)}{2K(k) - F(\phi_2, k) - F(\phi_1, k)}\right) - 1. \tag{11.55}$$

We observe that Eq. (11.55) only involves the single unknown k and can be numerically solved for some prescribed μ to determine the value for k. By further substitution into Eq. (11.54), the value for β can be obtained, and therefore, the attachment height y_0 may be calculated.

We also note that Eq. (11.49) of Model I coincides with Eq. (11.55) of Model II for the values $k = [(1 - \sin\psi_1)/2]^{1/2}$ and $k = [(1 - \sin\psi_2)/2]^{1/2}$. When $k = [(1 - \sin\psi_1)/2]^{1/2}$, we denote $k = k_1$, and $\mu = \mu_1$, so we may write

$$\mu_1 = 2\left(\frac{E(k_1) - E(\phi_3, k_1)}{K(k_1) - F(\phi_3, k_1)}\right) - 1,$$

where ϕ_3 is defined by $\phi_3 = \sin^{-1}(\sqrt{(1 - \sin\psi_2)/(1 - \sin\psi_1)})$. When $k = [(1 - \sin\psi_2)/2]^{1/2}$, we denote $k = k_2$ and $\mu = \mu_2$, so

$$\mu_2 = 2\left(\frac{E(k_2) - E(\phi_4, k_2)}{K(k_2) - F(\phi_4, k_2)}\right) - 1,$$

where ϕ_4 is defined by $\phi_4 = \sin^{-1}(\sqrt{(1 - \sin\psi_1)/(1 - \sin\psi_2)})$. We note that $\sin\phi_3 = \csc\phi_4$. A detailed numerical analysis of these equations is given in the following section.

11.5.3 RESULTS AND DISCUSSION

Some general features of the solutions for Model I and Model II are examined in this section. Firstly, the solution, which is characterised by the nondimensional parameter $\mu = (a\cos\psi_1 - b\cos\psi_2)/\ell$ subject to the constraint $-1 < \mu < 1$, is illustrated in Fig. 11.13, where we show the relationship between the parameter μ and a new parameter $B = 1/k^2$.

Fig. 11.13A can be divided into two main regions. The first region is $\mu < \mu_0$, which corresponds to $B < 2/(1 - \sin\psi_2)$, and the second region is $\mu > \mu_0$, which corresponds to $B > 2/(1 - \sin\psi_1)$. We note that these two values of B arise when Eqs (11.49), (11.55) coincide, as mentioned earlier. The parameter μ_0 is the asymptotic value for μ as k tends to zero, which can be analytically determined (see Exercise 7 at the end of this chapter), where it is shown to be

$$\mu_0 = 1 + 2\left\{\frac{\cos\omega_2 - \cos\omega_1}{\ln[\tan(\omega_2/2)/\tan(\omega_1/2)]}\right\},$$

where $\omega_1 = \pi/4 - \psi_1/2$ and $\omega_2 = \pi/4 - \psi_2/2$.

The region where $\mu < \mu_0$ can be further divided into three subregions, and the first of these is $\mu_3 < \mu < \mu_0$, where μ_3 is the asymptotic value of μ as k tends to infinity. We find that the solution asymptotes with the value $\mu = \mu_0$ and crosses the vertical axis at the point μ_3 where μ_3 can be analytically determined (see Exercise 4):

$$\mu_3 = \frac{\cos\psi_2 - \cos\psi_1}{\psi_1 - \psi_2}.$$

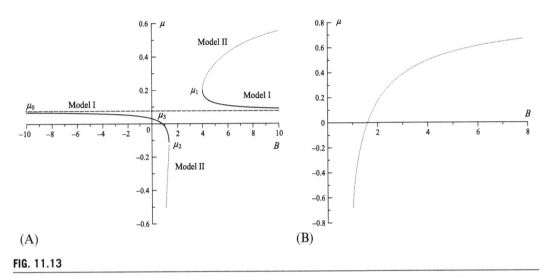

FIG. 11.13

The relationship between the parameter μ and B when ψ_1 and ψ_2 (A) have a different sign, (B) have the same sign.

In the first subregion, the parameter B is negative, which corresponds to a negative value of α and an imaginary modulus k for the elliptic functions. The second subregion is identified for $\mu_1 < \mu < \mu_3$, which corresponds to $0 < B \leq 2/(1 - \sin \psi_2)$. In this subregion, α is always positive and the modulus k is strictly real. The final subregion, which applies for Model II, is $-1 < \mu < \mu_1$, which corresponds to $1 < B \leq 2/(1 - \sin \psi_2)$. Again, α is always positive and k is strictly real.

The region for $\mu > \mu_0$ can be further divided into two subregions, which are $\mu_0 < \mu < \mu_2$ for Model I and $\mu_2 < \mu < 1$ for Model II. The value of B is always positive in both of these subregions, and it corresponds to a negative value of α and a complex value of angle ϕ of the form $\phi = -\pi/2 + i\varphi$. We note that in this region the curvature changes in sign from negative to positive.

We comment that the values of μ given in Eqs (11.49), (11.55) depend on the initial angles ψ_1 and ψ_2 of the two fullerenes, where we assume $\psi_1, \psi_2 \in (-\pi/2, \pi/2)$. When these two initial angles have the same sign (either both positive or both negative), the gradient at the attachment points of the two fullerenes are positive (negative), therefore the curvature of the arc has to change sign; and consequently, only Model II is applicable. This is shown in Fig. 11.13B, where μ_0 becomes zero and μ_1 and μ_2 meet at $B = 2/(1 - \sin \psi_2)$. Generally, the solution for μ may exist in either of the two main regions of Model I and Model II, as described previously. For the case when $\psi_1 = 0$ and $\psi_2 = \pi/2$, only the first region $\mu < \mu_0$ can be found with $\mu_0 = 1$. The points of attachment at the two fullerenes have gradients of ∞ and the solution is similar to the case for joining a carbon nanotube to a flat graphene sheet, as described in Section 11.3. By assuming $\psi_2 = \pi/2$ and varying the value of ψ_1, the point of attachment at the second fullerene above the origin has a gradient of ∞, so the solution is dependent on ψ_1, which is similar to the case for joining a carbon nanotube to a C_{60} fullerene, forming a nanobud as described in Section 11.4.

For a prescribed a, b, ψ_1, ψ_2, and ℓ, we can numerically determine values of $B = 1/k^2$, β and therefore y_0. However, it is not a simple task to choose a, b, ψ_1 and ψ_2 and vary ℓ to produce values

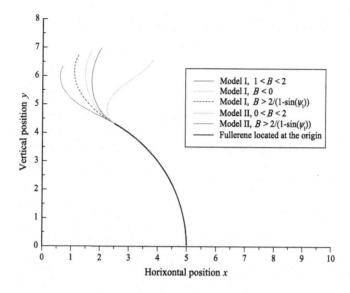

FIG. 11.14

Plots of joints $y = y(x)$ for the values of B corresponding to five regions for Model I and Model II, where $a = 5$, $\psi_1 = \pi/3$, $\psi_2 = -\pi/6$, and $\ell = 3$.

of B which fall in all five regions of the two models, as described previously. For convenience, we choose five different values of B corresponding to the five regions as shown in Fig. 11.13 from which we obtain values of β. As shown in Fig. 11.14, we assume $a = 5$, $\psi_1 = \pi/3$, $\psi_2 = -\pi/6$, and $\ell = 3$; consequently b can be determined by $(b\cos\psi_2, b\sin\psi_2) = (x(\phi_2), y(\phi_2))$, where $\phi_2 = \sin^{-1}(\sqrt{(1 - \sin\psi_2)/2k^2})$, $x(\phi)$ and $y(\phi)$ are given by Eqs (11.45), (11.44) for Model I, and are given by Eqs (11.53), (11.52) for Model II. In Fig. 11.15, the final joining structures of two fullerenes from the two models are illustrated, where $a = 5$, $\psi_1 = \pi/3$, $\psi_2 = -\pi/6$, and $\ell = 3$. We note that at present there is no experimental or simulation data which might confirm this theoretical study.

EXERCISES

11.1. Using the calculus of variations technique to derive the Euler-Lagrange equation, where we assume that

$$F = F(x, y, y', y'', y''').$$

11.2. Derive equations for $y(\phi)$ and $x(\phi)$, as given in Eqs (11.22), (11.23), for a joint between a carbon nanotube and a flat graphene sheet.

11.3. Use an algebraic package to plot the graph of relation between μ and B obtained from Eqs (11.20), (11.27).

11.4. For Model I of nanobuds, show that the equations for y_0 and a can be obtained as given in Eqs (11.31), (11.32).

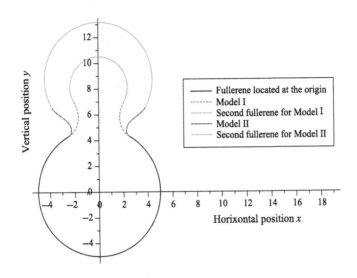

FIG. 11.15

Illustrations for connecting between two fullerenes by Model I and Model II where $a = 5$, $\psi_1 = \pi/3$, $\psi_2 = -\pi/6$, and $\ell = 3$.

11.5. For Model II of nanobuds, show that the arc length ℓ for the nanobuds can be derived as shown in Eq. (11.37).

11.6. From the derivation of nanobuds, use an algebraic package to plot the graph of relation between μ and B obtained from Eqs (11.34), (11.38) where $\psi = \pi/3$.

11.7. Derive the asymptotic formulae of μ_0 and μ_3 for nanopeanuts.

Hints and Solutions

CHAPTER 1

1.1. Hint
- See Section 1.2.1.

1.2. Hint
- See Section 1.2.2.

1.3. The chiral angle θ_0 is the angle between the chiral vector $\mathbf{C}$ and the zigzag line, as shown in Fig. E1. Rearranging the equation for the dot product of two nonzero vectors obtains

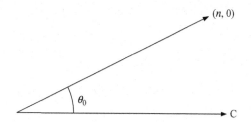

FIG. E1

Schematic of the chiral angle.

an equation for the angle between them, thus

$$\cos\theta = \frac{\mathbf{a}\cdot\mathbf{b}}{|\mathbf{a}|\cdot|\mathbf{b}|},$$

where in this case, the angle we are interested in is the chiral angle θ_0, and the vectors $\mathbf{a}$ and $\mathbf{b}$ are given by the chiral vector $\mathbf{C}$ and the zigzag line. Therefore the equation for the chiral angle becomes

$$\cos\theta_0 = \frac{\mathbf{C}\cdot n\mathbf{a}}{|\mathbf{C}|\cdot|n\mathbf{a}|}$$

$$= \frac{2n+m}{2(n^2+nm+m^2)^{1/2}},$$

where the magnitude of the chiral vector is given in the worked example for determining the nanotube radius.

1.4. Hint
- The translation vector $\mathbf{T}$ is parallel to the carbon nanotube axis and therefore perpendicular to the chiral vector $\mathbf{C}$.
- Due to the two vectors being perpendicular, we must have $\mathbf{C}\cdot\mathbf{T}=0$.

1.5. (i) $d_R = 3d = 30$, (ii) $d_R = d = 6$, (iii) $d_R = 3d = 9$.

1.6. The translation vector is $\mathbf{T} = t_1\mathbf{a} + t_2\mathbf{b}$. The magnitude of $\mathbf{T}$ is thus

$$|\mathbf{T}| = \sqrt{t_1^2\mathbf{a} \cdot \mathbf{a} + t_2^2\mathbf{b} \cdot \mathbf{b} + t_1 t_2\mathbf{a} \cdot \mathbf{b}}$$

$$= |\mathbf{a}|\sqrt{(2m+n)^2 + (2n+m)^2 - (2m+n)(2n+m)}/d_R$$

$$= |\mathbf{a}|\sqrt{3(n^2 + nm + m^2)}/d_R,$$

which is equal to $\sqrt{3}|\mathbf{C}|/d_R$. (i) $T = 2.46\,\text{Å}$, (ii) $T = 6.19\,\text{Å}$.

1.7. Hint
- See Fig. 1.7.
- Draw the chiral vector and the translation vector.

1.8. The MAPLE code below can be used to determine both d and d_R. This is commonly referred to as Euclid's algorithm:

```
> restart;
> G := proc(a,b)
        local i, tempA, tempB:
        if a = 0 then
            return(b):
        else
            tempA := a:
            tempB := b:
            for i while tempB <> 0 do
                if tempA > tempB then
                    tempA := tempA - tempB:
                else
                    tempB := tempB - tempA:
                end if:
            end do:
            return(tempA):
        end if:
    end proc:
```

1.9. See Worked Example 1.3 and Fig. E2 for the complete solution. This is the building block for a C_{260} fullerene.

1.10. Hint
- The area of a triangle is equal to $1/2 \times$ base $\times$ height.

1.11. First, we determine the surface density of carbon atoms on a graphene sheet. The surface area of one hexagon unit is given by

$$SA_h = \frac{3\sqrt{3}}{2}a_{C-C}^2.$$

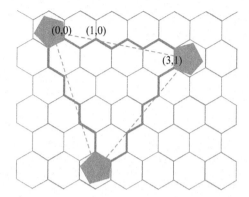

FIG. E2

Equilateral triangle building block for a (3, 1) fullerene.

Then the surface density of carbon atoms for one hexagon unit is simply

$$\eta_h = \frac{2}{SA_h} = \frac{4\sqrt{3}}{9a_{C-C}^2},$$

where we note that each atom contributes 1/3 to each hexagon unit; there are six in each hexagon. The number of carbon atoms per equilateral triangle is thus $n_C = \eta_h \times A_{nm}$, where A_{nm} is the surface area of one equilateral triangle given by Eq. (1.7). It then follows that because there are 20 equilateral triangle building blocks which make up one fullerene, the total number of carbon atoms in a fullerene is $N = 20(n^2 + nm + m^2)$.

1.12. Hint

- Project a polyhedron to a sphere (radial projection).
- Assume the sphere has radius equal to 1.
- The surface area of the sphere is also equal to the sum of the surface area of all faces of the network which now make up the sphere.
- The surface area of each face of a spherical polygon is equal to the sum of its angles minus $(n - 2)\pi$ for a polygon with n sides.

1.13. Hint

- Follow Worked Example 1.6.
- Allow shaped rings from triangles (n_3) to nonagon (n_9).
- Genus is the same as for a sphere.

1.14. A torus has genus 1; therefore Euler's theorem becomes

$$v - e + f = 2(1 - g) = 0.$$

If the torus is constructed from pentagons (n_5), hexagons (n_6), and heptagons (n_7) then we obtain the following equations for the faces, vertices, and edges of a torus:

$$f = n_5 + n_6 + n_7, \quad v = \frac{5n_5 + 6n_6 + 7n_7}{3}, \quad e = \frac{5n_5 + 6n_6 + 7n_7}{2}.$$

Substituting these into Euler's theorem above obtains

$$n_5 - n_7 = 0,$$

so that there must be equal numbers of pentagon and heptagon rings making up the nanotorus structure, but there can be any number of hexagon rings. Note that the edges and vertices equations are divided by 2 and 3, respectively, because for carbon atoms, every vertex of a polygon is shared by two others and every edge is shared by another.

1.15. Hint
- See Section 1.2.5.

1.16. The continuum approximation may be used when dealing with molecular structures which may be approximated to be symmetrical in shape. For example, the C_{60} fullerene is approximately equivalent to a sphere. By modelling the interaction in this way, time and computational effort are saved because the atoms are no longer represented discretely; rather, they are assumed to be smeared over the surface of each molecule.

1.17. Hint
- Use MAPLE or similar package.
- See Worked Example 1.10.

CHAPTER 2

2.1. Use the duplication formula (2.8).

2.2. The sketch should look like:

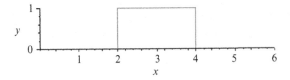

2.3.

$$\zeta(z) = \sum_{n=1}^{\infty} \frac{1}{\Gamma(z)} \int_0^{\infty} x^{z-1} e^{-nx} \, dx$$

$$= \frac{1}{\Gamma(z)} \int_0^{\infty} \sum_{n=1}^{\infty} x^{z-1} e^{-nx} \, dx$$

$$= \frac{1}{\Gamma(z)} \int_0^\infty \frac{x^{z-1}e^{-x}\,dx}{1-e^{-x}}$$

$$= \frac{1}{\Gamma(z)} \int_0^\infty \frac{x^{z-1}\,dx}{e^x-1},$$

where going from lines 2 to 3 uses the sum of a geometric series.

2.4.

$$\zeta(z,\alpha) = \sum_{n=0}^\infty \frac{1}{\Gamma(z)} \int_0^\infty x^{z-1}e^{-(n+\alpha)x}\,dx$$

$$= \frac{1}{\Gamma(z)} \int_0^\infty \sum_{n=0}^\infty x^{z-1}e^{-(n+\alpha)x}\,dx$$

$$= \frac{1}{\Gamma(z)} \int_0^\infty \frac{x^{z-1}e^{-\alpha x}}{1-e^{-x}}\,dx.$$

2.5. Expanding the left-hand side of

$$\frac{t}{e^t-1} = \sum_{n=0}^\infty \frac{B_n t^n}{n!}, \quad 0 < |t| < 2\pi,$$

obtains

$$\frac{t}{e^t-1} = \frac{1}{\ln(e)} - \frac{1}{2}t + \frac{1}{12}\ln(e)t^2 - \frac{1}{720}\ln(e)^3 t^4 + \frac{1}{30240}\ln(e)^5 t^6 + O(t^7),$$

where $\ln(e) = 1$. Similarly, expanding the right-hand side using the power series obtains

$$\sum_{n=0}^\infty \frac{B_n t^n}{n!} = B_0 + B_1 t + \frac{1}{2}B_2 t^2 + \frac{1}{6}B_3 t^3 + \frac{1}{24}B_4 t^4 + \frac{1}{120}B_5 t^5 + \frac{1}{720}B_6 t^6 + O(t^7).$$

By equating coefficients, we can deduce that $B_0 = 1$, $B_1 = -1/2$, $B_2 = 1/6$, $B_4 = -1/30$, $B_6 = 1/42$, etc., and that $B_n = 0$ when n is odd with the exception of $n = 1$.

2.6. This can be deduced from Eq. (2.18).

2.7. Expand $F(1,1;2;-z)$ using Eq. (2.20) to obtain

$$F(1,1;2;-z) = \sum_{n=0}^\infty \frac{(1)_n (1)_n}{(2)_n n!}(-z)^n = \sum_{n=0}^\infty \frac{n!}{(2)_n}(-z)^n,$$

on noting that $(1)_n = n!$. Since $n!/(2)_n = 1/(n+1)$, the equation becomes

$$F(1,1;2;-z) = \sum_{n=0}^\infty \frac{(-z)^n}{n+1}.$$

which on simplification becomes

$$F(1, 1; 2; -z) = \frac{\log(1 + z)}{z}.$$

2.8. Substitute Eq. (2.11) into Eq. (2.20).

2.9.

$$F = \sum_{n=0}^{\infty} \frac{(a)_n (b)_n}{(c)_n n!} z^n, \quad F' = \sum_{n=0}^{\infty} \frac{(a)_{n+1} (b)_{n+1}}{(c)_{n+1} n!} z^n, \quad F'' = \sum_{n=0}^{\infty} \frac{(a)_{n+2} (b)_{n+2}}{(c)_{n+2} n!} z^n,$$

Substitution into the differential equation gives

$$LHS = z(1 - z) \sum_{n=0}^{\infty} \frac{(a)_{n+2} (b)_{n+2}}{(c)_{n+2} n!} z^n + [c - (a + b + 1)z] \sum_{n=0}^{\infty} \frac{(a)_{n+1} (b)_{n+1}}{(c)_{n+1} n!} z^n$$

$$- ab \sum_{n=0}^{\infty} \frac{(a)_n (b)_n}{(c)_n n!} z^n$$

$$= \left(\frac{(a)_2 (b)_2}{(c)_2} + \frac{c(a)_2 (b)_2}{(c)_2} - \frac{(a + b + 1)(a)_1 (b)_1}{(c)_1} - \frac{ab(a)_1 (b)_1}{(c)_1} \right) z$$

$$+ \left(\frac{c(a)_1 (b)_1}{(c)_1} - \frac{ab(a)_0 (b)_0}{(c)_0} \right) + \sum_{n=2}^{\infty} \left\{ \frac{(a)_{n+1} (b)_{n+1}}{(c)_{n+1} (n - 1)!} - \frac{(a)_n (b)_n}{(c)_n (n - 2)!} \right.$$

$$+ \frac{c(a)_{n+1} (b)_{n+1}}{(c)_{n+1} n!} - \frac{(a + b + 1)(a)_n (b)_n}{(c)_n (n - 1)!} - \frac{ab(a)_n (b)_n}{(c)_n n!} \left. \right\} z^n$$

$$= [(a + 1)(b + 1) + c(a + 1)(b + 1) - (c + 1)(ab + a + b + 1)] \frac{abz}{c(c + 1)}$$

$$+ (ab - ab) + \sum_{n=2}^{\infty} \{ n(a + n)(b + n) - n(n - 1)(c + n) + c(a + n)(b + n)$$

$$- n(a + b + 1)(c + n) - ab(c + n) \} \frac{(a)_n (b)_n}{(c)_{n+1} n!} z^n$$

$$= 0 = RHS.$$

CHAPTER 4

4.1. (a) Using Eq. (4.6) and data from Table 4.5, we are able to determine A and B for each atom. This can be done using MAPLE as shown in the worksheet below.

```
>   restart;
>   A:= (sigma,epsilon) -> 4*epsilon*sigma^6:
>   B:= (sigma,epsilon) -> 4*epsilon*sigma^(12):

>   A_Li:= A(2.10,74.4*10^(-3)); B_Li := B(2.10,74.4*10^(-3));
                    A_Li := 25.52399761
                    B_Li := 2189.094267
```

```
>   A_Na:= A(2.50,43.6*10^(-3)); B_Na := B(2.50,43.6*10^(-3));
```
$$A\_Na := 42.57812500$$
$$B\_Na := 10395.05005$$
```
>   A_K:= A(2.93,28*10^(-3)); B_K := B(2.93,28*10^(-3));
```
$$A\_K := 70.86368700$$
$$B\_K := 44836.26908$$
```
>   A_He:= A(3.03,1.4*10^(-3)); B_He := B(3.03,1.4*10^(-3));
```
$$A\_He := 4.333549864$$
$$B\_He := 3353.509716$$
```
>   A_Ne:= A(3.18,2.2*10^(-3)); B_Ne := B(3.18,2.2*10^(-3));
```
$$A\_Ne := 9.100083812$$
$$B\_Ne := 9410.400604$$
```
>   A_Ar:= A(3.46,7.7*10^(-3)); B_Ar := B(3.46,7.7*10^(-3));
```
$$A\_Ar := 52.84541456$$
$$B\_Ar := 90670.05976$$
```
>   A_Kr:= A(3.55,5.3*10^(-3)); B_Kr := B(3.55,5.3*10^(-3));
```
$$A\_Kr := 42.43321904$$
$$B\_Kr := 84932.92824$$
```
>   A_Xe:= A(3.83,6.3*10^(-3)); B_Xe := B(3.83,6.3*10^(-3));
```
$$A\_Xe := 79.54139156$$
$$B\_Xe := 251064.8004$$

(b) See MAPLE worksheet below for the solution.

```
>   restart;
>   sigma_C := 3.5: e_C := 2.1*10^(-3):
>   sigma_Li := 2.10: e_Li := 74.4*10^(-3):
>   sigma_Na := 2.50: e_Na := 43.6*10^(-3):
>   sigma_K := 2.93: e_K := 28*10^(-3):
>   sigma_He := 3.03: e_He := 1.4*10^(-3):
>   sigma_Ne := 3.18: e_Ne := 2.2*10^(-3):
>   sigma_Ar:= 3.46: e_Ar := 7.7*10^(-3):
>   sigma_Kr:= 3.55: e_Kr := 5.3*10^(-3):
>   sigma_Xe:= 3.83: e_Xe := 6.3*10^(-3):

>   e := (e1,e2) -> sqrt(e1*e2):
>   sigma := (s1,s2) -> (s1+s2)/2:
>   sigma_C_Li := sigma(sigma_C,sigma_Li); e_C_Li := e(e_C,e_Li);
```

$$sigma\_C\_Li := 2.800000000$$
$$e\_C\_Li := 0.01249959999$$

```
> sigma_C_Na := sigma(sigma_C,sigma_Na); e_C_Na := e(e_C,e_Na);
```
$$sigma\_C\_Na := 3.000000000$$
$$e\_C\_Na := 0.009568698971$$

```
> sigma_C_K:= sigma(sigma_C,sigma_K); e_C_K := e(e_C,e_K);
```
$$sigma\_C\_K := 3.215000000$$
$$e\_C\_K := 0.007668115805$$

```
> sigma_C_He:= sigma(sigma_C,sigma_He); e_C_He := e(e_C,e_He);
```
$$sigma\_C\_He := 3.265000000$$
$$e\_C\_He := 0.001714642820$$

```
> sigma_C_Ne:= sigma(sigma_C,sigma_Ne); e_C_Ne := e(e_C,e_Ne);
```
$$sigma\_C\_Ne := 3.340000000$$
$$e\_C\_Ne := 0.002149418526$$

```
> sigma_C_Ar:= sigma(sigma_C,sigma_Ar); e_C_Ar := e(e_C,e_Ar);
```
$$sigma\_C\_Ar := 3.480000000$$
$$e\_C\_Ar := 0.004021193853$$

```
> sigma_C_Kr:= sigma(sigma_C,sigma_Kr); e_C_Kr := e(e_C,e_Kr);
```
$$sigma\_C\_Kr := 3.525000000$$
$$e\_C\_Kr := 0.003336165464$$

```
> sigma_C_Xe:= sigma(sigma_C,sigma_Xe); e_C_Xe := e(e_C,e_Xe);
```
$$sigma\_C\_Xe := 3.665000000$$
$$e\_C\_Xe := 0.003637306696$$

```
> A:= (sigma,epsilon) -> 4*epsilon*sigma^6:
> B:= (sigma,epsilon) -> 4*epsilon*sigma^(12):
> A_C_Li:= A(sigma_C_Li,e_C_Li); B_C_Li:= B(sigma_C_Li,e_C_Li);
```
$$A\_C\_Li := 24.09374416$$
$$B\_C\_Li := 11610.54170$$

```
> A_C_Na:= A(sigma_C_Na,e_C_Na); B_C_Na:= B(sigma_C_Na,e_C_Na);
```
$$A\_C\_Na := 27.90232620$$
$$B\_C\_Na := 20340.79580$$

```
> A_C_K:= A(sigma_C_K,e_C_K); B_C_K:= B(sigma_C_K,e_C_K);
```
$$A\_C\_K := 33.87150688$$
$$B\_C\_K := 37404.20096$$

```
> A_C_He:= A(sigma_C_He,e_C_He); B_C_He:= B(sigma_C_He,e_C_He);
```

$$A\_C\_He := 8.308694108$$
$$B\_C\_He := 10065.41959$$

> `A_C_Ne:= A(sigma_C_Ne,e_C_Ne); B_C_Ne:= B(sigma_C_Ne,e_C_Ne);`

$$A\_C\_Ne := 11.93602665$$
$$B\_C\_Ne := 16570.61324$$

> `A_C_Ar:= A(sigma_C_Ar,e_C_Ar); B_C_Ar:= B(sigma_C_Ar,e_C_Ar);`

$$A\_C\_Ar := 28.56869910$$
$$B\_C\_Ar := 50741.80696$$

> `A_C_Kr:= A(sigma_C_Kr,e_C_Kr); B_C_Kr:= B(sigma_C_Kr,e_C_Kr);`

$$A\_C\_Kr := 25.60131667$$
$$B\_C\_Kr := 49115.32584$$

> `A_C_Xe:= A(sigma_C_Xe,e_C_Xe); B_C_Xe:= B(sigma_C_Xe,e_C_Xe);`

$$A\_C\_Xe := 35.26012010$$
$$B\_C\_Xe := 85453.06828$$

(c) By substituting the values of A and B as obtained in (b) for each type of interaction into Eq. (4.7), we can determine the critical radius of a spherical fullerene for which the minimum energy configuration of the atom is at $r = 0$. In Table E1, we present these critical radii of fullerenes for those atoms shown.

Table E1 Critical Radius of a Fullerene for a Particular Atom Having Minimum Energy at Its Centre

X	Li+	Na+	K+	He	Ne	Ar	Kr	Xe
a (Å)	3.26	3.49	3.75	3.80	3.89	4.05	4.11	4.27

4.2. (a) Using a cylindrical coordinate system, we assume that the carbon atom is located at $(\varepsilon, 0, 0)$ with respect to the centre of a carbon nanotube of infinite extent to a parametric equation $(b\cos\theta, b\sin\theta, z)$. The potential energy can be determined from performing a surface integral of the Lennard-Jones potential over the carbon nanotube, thus

$$E = \eta_g b \int_{-\pi}^{\pi} \int_{-\infty}^{\infty} \left(-\frac{A}{\rho^6} + \frac{B}{\rho^{12}} \right) dz d\theta, \tag{E4.1}$$

where η_g denotes the mean atomic surface density of the carbon nanotube and ρ is the typical distance between the atom and the nanotube, which is given by

$$\rho^2 = (b\cos\theta - \varepsilon)^2 + b^2 \sin^2\theta + z^2 = \lambda^2 + z^2,$$

where $\lambda^2 = b^2 - 2\varepsilon b \cos\theta + \varepsilon^2$. Thus from Eq. (E4.1) we have

$$E = \eta_g b \int_{-\pi}^{\pi} \int_{-\infty}^{\infty} \left(-\frac{A}{(\lambda^2 + z^2)^3} + \frac{B}{(\lambda^2 + z^2)^6} \right) dz d\theta. \tag{E4.2}$$

By introducing $z = \lambda \tan \psi$, we have $dz = \lambda \sec^2 \psi \, d\psi$. The fact that $1 + \tan^2 \psi = \sec^2 \psi$, Eq. (E4.2) reduces to

$$
\begin{aligned}
E &= \eta_g b \int_{-\pi}^{\pi} \left\{ -\frac{A}{\lambda^5} \int_{-\pi/2}^{\pi/2} \cos^4 \psi \, d\psi + \frac{B}{\lambda^{11}} \int_{-\pi/2}^{\pi/2} \cos^{10} \psi \, d\psi \right\} d\theta \\
&= \eta_g b \int_{-\pi}^{\pi} \left\{ -\frac{3\pi A}{8} \frac{1}{\lambda^5} + \frac{63\pi B}{256} \frac{1}{\lambda^{11}} \right\} d\theta \\
&= \eta_g b \left\{ -\frac{3\pi A}{8} \int_{-\pi}^{\pi} \frac{1}{\lambda^5} d\theta + \frac{63\pi B}{256} \int_{-\pi}^{\pi} \frac{1}{\lambda^{11}} d\theta \right\}.
\end{aligned} \tag{E4.3}
$$

By introducing $\alpha = b^2 + \varepsilon^2$ and $\beta = 2\varepsilon b$, we have $\lambda = (\alpha - \beta \cos \theta)^{1/2}$ so that Eq. (E4.3) becomes

$$
E = \eta_g b \left\{ -\frac{3\pi A}{8} \int_{-\pi}^{\pi} \frac{1}{(\alpha - \beta \cos \theta)^{5/2}} d\theta + \frac{63\pi B}{256} \int_{-\pi}^{\pi} \frac{1}{(\alpha - \beta \cos \theta)^{11/2}} d\theta \right\}.
$$

Finally, from Eqs (4.13), (4.15), the potential energy for an offset atom inside a carbon nanotube is given by

$$
E = \eta_g b \left(-\frac{3\pi A}{8} J_2 + \frac{63\pi B}{256} J_5 \right),
$$

where here J_n is given by

$$
J_n = \frac{2\pi}{(b - \varepsilon)^{2n+1}} F\left(n + \frac{1}{2}, \frac{1}{2}; 1; -\frac{4\varepsilon b}{(b - \varepsilon)^2} \right),
$$

where $F(a, b; c, z)$ is the usual hypergeometric function.

(b) Using Eq. (1.1) from Chapter 1, we find that the radius of a $(15, 15)$ carbon nanotube is $b = 10.17$ Å. Here, we use $\eta_g = 0.3812$ Å^{-2}, $A = 15.41$ eV Å^6, and $B = 22534.75$ eV Å^{12}. By running the MAPLE worksheet attached below, we find that in a $(15, 15)$ carbon nanotube, the equilibrium distance of a carbon atom with respect to the centre of the carbon nanotube is at $\varepsilon = 6.78$ Å.

```
>  restart;
>  with(plots):
>  r := (n,m) -> evalf(1.42*sqrt(3*(n^2 + n*m + m^2))/(2*Pi)):
```
Define the radius of a $(15, 15)$ carbon nanotube.
```
>  b:= r(15,15);
```
$$
b := 10.17000086
$$
Define the attractive and repulsive constants for C–C interactions.
```
>  A:= 15.41: B:= 22534.75:
```
Define the mean atomic density.
```
>  n_g := 0.3812:
```

```
>   J := (n) ->
>   2*Pi*hypergeom([n+1/2,1/2],[1],-4*epsilon*b/(b-epsilon)^2)/
>   (b-epsilon)^(2*n +1):

>   E := b*n_g*(-A*3*Pi*J(2)/8 + B*63*Pi*J(5)/256):


>   plot(E,epsilon =0.1..7.5);
```

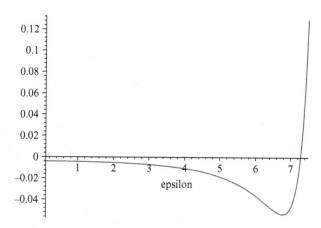

```
>   eq_minimise:= diff(E,epsilon):


>   epsilon_min:= fsolve(eq_minimise,epsilon=7);
```
$$epsilon\_min := 6.779731374$$
```
>   E_min := evalf(subs(epsilon = epsilon_min,E));
```
$$E\_min := -0.05347493396$$

4.3. Hint

- Use the same procedure as shown in Solution 1(b), but change the attractive and repulsive constants A and B for each type of atom.

4.4. See MAPLE worksheet below:

```
>   restart;
>   A := 17.4: B:= 29*10^3:
>   ng:= 0.3812: nf := 0.3767:
>   a := 7.12:
>   epsilon := 1:
>   alpha:= b^2 + epsilon^2 - a^2: beta := 2*b*epsilon:
>   J := n ->
>   2*Pi*hypergeom([n+1/2,1/2],[1],-2*beta/(alpha-beta))/(alpha-beta)^
>   (n + 1/2):
```

```
>  E := 4*Pi^2*a^2*b*ng*nf*((B/5)*(315*J(5)/256 + 1155*a^2*J(6)/64 +
>  9009*a^4*J(7)/128 + 6435*a^6*J(8)/64 + 12155*a^8*J(9)/256) -
>  (A/8)*(3*J(2) + 5*a^2*J(3))):
```

```
>  plot(E,b=10..16);
```

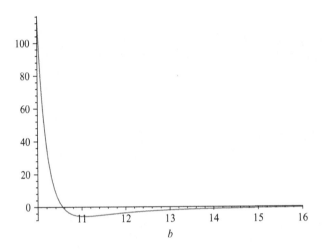

```
>  fsolve(diff(E,b),b=11);
```
$$11.06029952$$

4.5. Since the van der Waals force is a short range force, we consider only the nearest neighbour interactions. Therefore we rewrite Eq. (4.23) as

$$E^* = E^{tot}/L_1 = ab\eta_g^2\pi \int_0^{2\pi} \int_0^{2\pi} \left(-\frac{3A}{4}\frac{1}{\lambda^5} + \frac{63B}{128}\frac{1}{\lambda^{11}} \right) d\theta_1 d\theta_2,$$

where $\lambda^2 = a^2 + b^2 - 2ab\cos(\theta_1 - \theta_2)$, and noting that $\varepsilon = 0$ because it is a concentric multiwalled nanotube. Thus from Worked Example 4.6, E^* is given by

$$E^* = ab\eta_g^2\pi \left[-\frac{3A\pi^2}{(a+b)^5} F\left(\frac{5}{2},\frac{1}{2};1;\frac{4ab}{(a+b)^2}\right) + \frac{63B\pi^2}{32(a+b)^{11}} F\left(\frac{11}{2},\frac{1}{2};1;\frac{4ab}{(a+b)^2}\right) \right].$$

The first step is to determine the radius b, which results in the minimum energy occurring when the inner tube is a $(5,5)$ nanotube with $a = 3.392$ Å. As such, we need to solve for b, which satisfies $dE^*/db = 0$; this is shown in the MAPLE worksheet below. Repeatedly, we use b as a to find the radius of the next shell of the multiwalled carbon nanotube. In Table E2, radii for each shell of the multiwalled carbon nanotube are presented for the case of a $(5,5)$ as the innermost nanotube.

Table E2 Radii for the Next Five Shells of a Multiwalled Carbon Nanotube, Given That a $(5, 5)$ With $a = 3.392$ Å Is the Innermost Nanotube					
Shell	**2nd**	**3rd**	**4th**	**5th**	**6th**
b(Å)	6.821	10.248	13.673	17.096	20.517

```
>   restart;


>   E :=
>   a*b*ng^2*Pi*(-3*A*Pi^2*hypergeom([5/2,1/2],[1],4*a*b/(a+b)^2)/(a+b)
>   ^5 +
>   63*B*Pi^2*hypergeom([11/2,1/2],[1],4*a*b/(a+b)^2)/(32*(a+b)^(11))):


>   eq_min:= diff(E,b):


>   ng := 0.3812:

>   A := 15.2: B := 24.1*10^3:

>   a1 := 3.392:

>   plot(subs(a = a1, E),b=6.2..12):

>   b1:= fsolve(subs(a=a1,eq_min),b=7);
```
$$b1 := 6.821078479$$
```
>   plot(subs(a = b1, E),b=9.6..12):

>   b2:= fsolve(subs(a=b1,eq_min),b=10);
```
$$b2 := 10.24826287$$
```
>   plot(subs(a = b2, E),b=13..15):

>   b3:= fsolve(subs(a=b2,eq_min),b=13);
```
$$b3 := 13.67291718$$
```
>   plot(subs(a = b3, E),b=16.5..19):

>   b4:= fsolve(subs(a=b3,eq_min),b=17);
```
$$b4 := 17.09578972$$
```
>   plot(subs(a = b4, E),b=20..23):

>   b5:= fsolve(subs(a=b4,eq_min),b=20);
```
$$b5 := 20.51739980$$

4.6. Using a cylindrical polar coordinate system, we assume that an atom is located at $(\varepsilon \cos \phi, \varepsilon \sin \phi, 0)$ relative to the centre of a nanotube bundle. Thus the distance from the atom to the axis of kth tube in the bundle is given by

$$d_k^2 = \left(R \cos\left(\frac{2k\pi}{N}\right) - \varepsilon \cos \phi\right)^2 + \left(R \sin\left(\frac{2k\pi}{N}\right) - \varepsilon \sin \phi\right)^2$$

$$= (R - \varepsilon)^2 + 4R\varepsilon \sin^2\left(\frac{k\pi}{N} - \frac{\phi}{2}\right).$$

If we denote the interaction energy between an atom and a single nanotube a distance d_k apart by E, then the total interaction energy between an atom and a bundle of N-fold symmetry is given by

$$E_{tot} = \sum_{k=1}^{N} E\left(\left[(R - \varepsilon)^2 + 4R\varepsilon \sin^2\left(\frac{k\pi}{N} - \frac{\phi}{2}\right)\right]^{1/2}\right),$$

where E is given in Chapter 3 as $E(\delta) = \eta_g(-AI_3 + BI_6)$, where

$$I_n = \frac{2\pi b}{\delta^{2n-1}} B\left(n - \frac{1}{2}, \frac{1}{2}\right) F\left(n - \frac{1}{2}, n - \frac{1}{2}; 1; \frac{b^2}{\delta^2}\right).$$

4.7. The total interaction energy of a fullerene at the centre of a nanotube bundle is given by $W_f = -NE_{ft}(R)$, where E_{ft} is defined by Eq. (4.37), noting that in Eqs (4.37), (4.38) that $d_k = R$. We note that the minimum for any value of N is given by the minimum of the function E_{ft}. We model the C_{240} fullerene as a sphere with radius $r_0 = 7.12$ Å and mean surface density $\eta_f = 0.3767$ Å^{-2}, and the $(16, 16)$ nanotube as a cylinder with radius $r = 10.856$ Å and mean surface density $\eta_t = 0.3812$ Å^{-2}. Substituting these values into Eq. (4.37), we deduce $E_{ft}(R)$, which upon minimising gives rise to the critical bundle radius size. Here, we use $A = 17.4$ eV Å^6 and $B = 29 \times 10^3$ eV Å^{12}. From the MAPLE worksheet attached below, we find for a bundle of $(16, 16)$ nanotubes that we need the bundle radius $R = 20.92$ Å such that the minimum energy configuration of the C_{240} fullerene is where its centre is on the tube central axis.

```
>   restart;
>   r0 := 7.12: r:= 10.856:
>   nf := 0.3767: nt := 0.3812:
>   A := 17.4:  B := 29*10^3:
>   J := (n) -> 2*Pi*hypergeom([1/2,
    n+1/2],[1],-4*r*R/((r - R)^2 -
    r0^2))/((r - R)^2 - r0^2)^(n+1/2):
>   E_ft := 4*Pi^2*r0^2*r*nf*nt*(
    B*(315*J(5)/256 + 1155*r0^2*J(6)/64 + 9009*r0^4*J(7)/128 +
    6435*r0^6*J(8)/64 + 12155*r0^8*J(9)/256 )/5 - A*(
    3*J(2) +5*r0^2*J(3) )/8 ):

>   plot(E_ft, R = 20..25);
```

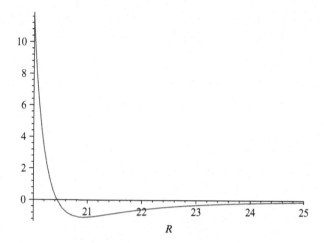

```
>  Eq_min := diff(E_ft,R):
>  R_min:= fsolve(Eq_min,R=21);
```
$$R\_min := 20.92235196$$

CHAPTER 5

5.1. We refer the reader to Sections 5.1.1 and 5.1.2 for the definition and explanation of the acceptance condition and the suction energy.

5.2. Firstly, we use the algebraic package MAPLE to plot the graph of the relation between the suction energy and the tube radius b.

```
> restart;
> A:=17.4:
> B:=29000:
> n_f:=0.3789:
> n_g:=0.3812:
> a:=3.55:
```

We define μ, which is $\mu = a^2/(b^2 - a^2)$.

```
> mu := b -> a^2/(b^2-a^2):
```

The suction energy is given by Eq. (5.12).

```
> W := b -> Pi^3*n_f*n_g*a^2*b*(A*(3+5*mu(b))
  - B*(315+4620*mu(b)+18018*mu(b)^2+25740*mu(b)^3+12155*mu(b)^4)
  /(160*(b^2-a^2)^3))/(b^2-a^2)^(5/2):
```

We comment that the reader should check this program by plotting the relation between the suction energy W and the tube radius b and compare with Fig. 5.8. Here, we use `fsolve()` to find a point which crosses the x-axis.

> `fsolve(W(b),b=6.5);`

`6.270361785`

The maximum point is determined by differentiating `W(b)` with respect to b and solve for the solution.

> `fsolve(diff(W(b),b),b=6.5);`

`6.782835883`

5.3. In an axially symmetric cylindrical polar coordinate system (r, z), an atom is assumed located at $(0, Z)$, which might be inside or outside of the carbon nanotube that is assumed to be of semiinfinite length, centred around the positive z-axis and of radius b. The parametric form of the equation for the surface of the carbon nanotube is (b, z), where $z \geqslant 0$. As shown in Fig. E3, the distance ρ between the atom and a typical surface element of the tube is given by $\rho^2 = b^2 + (Z - z)^2$. Due to the symmetry of the problem, only the force in the axial direction needs to be considered. That is, $F_z = F_{vdW}(Z - z)/\rho$, where F_{vdW} is the van der Waals interaction force defined by

$$F_{vdW} = -\frac{d\Phi}{d\rho} = -\frac{6A}{\rho^7} + \frac{12B}{\rho^{13}}.$$

Consequently, the interaction force between an atom located on the z-axis and all the atoms of the carbon nanotube is given by

$$F_z^{tot}(Z) = -2\pi b\eta_g \int_0^\infty \frac{d\Phi}{d\rho} \frac{(Z - z)}{\rho} dz, \tag{E5.1}$$

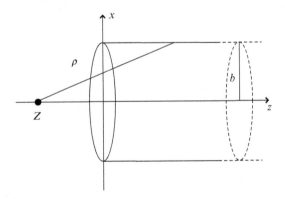

FIG. E3

Geometry of the single atom entering a carbon nanotube.

where η_g is the uniform surface density of carbon atoms in a graphene structure, such as a carbon nanotube. As $\rho^2 = a^2 + (Z - z)^2$ and therefore $d\rho = -((Z - z)/\rho)dz$, Eq. (E5.1) becomes

$$F_z^{tot}(Z) = 2\pi b\eta_g \int_{\sqrt{b^2+Z^2}}^{\infty} \frac{d\Phi}{d\rho} d\rho,$$

$$= 2\pi b\eta_g \left[\frac{A}{(b^2 + Z^2)^3} - \frac{B}{(b^2 + Z^2)^6} \right], \quad \text{(E5.2)}$$

where $F_z^{tot}(Z)$ is a continuous function with zeros at $Z = \pm Z_o$; thus

$$Z_o = b \left[\left(\frac{B}{Ab^6} \right)^{1/3} - 1 \right]^{1/2},$$

and Z_o is real only when $b \leqslant b_o$, where $b_o = \sqrt[6]{B/A}$.

The following is the MAPLE code to plot the force experienced by an atom due to van der Waals interaction profiles for three distinct carbon nanotube radii

```
> restart;
> A:=17.4:
> B:=29000:
> eta:=3.812:
> F:=(Z,b)->2*Pi*b*n*(A/(b^2+Z^2)^3 - B/(b^2+Z^2)^6):
> plot([F(Z,3.32),F(Z,3.443),F(Z,3.739)],Z=-10..10);
```

In Fig. E4, $F_z^{tot}(Z)$ is plotted for carbon nanotubes of various radii, which illustrates that as the radius of the nanotube increases beyond a_o (in this case $b_o \approx 3.443$), the value of $F_z^{tot}(Z)$ remains positive for all values of Z.

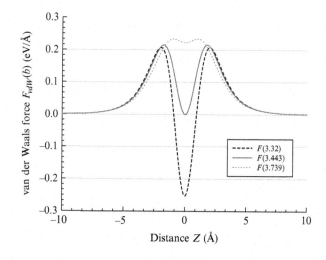

FIG. E4

Force experienced by an atom due to van der Waals interaction with a semiinfinite carbon nanotube.

The integral of $F_z^{tot}(Z)$ represents the work done by the van der Waals forces, which are imparted onto the atom in the form of kinetic energy. For the atom to be accepted into the nanotube, the sum of its initial kinetic energy and that received by moving from $-\infty$ to $-Z_o$ needs to be greater than that which is lost when the van der Waals force is negative (i.e. in the region $-Z_o < Z < Z_o$). This is termed the acceptance energy (W_a), which allows the acceptance condition to be written as

$$\frac{Mv_o^2}{2} + W_a > 0,$$

where M is the mass of the atom and v_o is its initial velocity, and

$$W_a = 2\pi b\eta_g \int_{-\infty}^{Z_o} \left[\frac{A}{(b^2 + Z^2)^3} - \frac{B}{(b^2 + Z^2)^6} \right] dZ.$$

Employing the substitution $Z = b \tan \psi$, this integral becomes

$$W_a = \frac{2\pi \eta_g}{a^4} \int_{-\pi/2}^{\psi_o} \left(A \cos^4 \psi - \frac{B}{a^6} \cos^{10} \psi \right) d\psi, \qquad \text{(E5.3)}$$

where $\psi_o = \tan^{-1}\left\{ \left[B/\left(Ab^6\right) \right]^{1/3} - 1 \right\}^{1/2}$. Evaluation of this integral gives the acceptance energy in explicit form as

$$W_a = \frac{\pi \eta_g}{128b^4} \left\{ 32A \left[\sin \psi_o \left(2\cos^3 \psi_o + 3 \cos \psi_o \right) + 3 \left(\psi_o + \frac{\pi}{2} \right) \right] \right.$$
$$- \frac{B}{b^6} \left[\frac{\sin \psi_o}{5} \left(128 \cos^9 \psi_o + 144 \cos^7 \psi_o + 168 \cos^5 \psi_o \right.\right.$$
$$\left.\left.\left. + 210 \cos^3 \psi_o + 315 \cos \psi_o \right) + 63 \left(\psi_o + \frac{\pi}{2} \right) \right] \right\}.$$

Assuming that the atom is initially at rest, then the acceptance condition becomes simply $W_a > 0$. Using the values from Tables 1.5–1.7, the acceptance energy is calculated for nanotubes of various radii and graphed in Fig. E5, using the values of Z_o as graphed in Fig. E6 with the MAPLE code

```
> restart:
> A:=17.4:
> B:=29000:
> n:=evalf(4*sqrt(3)/(9*(1.42)^2)):
  # For the acceptance energy shown in Fig. E5
> psi:= b -> arctan(sqrt((B/(A*b^6))^(1/3) - 1)):
> Wa:= b -> Pi*n/(128*b^4)*( 32*A*(sin(psi(b))*
  (2*cos(psi(b))^3+ 3*cos(psi(b)))+3*(psi(b)+Pi/2)) -B/b^6*(
  sin(psi(b))/5*(128*cos(psi(b))^9 + 144*cos(psi(b))^7 +
  168*cos(psi(b))^5 + 210*cos(psi(b))^3 + 315*cos(psi(b))) +
  63*(psi(b)+Pi/2) )):
> plot(Wa(b),b=3.15..3.45);
  # For the upper limit of integration Zo as shown in Fig. E6
```

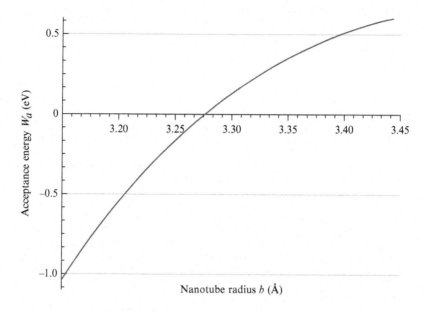

FIG. E5

Acceptance energy threshold for an atom to be sucked into a carbon nanotube.

```
> Zo:= b -> b*sqrt((B/(A*b^6))^(1/3)-1):
> plot(Zo(b),b=3.15..3.45);
```
Worthy of comment is that the acceptance energy is positive for tubes of radius $a > 3.276$ Å. This radius value is smaller than that of a $(5, 5)$ carbon nanotube, using the usual notation (n, m), where n and m are positive integers representing the helicity of a carbon nanotube. As $(5, 5)$ is the smallest carbon nanotube expected to be physical, therefore all physical carbon nanotubes will accept a single atom from rest. However, for a nanotube with radius less than this size (e.g. a $(7, 2)$ nanotube with a radius of $b = 3.206$ Å), then this model predicts that it would not accept an atom by suction force alone and that the atom would need to possess an initial velocity for it to overcome the negative acceptance energy. Also worth mentioning is that when $b > b_o$, the force graph does not cross the axis; therefore Z_o is not real. In this case the atom will always be accepted by the nanotube.

5.4. The suction energy (W) can be calculated as the total integral of $F_z^{tot}(Z)$ from $-\infty$ to ∞, which is a good approximation where the atom starts more than 10 Å outside of the tube end and moves to a point more than 10 Å within the nanotube. It can be seen that this is just Eq. (E5.3) with the upper limit of the integral (ψ_o) replaced with $\pi/2$; therefore evaluation of this integral yields

$$W = \frac{3\pi^2 \eta_g}{128 b^4}\left(32A - \frac{21B}{b^6}\right). \qquad (E5.4)$$

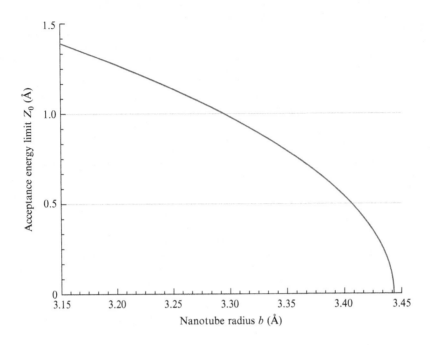

FIG. E6

Upper limit of integration Z_o used to determine the acceptance energy for an atom and carbon nanotube.

Assuming that the atom is initially at rest, the increase in its velocity (v) can be calculated directly from the kinetic energy formula and is explicitly given by

$$v = \frac{\pi}{8b^2} \sqrt{\frac{3\eta_g}{M} \left(32A - \frac{21B}{b^6} \right)}.$$

Note that care must be taken when calculating v, as the value inside of the parentheses may be negative. In this case, the atom loses energy when entering the tube and will decelerate upon entering the tube. By differentiating Eq. (E5.4), it is possible to calculate the tube radius b_{max}, which provides the maximum suction energy and therefore the maximum velocity on entering the tube. This occurs for a value of radius b_{max}, which is given by

$$b_{max} = \sqrt[6]{\frac{105B}{64A}},$$

and for the values of A and B in Table 1.7, $b_{max} \approx 3.739$ Å. Both (6, 5) and (9, 1) nanotubes have a radius of $b = 3.737$ Å, which is very close to b_{max}.

In Fig. E7, the suction energy is graphed for various carbon nanotubes illustrating a maximum value occurring at $b = b_{max}$. The MAPLE code is

```
> restart:
> A:=17.4:
```

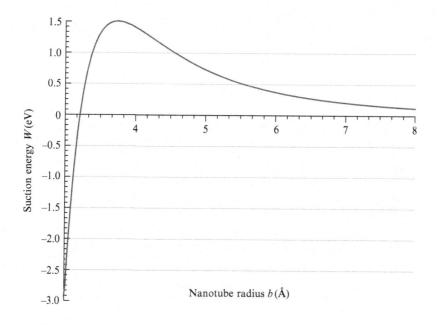

FIG. E7

Suction energy for an atom entering a carbon nanotube.

```
> B:=29000:
> n:=evalf(4*sqrt(3)/(9*(1.42)^2)):
> W:=b->Pi^2*n/(4*b^4)*(3*A-63*B/(32*b^6)):
> plot(W(b),b=3..8);
```

The suction energy W is positive for any value of radius $b > 3.210\,\text{Å}$, which means there is a range of nanotube radii $3.210 < b < 3.276\,\text{Å}$ for which W is positive but W_a is negative. In other words, an atom accepted into a nanotube with a radius in this range would experience an increase in velocity. However, the atom would not be sucked in from rest due to the magnitude of the repulsive component of the van der Waals force experienced as it crosses the tube opening. It is worth mentioning that it is not expected that physical nanotubes with radii falling within this range exist, as it is assumed that $(5,5)$ is the smallest physical carbon nanotube.

5.5. The suction energy (W) for a fullerene can be determined using the same method as was used to obtain the acceptance energy (W_a). Using Eq. (5.3), where $F(Z)$ is given by Eq. (5.5), we make the substitution $Z = \sqrt{b^2 - a^2}\tan\psi$ so that $\lambda = (b^2 - a^2)\sec^2\psi/a^2$ and $dZ = \sqrt{b^2 - a^2}\sec^2\psi\,d\psi$. Thus we obtain the equation given by Eq. (5.11) except with the upper limit changed from ψ_o to $\pi/2$. In other words, we have

$$W = \frac{8\pi^2\eta_f\eta_g b}{a^2\sqrt{b^2 - a^2}}\left[A\left(J_2 + 2J_3\right) - \frac{B}{5a^6}\left(5J_5 + 80J_6 + 336J_7 + 512J_8 + 256J_9\right)\right],$$

where $J_n = a^{2n}(b^2 - a^2)^{-n} \int_{-\pi/2}^{\pi/2} \cos^{2n} \psi \, d\psi$. Thus from Gradshteyn and Ryzhik (2000) (3.621(3)), we obtain

$$J_n = \frac{a^{2n}}{(b^2 - a^2)^n} \frac{(2n-1)!!}{(2n)!!} \pi,$$

where !! represents the double factorial notation such that $(2n - 1)!! = (2n - 1)(2n - 3)\ldots 3 \cdot 1$ and $(2n)!! = (2n)(2n - 2)\ldots 4 \cdot 2$. Substitution and simplification gives Eq. (5.12).

5.6. Refer to Fig. 5.16. Approximately (a) 6.75 Å, (b) 5.5 Å, and (c) 3.75 Å.

5.7. Due to the short-range nature of the van der Waals interaction, when the fullerene is more than 10 Å from the bundle end, either inside or outside of the bundle, the force may be considered negligible. Using this observation, we can use the expressions given in Section 4.9 to characterise the force at both ends of a finite tube, provided that the tube half-length L is greater than 10 Å. For the optimised bundles given in Section 4.9, the equilibrium position for the fullerene is on the bundle axis, and therefore the total van der Waals force is given by $F_{tot}^z = NF_{vdW}^z$, which is given by Eqs (4.37), (4.38). In Fig. E8, we show the force profiles for the four configurations. We note that the force is always positive, and therefore the C_{60} fullerene is sucked into the bundle.

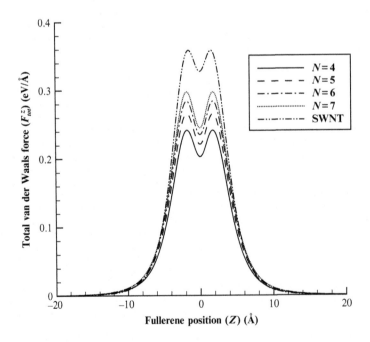

FIG. E8

Total van der Waals force for a fullerene entering a nanotube bundle for various configurations of ideal oscillators.

CHAPTER 6

6.1. Use the same technique as described in Section 5.2.1, instead of integrating z from zero to ∞ of Eq. (5.6), in this case, we have to integrate z from $Z - L$ to $Z + L$:

6.2.
```
> restart;
> A:=17.4:
  > B:=29000:
  > n_f:=0.3789:
  > n_g:=0.3812:
  > a:=3.55:
  > rho1:= (b,L,Z) -> sqrt(b^2 +(Z+L)^2):
  > rho2:= (b,L,Z) -> sqrt(b^2+(Z-L)^2):
  > Energy := (rho,b,L,Z) -> n_f*Pi*a*( A*(1/(rho(b,L,Z)+a)^4
    - 1/(rho(b,L,Z)-a)^4)/2- B*(1/(rho(b,L,Z)+a)^10
    - 1/(rho(b,L,Z)-a)^10)/5)/rho(b,L,Z):
  > Force:= (b,L,Z) -> 2*Pi*b*n_g*(Energy(rho2(b,L,Z),b,L,Z)
    - Energy(rho1(b,L,Z),b,L,Z)):
```
For a $(9, 9)$ carbon nanotube
```
> plot(Force(6.106,50,Z),Z=-80..80);
```

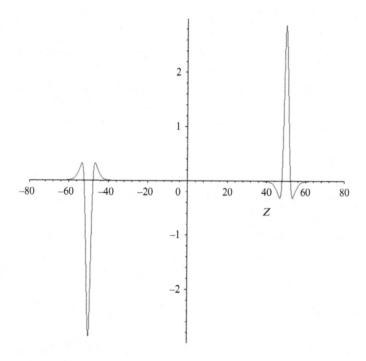

For a $(11, 11)$ carbon nanotube:

```
> plot([Force(7.463,50,Z)],Z=-80..80);
```

From this exercise, you will see that a $(9, 9)$ carbon nanotube does not accept the C_{60} fullerene by suction forces alone due to the strong repulsive forces at both ends of the carbon nanotube. However, the C_{60} fullerene is easily accepted into a $(11, 11)$ carbon nanotube by suction force alone.

6.3.

```
> restart;
```

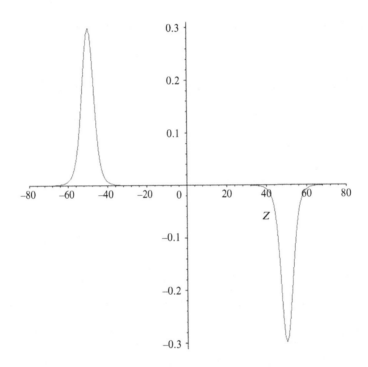

```
> A:=17.4:
> B:=29000:
> n_f:=0.3789:
> n_g:=0.3812:
> a:=3.55:
> M:=1.196*10^(-24):
> mu := b -> a^2/(b^2-a^2):
> W := b -> Pi^3*n_f*n_g*a^2*b*(A*(3+5*mu(b)) - B*(315+4620*mu(b)
  +18018*mu(b)^2+25740*mu(b)^3+12155*mu(b)^4)/(160*(b^2-a^2)^3))
  /(b^2-a^2)^(5/2):
> v:= (b,v0) -> sqrt(2*W(b)*1.602*10^(-19)/M +v0^2):
```

```
> f:= (b,L,v0) -> 10^(-9)*v(b,v0)/(4*L):
> plot(f(b,50*10^(-10),0),b=6.5..8);
```

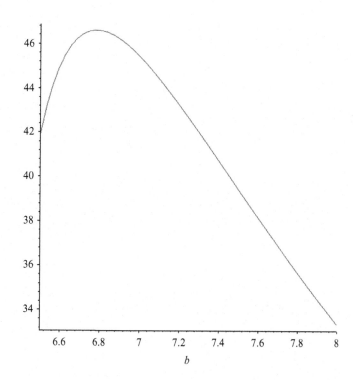

We can find the maximum frequency by differentiating the frequency equation with respect to the radius b and use `fsolve()` to find the solution.

```
> fsolve(diff(f(b,50*10^(-10),0),b),b=6.6);
```

The carbon nanotube with radius $b = 6.782835883$ Å will give the highest oscillation frequency for the C_{60} fullerene oscillating inside a carbon nanotube.

6.4. In this exercise, we utilise the same MAPLE code as in the previous exercise for the oscillation frequency equation. Assuming that $v_0 = 1152\,\text{ms}^{-1}$ and $L = 50 \times 10^{-10}$ m, we use the `fsolve()` command to find the smallest nanotube radius b for which the C_{60} molecule starts to oscillate. The solution is obtained as $b = 6.106$ Å, which is approximately equal to the radius of the $(9, 9)$ carbon nanotube.

```
> fsolve(f(b,50*10^(-10),1152),b=6.2);
```

6.5. To find the maximum frequency for the double-walled carbon nanotube oscillator, we differentiate Eq. (6.27) with respect to d and solve $\partial f / \partial d = 0$ for d. You will obtain $d = (\alpha^2 \ell_4 - 2v_0^2)/(2\alpha^2)$, which will give rise to the maximum frequency. Assuming $v_0 = 0$, we differentiate Eq. (6.27) with respect to d and again solving $\partial f / \partial d = 0$ for d, we obtain $d = \ell_4/2$ for which the system gives rise to the maximum frequency.

6.6. We utilise the MAPLE code, which is given in Worked Example 6.5, to plot the relation between the oscillation frequency and the inner tube half-length L_1.

```
> plot(fre(L1,(L2-L1)/2,0),L1=3.55..95);
```

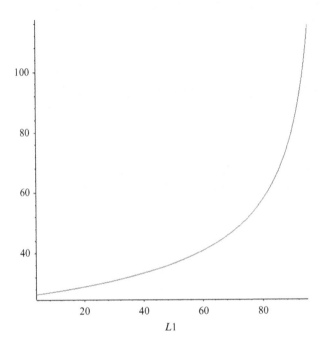

6.7. In an axially symmetric cylindrical polar coordinate system (r, z), an atom is assumed located at $(0, Z)$ inside a carbon nanotube of length $2L$, centred around the z-axis and of radius a, as depicted in Fig. E9. Here, it is assumed that the atom oscillates along the z-axis. This assumption is valid for the carbon nanotube $(6, 6)$, where the atom is likely to be on the z-axis due to the minimum potential energy, which can be shown using Eq. (4.33). From the symmetry of the problem, only the force in the axial direction is of concern, and neglecting the frictional force $F_r(Z)$, from Eq. (6.1) gives

$$M\frac{d^2Z}{dt^2} = F_z^{tot}(Z),$$

where M denotes the mass of a single carbon atom and $F_z^{tot}(Z)$ is the total axial van der Waals interaction force between the atom and the carbon nanotube length $2L$, given by

$$F_z^{tot}(Z) = 2\pi b\eta g \left[-\frac{A}{(b^2 + (Z - L)^2)^3} + \frac{B}{(b^2 + (Z - L)^2)^6} \right.$$
$$\left. + \frac{A}{(b^2 + (Z + L)^2)^3} - \frac{B}{(b^2 + (Z + L)^2)^6} \right]. \tag{E6.1}$$

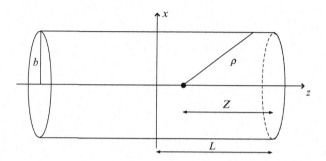

FIG. E9

Geometry of the single-atom oscillation.

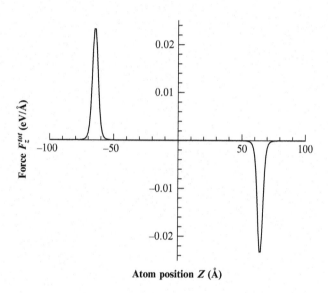

FIG. E10

Plot of $F_z^{tot}(Z)$ as given by Eq. (E6.1) for the atom oscillating inside the carbon nanotube $(6,6)$.

The reader is referred to Chapter 5 for details of this derivation. In Fig. E10, $F_z^{tot}(Z)$ is plotted as given by Eq. (E6.1) for the case of the carbon nanotube $(6,6)$ ($a = 4.071$ Å). Noting that $(6,6)$ with $a = 4.071$ Å satisfies the condition stated in Chapter 5, where both ends can generate the necessary attractive force to suck the atom inside, the atom oscillates and never escapes the carbon nanotube.

6.8. Given the similarities between the force and energy profiles we find in Exercise 7 of Chapter 5 and those for the oscillation of a C_{60} fullerene inside a carbon nanotube, we adopt exactly the same model and use the equation for frequency of $f = (2W/m_f)^{1/2}/4L$, where M_f

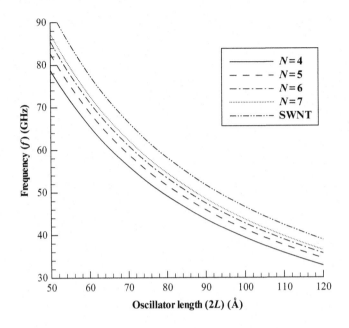

FIG. E11

Frequency for a fullerene nanotube bundle oscillator for various configurations of ideal oscillators varying the bundle length $2L$.

is the mass of the fullerene. In Fig. E11, we plot the oscillation frequency for the various ideal oscillators, varying the bundle length $2L$. We note that the details for the oscillation of a C_{60} molecule inside a nanotube bundle can be found in Cox et al. (2007b).

CHAPTER 7

7.1. We refer the reader to Worked Example 7.3 for the details of the calculation. The following is the program in MAPLE:

```
> restart:
> a:= 3.55:
> nf:= 0.3789:
    > ng:= 0.3812:
    > A:= 17.4:
    > B:= 29000:
    > d := (b,x) -> sqrt((b-x)^2 + 4*b*x*sin(theta/2)^2+(z-Z)^2):
    > P:= r -> 4*Pi*a^2*nf*((B*( 5/(r^2-a^2)^6 + 80*a^2/(r^2-a^2)^7
      + 336*a^4/(r^2-a^2)^8 + 512*a^6/(r^2-a^2)^9 + 256*a^8/(r^2-a^2)
      ^(10))/5 - A*(1/(r^2-a^2)^3 + 2*a^2/(r^2-a^2)^4 ))):
```

```
> Energy:= (b,x) -> b*ng*Int(Int(P(d(b,x)),theta=-Pi..Pi),
  z=-infinity..infinity):
> plot(Energy(6.784,13.034),Z=-10..10);
```

7.2. We refer the reader to Worked Example 7.5 for the detailed calculation, and the following is the program in MAPLE:

```
> restart;
> a:= 3.55:
> nf:= 0.3789:
> ng:= 0.3812:
> Agc:= 17.4:
> Bgc:= 29000:
> Acc:= 20:
> Bcc:= 34800:
> d1:=2*Z:
> d2:=sqrt(4*e^2+Z^2):
> P_ball:= (r,n) -> 4*Pi^2*a^2*nf^2/(r*(2-n)*(3-n)) *
  (1/(2*a+r)^(n-3) + 1/(-r)^(n-3) - 1/(2*a-r)^(n-3) -1/r^(n-3)):
> Un:= r -> -Acc*P_ball(r,6)+Bcc*P_ball(r,12):
> E_ball:= k -> (2*k-2)*Un(d1) + (2*k-1)*Un(d2):
> P_tube:= r -> 4*Pi*a^2*nf*((Bgc*(5/(r^2-a^2)^6 +
  80*a^2/(r^2-a^2)^7+336*a^4/(r^2-a^2)^8+512*a^6/(r^2-a^2)^9
  + 256*a^8/(r^2-a^2)^(10))/5 - Agc*(1/(r^2-a^2)^3+2*a^2/(r^2-a^2)^4))):
> delta_i:= (b,i)->sqrt((b+e)^2-4*b*e*sin(theta/2)^2+(z-Z*(i-1))^2):
> delta_j:= (b,j)->sqrt((b-e)^2+4*b*e*sin(theta/2)^2+(z-2*Z*(j-1))^2):
> E_i:= (b,k)->sum(b*ng*Int(Int(P_tube(delta_i(b,i)),theta=-Pi..Pi),
  z=-infinity..infinity),i=1..k):
> E_j:= (b,k)->sum(b*ng*Int(Int(P_tube(delta_j(b,j)),theta=-Pi..Pi),
  z=-infinity..infinity),j=1..k+1):
> E_tube:= (b,k) -> E_i(b,k) + E_j(b,k):
> E_tot:= (k,b) -> E_ball(k) + E_tube(b,k):
> e:=0:
> plot([E_tot(1,6.784),E(5,6.784)],Z=9..11);
> fsolve(diff(E(1,6.784),Z),Z=10);
```

7.3. In this exercise the case of a spheroidal fullerene is considered centred on and rotationally symmetric around the z-axis, as well as enclosed in a cylinder of infinite extent and centred around the z-axis. In this case the form of the integral is equivalent to Eq. (7.23), except that the lower z limit is changed to $-\infty$ and the parameter defining the location of the spheroid, $Z = 0$. That is,

$$E = ab\eta_c\eta_s \int_0^{2\pi} \int_0^\pi \int_{-\infty}^\infty \int_0^{2\pi} \left(-\frac{A}{\rho^6} + \frac{B}{\rho^{12}}\right) \sin\phi\sqrt{a^2\cos^2\phi + c^2\sin^2\phi}\,d\psi\,dz\,d\phi\,d\theta, \quad (E7.1)$$

where ρ is defined in Eq. (7.22). In this derivation of the expression, the infinite extent of z in both directions makes it possible to consider this integration first. To this end the substitution κ, which is defined in Eq. (7.29), is made, then consider I_z, namely

$$I_z = \int_{-\infty}^{\infty} \left\{ -A[\kappa^2 + (z - c \cos \phi - Z)^2]^{-3} + B[\kappa^2 + (z - c \cos \phi - Z)^2]^{-6} \right\} dz.$$

Now as before, substituting $\gamma = z - c \cos \phi - Z$ gives

$$I_z = \int_{-\infty}^{\infty} \left[-A(\kappa^2 + \gamma^2)^{-3} + B(\kappa^2 + \gamma^2)^{-6} \right] d\gamma.$$

A further substitution $\gamma = \kappa \tan \tau$ gives

$$I_z = \int_{-\pi/2}^{\pi/2} \left(-A\kappa^{-5} \cos^4 \tau + B\kappa^{-11} \cos^{10} \tau \right) d\tau,$$

which upon integrating, produces

$$I_z = (3\pi/8) \left[-A\kappa^{-5} + (21B/32)\kappa^{-11} \right].$$

Next, the integral of I_z is performed over the angles θ and ψ, namely

$$I_{\theta,\psi} = \frac{3\pi}{8} \int_0^{2\pi} \int_0^{2\pi} \left\{ -A \left[\beta_1^2 - (\beta_1^2 - \beta_2^2) \sin^2 ((\theta - \psi)/2) \right]^{-\frac{5}{2}} \right.$$
$$\left. + \frac{21B}{32} \left[\beta_1^2 - (\beta_1^2 - \beta_2^2) \sin^2 ((\theta - \psi)/2) \right]^{-\frac{11}{2}} \right\} d\psi \, d\theta. \tag{E7.2}$$

Appealing to the same arguments of rotational symmetry presented in Section 7.3.1.1, the double integral in Eq. (E7.2) can be reduced to the single integral

$$I_{\theta,\psi} = \frac{3\pi^2}{4} \int_0^{2\pi} \left\{ -A \left[\beta_1^2 - \left(\beta_1^2 - \beta_2^2 \right) \sin^2 \frac{\theta}{2} \right]^{-\frac{5}{2}} \right.$$
$$\left. + \frac{21B}{32} \left[\beta_1^2 - \left(\beta_1^2 - \beta_2^2 \right) \sin^2 \frac{\theta}{2} \right]^{-\frac{11}{2}} \right\} d\theta, \tag{E7.3}$$

and by bisecting the interval and making the substitution $t = \sin^2(\theta/2)$, Eq. (E7.3) becomes

$$I_{\theta,\psi} = \frac{3\pi^2}{2} \int_0^1 \left\{ -\frac{A}{\beta_1^5} \left[1 - \left(1 - \frac{\beta_2^2}{\beta_1^2} \right) t \right]^{-\frac{5}{2}} + \frac{21B}{32\beta_1^{11}} \left[1 - \left(1 - \frac{\beta_2^2}{\beta_1^2} \right) t \right]^{-\frac{11}{2}} \right\}$$
$$t^{-\frac{1}{2}} (1 - t)^{-\frac{1}{2}} dt,$$

which involves the fundamental integral representation for the hypergeometric function, as given in Section 2.1.3 of Erdélyi et al. (1953), to become

$$I_{\theta,\psi} = \frac{3\pi^3}{2} \left[-\frac{A}{\beta_1^5} F\left(\frac{5}{2}, \frac{1}{2}; 1; 1 - \frac{\beta_2^2}{\beta_1^2} \right) + \frac{21B}{32\beta_1^{11}} F\left(\frac{11}{2}, \frac{1}{2}; 1; 1 - \frac{\beta_2^2}{\beta_1^2} \right) \right],$$

where $F(a, b; c; z)$ is the usual hypergeometric function. With the use of Eq. (7.33), the integral $I_{\theta,\psi}$ becomes

$$I_{\theta,\psi} = (3\pi^3/2)\left[-A(\beta_1\beta_2)^{-5/2}P_{3/2}(\epsilon) + (21B/32)(\beta_1\beta_2)^{-11/2}P_{9/2}(\epsilon)\right].$$

Thus the potential E for an ellipsoidal fullerene inside an infinite carbon nanotube can now be written in the form

$$E = 3\pi^3 ab\eta_c\eta_s\left[-AI_{5/2} + (21B/32)I_{11/2}\right], \tag{E7.4}$$

where the integral I_n is defined by

$$I_n = \int_0^{\pi/2} P_{n-1}\left(\frac{b^2 + a^2\sin^2\phi}{b^2 - a^2\sin^2\phi}\right)\frac{\sin\phi\sqrt{c^2\sin^2\phi + a^2\cos^2\phi}}{(b^2 - a^2\sin^2\phi)^n}d\phi.$$

In Fig. E12, the potential E, as given by Eq. (E7.4), is plotted with respect to the tube radius b for different values of spheroid major and minor radii a and c. It is evident from this graph that the value of a is the dominant feature for determining the radius of nanotube b, suitable for accepting the fullerene in question.

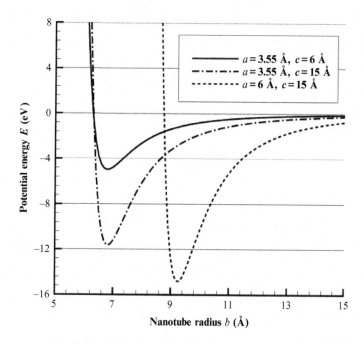

FIG. E12

Variation of the interaction potential for ellipsoidal fullerenes with radii of carbon nanotube.

CHAPTER 8

8.1. A carbon nanotube has: a larger inner volume, distinct inner and outer surfaces for functionalisation and readily removable end caps which make the inner volume and surfaces more accessible. In addition, nanotubes have been shown to enter the cell nuclei, which suggests the possibility of gene delivery.

8.2. Refer to Section 8.2.

8.3. Refer to Section 8.2.4.

8.4. Make use of the empirical combining rules given by Eq. (8.9).

8.5. The bond distances between boron and carbon atoms is 1.55 Å, while carbon and carbon atoms is 1.42 Å. Thus following Worked Example 8.5, we can determine the lengths of the edges of both the hexagonal boron array and the triangular carbon array which are illustrated in Fig. E13. Note that we have assumed that all angles in the lattice are 120°. With reference to Fig. E13, the length of each edge for the boron hexagonal array x is given by

$$x = 1.42 + 2(1.55 \cos 60°) = 2.97 \text{ Å}.$$

Similarly, the length of each edge for the carbon triangular array y is given by

$$y = 2 \times 1.42 \sin 60 + 2 \times 1.55 \sin 60° = 5.144 \text{ Å}.$$

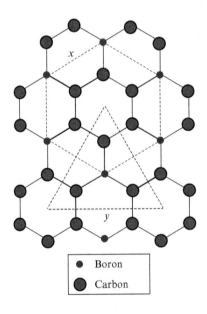

FIG. E13

Surface density for the boron carbide nanotube.

Therefore the surface density for the boron atoms is

$$\eta_B = \frac{4\sqrt{3}}{9x^2} = 0.08727 \, \text{atoms}/\text{Å}^2,$$

and for the carbon atoms is

$$\eta_C = 6\frac{4\sqrt{3}}{9y^2} = 0.1745 \, \text{atoms}/\text{Å}^2$$

Note that the surface density of carbon atoms η_C is multiplied by six because six carbon atoms are weighted at each vertex of the triangular array.

8.6. Below are the steps required to determine the interaction force for a single water molecule:

a. Determine the coordinate positions of each atom of the water molecule.

b. Determine the centre of mass of the water molecule and assume that this lies on the nanotube axis.

c. Convert the values of ϵ and σ into the Lennard-Jones constants A and B.

d. Use MAPLE to plot the force. To start, refresh the memory and use the plots package.

```
> restart; with(plots):
```

Define the coordinate positions.

```
> Ox := 0: Oy := 0: Oz := Z+0.0646:
> H1x := 0.8165: H1y := 0: H1z := Z-0.5128:
> H2x := -0.8165: H2y := 0: H2z := Z-0.5128:
```

Then define the Lennard-Jones parameters for the interaction between the water molecule atoms and the carbon nanotube.

```
> COA := 35.72: COB := 62136: CHA := 14.94: CHB := 14544:
```

Determine the surface density of carbon atoms.

```
> n := evalf(4*sqrt(3)/(9*(1.42)^2)):
```

Calculate the force, as given in Worked Example 8.3.

```
> alpha1 := (a,epsilon,z,zeta) -> sqrt((a+epsilon)^2 + (zeta-z)^2):
> alpha2 := (a,epsilon,z,zeta) -> sqrt((a-epsilon)^2 + (zeta-z)^2):
> Force := (a,epsilon,z,zeta,A,B) ->
  evalf(2*Pi*n*a*((-A/alpha1(a,epsilon,z,zeta)^6)
  *hypergeom([3,1/2],[1],(alpha1(a,epsilon,z,zeta)^2
  -alpha2(a,epsilon,z,zeta)^2)/alpha1(a,epsilon,z,zeta)^2)
  + (B/alpha1(a,epsilon,z,zeta)^(12))
  *hypergeom([6,1/2],[1],(alpha1(a,epsilon,z,zeta)^2
  -alpha2(a,epsilon,z,zeta)^2)/alpha1(a,epsilon,z,zeta)^2))):
> FT := (a,Z) -> -Re(Force(a,sqrt(Ox^2+Oy^2),Z+0.0646,0,COA,COB)
  + Force(a,sqrt(H1x^2+H1y^2),Z-0.5128,0,CHA,CHB)
  + Force(a,sqrt(H2x^2+H2y^2),Z-0.5128,0,CHA,CHB)):
```

Plot the force.

```
> plot([FT(3.39,Z),FT(4.9,Z),FT(5.5,Z)],Z=-15..15,
  linestyle=[SOLID,DASH,DOT],color=BLACK);
```

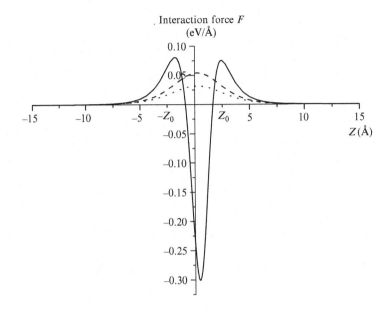

CHAPTER 9

9.1. The new 'polyhedral' model is derived by ensuring that all bonds are equal in length, all angles between adjacent bonds are equal and all atoms lie equidistant from a central cylindrical axis. In the 'rolled-up' model, it is assumed that a graphene sheet is rolled-up to form a cylinder and that all atoms lie on this cylinder. In this conventional 'rolled-up' model, to satisfy the constraints of curvature, the bond angles must become smaller than the 120° found in a graphene sheet. In fact, the conventional model can be determined from the new 'polyhedral' model by taking first-order terms.

These two models exhibit the greatest difference for nanotubes of small radii because this is when the effect of curvature is the most pronounced.

9.2. Considering Eq. (9.5), substitute the expansion for ψ given in Worked Example 9.1 produces the following:

$$a_0(h) = \frac{(2+h)^2}{4(1+h+h^2)},$$

(E9.1)

$$a_1(h) = \frac{\pi^2 h^2 (2+h)^2 (1+2h)^2 (h-1)^2}{64(1+h+h^2)^5},$$

(E9.2)

and from $h = m/n$ Eqs (E9.1), (E9.2), give the expansion for $\cos^2 \theta$ given by Eq. (9.21).

CHAPTER 10

10.1. We refer the reader to Sections 10.1.1 and 10.1.2 for the full details of the two approaches. In short, we minimise the distance between two atoms at the junction for the variation in bond length approach, and we minimise the bond angle by fixing all the bond lengths for the variation in the bond angle approach.

10.2. In terms of the variation in bond length, all bond lengths on two nanostructures are fixed. However, the bond lengths which connect between the two atoms may vary from the assumed bond length, and the bond angles at the connection sites may also vary. This means that the variation in bond length approach affects both the bond stretching and the bending angle terms in the energy equation. On the other hand, all the bond lengths fixed for the variation in the bond angle approach and only the bond angle varies; therefore the bond stretching term will always be zero. Thus the variation in bond angle is a better approach.

10.3. The terminology $N(5,5)_p$ refers to N sections of a $(5,5)$ nanotube where each section consists of p atoms. Therefore $5(5,5)_{18}5(9,0)_{30}$ refers to an elbow constructed from five sections of a $(5,5)$ nanotube and five sections of a $(9,0)$ nanotube. The total number of atoms for this elbow is obtained as $(5 \times 18) + (5 \times 30) = 240$ atoms.

10.4. We refer reader to Section 10.2.1 for the derivation of Eqs (10.6), (10.8).

10.5. We refer reader to Worked Example 10.4 for the detail of MAPLE program for the calculation:

```
> restart;
> with(LinearAlgebra):
> with(geom3d):
> sigma:=1:
> point(A1,evalf(sigma/2),evalf(-sigma*sin(Pi*72/180)),0):
> point(A2,0,evalf(sigma*cos(Pi*54/180)),0):
> point(A3,evalf(sigma*sin(Pi*4/180)),0,0):
> midpoint(M,A1,A2):
> point(U,xcoord(A3)-xcoord(M),ycoord(A3)-ycoord(M),zcoord(A3)
  -zcoord(M)):
> point(V,(xcoord(A2)-xcoord(A1))/distance(A1,A2),(ycoord(A2)
  -ycoord(A1))/distance(A1,A2),(zcoord(A2)-zcoord(A1))
  /distance(A1,A2)):
> W:=Matrix([[i,j,k],[xcoord(U),ycoord(U),zcoord(U)],
  [xcoord(V),ycoord(V),zcoord(V)]]):
> Determinant(W):
```
A circular path of atom A_3 is obtained as

$$(0.250 - 0.180\cos\phi, -0.182 - 0.182\cos\phi, -0.115\sin\phi).$$

10.6. See Baowan et al. Nanotechnology **19** (2008) 075704.

10.7. See Baowan et al. Nanotechnology **19** (2008) 075704.

CHAPTER 11

11.1. See Worked Example 11.1

11.2. We refer the reader to Section 11.3.1 for the derivation of Eqs (11.22), (11.23), where in this exercise we have

$$\frac{dy}{d\phi} = -2\beta k \sin \phi,$$

and

$$\frac{dx}{d\phi} = -\beta \frac{1 - 2k^2 \sin^2 \phi}{\left(1 - k^2 \sin^2 \phi\right)^{1/2}}.$$

11.3. On using MAPLE, we have:

```
> restart;
> with(plots):
  > mu1:= 2*((EllipticE(sqrt(B/2),1/sqrt(B)))/(EllipticF
    (sqrt(B/2),1/sqrt(B))) )-1:
  > mu2:= 2*((2*EllipticE(1/sqrt(B))-EllipticE(sqrt(B/2),1/sqrt(B)))/
    (2*EllipticK(1/sqrt(B))-EllipticF(sqrt(B/2),1/sqrt(B))) )-1:
  > p1:=plot(eval(mu1),B=-4..2):
  > p2:=plot(eval(mu2),B=1..2):
  > display([p1,p2]);
```

We note that on using a normal elliptic function in MAPLE, in the argument $E(t, k)$, we have to put $E(\arcsin(t), k)$ to calculate the elliptic integral in trigonometric form.

11.4. Only a positive curvature is considered in Model I, so that for the parametric equation form of y, we have

$$\frac{dy}{d\phi} = 2\beta k \sin \phi.$$

Upon integrating this equation, we may deduce

$$y(\phi) = -2\beta k \cos \phi + A,$$

where A is an arbitrary constant. By using the boundary condition $y(\phi_f) = b \sin \psi$, we obtain

$$y(\phi) = b \sin \psi + 2\beta k(\cos \phi_f - \cos \phi).$$

At the attachment point of the tube, $y(\phi_t) = y_c$, we may deduce

$$y_c = b \sin \psi + 2\beta k(\cos \phi_f - \cos \phi_t).$$

Now, we determine the corresponding parametric equation for x, and after changing to the new parameter ϕ, we obtain

$$\frac{dx}{d\phi} = \beta \left[2\left(1 + k^2 \sin^2 \phi\right)^{1/2} - \frac{1}{\left(1 + k^2 \sin^2 \phi\right)^{1/2}} \right],$$

which upon integration yields

$$x(\phi) = \beta \left[2E(\phi,k) - F(\phi,k) \right] + A,$$

where again A denotes the constant of integration. By using the boundary condition $x(\phi_f) = b\cos\psi$, we obtain

$$x(\phi) = b\cos\psi + \beta \left\{ 2[E(\phi,k) - E(\phi_f,k)] - [F(\phi,k) - F(\phi_f,k)] \right\},$$

At the attachment point of the tube, $x(\phi_t) = a$, we may deduce

$$a = b\cos\psi + \beta \left\{ 2[E(\phi_t,k) - E(\phi_f,k)] - [F(\phi_t,k) - F(\phi_f,k)] \right\}.$$

11.5. The arc length constraint is obtained from the two regions, and we have

$$\ell = \int_{b\cos\psi}^{x_c} (1+y'^2)^{1/2}dx + \int_{x_c}^{a} (1+y'^2)^{1/2}dx.$$

We note that in the above equation, the curvature of the first term is positive and the curvature of the second term in negative. On making the substitution $y' = \tan\theta$, $dx/d\theta = \cos\theta/\sqrt{\lambda + \alpha\cos\theta}$ for the first term and $dx/d\theta = -\cos\theta/\sqrt{\lambda + \alpha\cos\theta}$ for the second term, we my deduce

$$\ell = \int_{\psi-\pi/2}^{\theta_c} \frac{d\theta}{(\lambda + \alpha\cos\theta)^{1/2}} - \int_{\theta_c}^{-\pi/2} \frac{d\theta}{(\lambda + \alpha\cos\theta)^{1/2}}.$$

Now, we change to the parameter ϕ, and the above equation can be written as

$$\ell = \beta \left[\int_{\phi_f}^{-\pi/2} \frac{d\phi}{(1 - k^2\sin^2\phi)^{1/2}} - \int_{-\pi/2}^{\phi_t} \frac{d\phi}{(1 - k^2\sin^2\phi)^{1/2}} \right]$$
$$= \beta[2K(k) - F(\phi_t,k) - F(\phi_f,k)].$$

11.6. On using MAPLE, we have

```
> restart;
> with(plots):
> mu1:= x -> 2*((EllipticE(sqrt(B/2),1/sqrt(B))-EllipticE(sqrt(B*(1-x)/2),
  1/sqrt(B)))
  /(EllipticF(sqrt(B/2),1/sqrt(B))-EllipticF(sqrt(B*(1-x)/2),1/sqrt(B))))-1:
> mu2:= x -> 2*( (2*EllipticE(1/sqrt(B))-EllipticE(sqrt(B/2),1/sqrt(B))-
  EllipticE(sqrt(B*(1-x)/2),1/sqrt(B)))/ (2*EllipticK(1/sqrt(B))-
  EllipticF(sqrt(B/2),1/sqrt(B))-EllipticF(sqrt(B*(1-x)/2),1/sqrt(B))))-1:
> p1:=plot(mu1(sin(Pi/3)),B=-10..20):
> p2:=plot(mu2(sin(Pi/3)),B=-10..20):
> display([p1,p2]);
```

11.7. These formulae are not immediately apparent from Eq. (11.49) and follow only after rearrangement of Eq. (11.49) to the particular form Eq. (E11.3) given below. Firstly, we consider the usual Legendre incomplete elliptic integral of the first kind $F(\phi, k)$, as defined by Byrd and Friedman, and for the value of ϕ when $\phi \in (\phi_2, \phi_1)$, we may deduce

$$F(\phi_2, k) - F(\phi_1, k) = \int_{\phi_1}^{\phi_2} \frac{d\phi}{\sqrt{1 - k^2 \sin^2 \phi}},$$

where $\phi_2 = \sin^{-1}(\sqrt{(1 - \sin \psi_2)/2k^2})$ and $\phi_1 = \sin^{-1}(\sqrt{(1 - \sin \psi_1)/2k^2})$, and ψ_1 and ψ_2 are the initial defect angles of two fullerenes. On making the substitution $k \sin \phi = \sin \lambda$, we may derive

$$F(\phi_2, k) - F(\phi_1, k) = \int_{\omega_1}^{\omega_2} \frac{d\lambda}{\sqrt{k^2 - \sin^2 \lambda}}, \tag{E11.1}$$

where $\omega_1 = \pi/4 - \psi_1/2$ and $\omega_2 = \pi/4 - \psi_2/2$. By precisely the same considerations for the incomplete elliptic integral of the second kind $E(\phi, k)$ for $\phi \in (\phi_2, \phi_1)$, we obtain

$$E(\phi_2, k) - E(\phi_1, k) = \int_{\omega_1}^{\omega_2} \frac{\cos^2 \lambda}{\sqrt{k^2 - \sin^2 \lambda}} d\lambda. \tag{E11.2}$$

By substituting Eqs (E11.1), (E11.2) into Eq. (11.49), we may derive

$$\mu = \left(2 \int_{\omega_1}^{\omega_2} \frac{\cos^2 \lambda \, d\lambda}{\sqrt{k^2 - \sin^2 \lambda}} - \int_{\omega_1}^{\omega_2} \frac{d\lambda}{\sqrt{k^2 - \sin^2 \lambda}} \right) \Big/ \int_{\omega_1}^{\omega_2} \frac{d\lambda}{\sqrt{k^2 - \sin^2 \lambda}}. \tag{E11.3}$$

For μ_0, which is the asymptotic value as k tends to zero, by using $\cos 2\theta = 2 \cos^2 \theta - 1 = 1 - 2 \sin^2 \theta$, Eq. (E11.3) can be formally reduced, and μ_0 becomes

$$\mu_0 = 1 - 2 \int_{\omega_1}^{\omega_2} \sin \lambda \, d\lambda \Big/ \int_{\omega_1}^{\omega_2} \frac{d\lambda}{\sin \lambda}.$$

On evaluating the above equation, we obtain

$$\mu_0 = 1 + 2 \left\{ \frac{\cos \omega_2 - \cos \omega_1}{\ln \left[\tan(\omega_2/2)/ \tan(\omega_1/2) \right]} \right\}.$$

For μ_3, which is the asymptotic value as k tends to infinity, from Eq. (E11.3) we have

$$\mu_3 = \left(2 \int_{\omega_1}^{\omega_2} \cos^2 \lambda \, d\lambda - \int_{\omega_1}^{\omega_2} d\lambda \right) \Big/ \int_{\omega_1}^{\omega_2} d\lambda,$$

and from this we may deduce

$$\mu_3 = \frac{\cos \psi_2 - \cos \psi_1}{\psi_1 - \psi_2}.$$

Bibliography

Books

Andrews, G.E., Askey, R., Roy, R., 1999. Special Functions. Cambridge University Press, Cambridge.

Bailey, W.N., 1972. Generalized Hypergeometric Series. Hafner Publishing Co., New York.

Byrd, P.F., Friedman, M.D., 1971. Handbook of Elliptic Integrals for Engineers and Scientists, second ed. Springer-Verlag, Berlin.

Companion, A.L., 1979. Chemical Bonding, second ed. McGraw-Hill, New York.

Coxeter, H.S.M., 1969. Introduction to Geometry. Wiley, New York.

Cromwell, P., 1996. Polyhedra. Cambridge University Press, New York.

Dresselhaus, M.S., Dresselhaus, G., Eklund, P.C., 1996. Science of Fullerenes and Carbon Nanotubes. Academic Press, San Diego.

Elsgolc, L.E., 1961. Calculus of Variations. Pergamon Press, London.

Erdélyi, A., Magnus, W., Oberhettinger, F., Tricomi, F.G., 1953. Higher Transcendental Functions, vol. 1. McGraw-Hill, New York.

Gradshteyn, I.S., Ryzhik, I.M., 2000. Table of Integrals, Series and Products, sixth ed. Academic Press, San Diego, USA.

Harris, P.J.F., 1999. Carbon Nanotubes and Related Structures. Cambridge University Press, Cambridge. (Hardcover).

Harris, P.J.F., 2002 Carbon Nanotubes and Related Structures. Cambridge University Press, Cambridge. (Paperback).

Hirschfelder, J.O., Curtiss, C.F., Bird, R.B., 1954. Molecular Theory of Gases and Liquids. John Wiley and Sons, New York.

Hunter, R.J., 2001. Foundations of Colloid Science, second ed. Oxford University press, New York.

Israelachvili, J., 1992. Intermolecular and Surface Forces, second ed. Academic Press, San Diego.

Magnus, W., Oberhettinger, F., Soni, R.P., 1966. Formulas and Theorems for the Special Functions of Mathematical Physics, third ed. Springer-Verlag, Berlin.

Saito, R., Dresselhaus, G., Dresselhaus, M., 1998. Physical Properties of Carbon Nanotubes. Imperial College Press, London.

Journal Articles

Banhart, F., Ajayan, P.M., 1996. Carbon onions as nanoscopic pressure cells for diamond formation. Nature 382, 433–435.

Banhart, F., Füller, T., Ph., R., Ajayan, P.M., 1997. The formation, anealing and self-compression of carbon onions under electron irradiation. Chem. Phys. Lett. 269, 349–355.

Baowan, D., Thamwattana, N., Hill, J.M., 2007. Encapsulation of C_{60} fullerenes into single-walled nanotubes: Fundamental mechanical principles and conventional applied mathematical modeling. Phys. Rev. B 76, 155411.

Baowan, D., Thamwattana, N., Hill, J.M., 2008. Suction energy and offset configuration for double-walled carbon nanotubes. Commun. Nonlinear Sci. Numer. Simul. 13, 1431–1447.

Baughman, R.H., Cui, C., Zakhidov, A.A., Iqbal, Z., Barisci, J.N., Spinks, G.M., Wallace, G.G., Mazzoldi, A., De Rossi, D., Rinzler, A.G., Jaschinski, O., Roth, A., Kertesz, M., 1999. Carbon nanotube actuators. Science 284, 1340–1344.

Baughman, R.H., Zakhidov, A.A., de Heer, W.A., 2002. Carbon nanotubes–the route toward applications. Science 297, 787–792.

Bianco, A., Kostarelos, K., Prato, M., 2005. Applications of carbon nanotubes in drug delivery. Curr. Opin. Chem. Biol. 9, 674–679.

Breton, J., Gonzalez-Platas, J., Girardet, C., 1994. Endohedral adsorption in graphitic nanotubules. J. Chem. Phys. 101, 3334–3340.

Burchnall, J.L., Chaundy, T.W., 1940. Expansions of Appell's double hypergeometric functions. Q. J. Math. os-11, 249–270.

Cabria, I., Mintmire, J.W., White, C.T., 2003. Metallic and semiconducting narrow carbon nanotubes. Phys. Rev. B 67, 121406.

Colavecchia, F.D., Gasaneo, G., Miraglia, J.E., 2001. Numerical evaluation of Appell's F_1 hypergeometric function. Comput. Phys. Commun. 138, 29–43.

Cox, B.J., Thamwattana, N., Hill, J.M., 2007a. Mechanics of atoms and fullerenes in single-walled carbon nanotubes. II. Oscillatory behaviour. Proc. R. Soc. A 463, 477–494.

Cox, B.J., Thamwattana, N., Hill, J.M., 2007b. Mechanics of fullerenes oscillating in carbon nanotube bundles. J. Phys. A 40, 13197–13208.

Cumings, J., Zettl, A., 2000. Low-friction nanoscale linear bearing realized from multi-walled carbon nanotubes. Science 289, 602–604.

De Crescenzi, M., Castrucci, P., Scarselli, M., Diociaiuti, M., Chaudhari, P.S., Balasubramanian, C., Bhave, T.M., Bhoraskar, S.V., 2005. Experimental imaging of silicon nanotubes. Appl. Phys. Lett. 86, 231901.

Dresselhaus, M., Dresselhaus, G., Saito, R., 1995. Physics of carbon nanotubes. Carbon 33, 883–891.

Dunlap, B.I., 1992. Connecting carbon tubules. Phys. Rev. B 46, 1933–1936.

Dunlap, B.I., 1994. Relating carbon tubules. Phys. Rev. B 49, 5643–5651.

Dunlap, B.I., Zope, R.R., 2006. Efficient quantum-chemical geometry optimization and the structure of large icosahedral fullerenes. Chem. Phys. Lett. 422, 451–454.

Ferrari, M., 2005. Cancer nanotechnology: opportunities and challenges. Nat. Rev. Cancer 5, 161–171.

Fonseca, A., Hernadi, K., Nagy, J.b., Lambin, P., Lucas, A.A., 1995. Model structure of perfectly graphitizable coiled carbon nanotubes. Carbon 33, 1759–1775.

Ge, M., Sattler, K., 1994. Observation of fullerene cones. Chem. Phys. Lett. 220, 192–196.

Girifalco, L.A., 1992. Molecular properties of C_{60} in the gas and solid phases. J. Phys. Chem. 96, 858–861.

Girifalco, L.A., Hodak, M., Lee, R.S., 2000. Carbon nanotubes, buckyballs, ropes, and a universal graphitic potential. Phys. Rev. B 62, 13104–13110.

Hilder, T.A., Hill, J.M., 2007. Continuous versus discrete for interacting carbon nanostructures. J. Phys. A 40, 3851–3868.

Hillebrenner, H., Buyukserin, F., Stewart, J.D., Martin, C.R., 2006. Template synthesized nanotubes for biomedical delivery applications. Nanomed 1, 39–50.

Hodak, M., Girifalco, L.A., 2001. Fullerenes inside carbon nanotubes and multi-walled carbon nanotubes: optimum and maximum sizes. Chem. Phys. Lett. 350, 405–411.

Hodak, M., Girifalco, L.A., 2003. Ordered phases of fullerene molecules formed inside carbon nanotubes. Phys. Rev. B 67, 075419.

Iijima, S., 1991. Helical microtubules of graphitic carbon. Nature 354, 56–58.

Ishigami, M., Aloni, S., Zettl, A., 2003. Properties of boron nitride nanotubes. In: AIP Conference Proceedings for 12th International Conference STM'03, vol. 696, pp. 94–99.

Jiang, H., Zhang, P., Liu, B., Huang, Y., Geubelle, P.H., Gao, H., Hwang, K.C., 2003. The effect of nanotube radius on the constitutive model for carbon nanotubes. Comput. Mater. Sci. 28, 429–442.

Jorio, A., Fantinin, C., Pimenta, M.A., Capaz, R.B., Samsonidze, G.G., Dresselhaus, G., Dresselhaus, M.S., Jiang, J., Kobayashi, N., Grüneis, A., Saito, R., 2005. Resonance Raman spectroscopy (n, m)-dependent effects in small-diameter single-wall carbon nanotubes. Phys. Rev. B 075401.

Kang, J.W., Song, K.O., Hwang, H.J., Jiang, Q., 2006. Nanotube oscillator based on a short single-walled carbon nanotube bundle. Nanotechnology 17, 2250–2258.

Krishnan, A., Dujardin, E., Treacy, M.M.J., Hugdahl, J., Lynum, S., Ebbesen, T.W., 1997. Nature 388, 451–454.

Kroto, H.W., 1987. The stability of the fullerenes C_n ($n = 24$, 28, 32, 50, 60 and 70). Nature 329, 529–531.

Kroto, H.W., McKay, K., 1988. The formation of quasi-icosahedral spiral shell carbon particles. Nature 331, 328–331.

Lee, J.H., 2006. A study on a boron-nitride nanotube as a gigahertz oscillator. J. Korean Phys. Soc. 49 (1), 172–176.

Lennard-Jones, J.E., 1931. Cohesion. Proc. Phys. Soc. 43, 461–482.

Liu, P., W., Z.Y., Lu, C., 2005. Oscillatory behaviour of C60-nanotube oscillators: a molecular-dynamics study. J. Appl. Phys. 97, 094313.

Lu, J.P., Yang, W., 1994. The shape of large single- and multiple-shell fullerenes. Phys. Rev. B 49, 11421–11424.

Machón, M., Reich, S. Thomsen, C., Sánchez-Portal, D., Ordejón, P., 2002. Ab initio calculations of the optical properties of 4-å-diameter single walled nanotubes. Phys. Rev. B 155410.

Martin, C.R., Kohli, P., 2003. The emerging field of nanotube biotechnology. Nat. Rev. Drug Discov. 2, 29–37.

Miyamoto, Y., Rubio, A., Louie, S.G., Cohen, M.L., 1994. Electronic properties of tubule forms of hexagonal BC_3. Phys. Rev. B 50, 18360–18366.

Mu, C., Zhao, Q. Xu, D., Zhuang, Q., Shao, Y., 2007. Silicon nanotube array/gold electrode for direct electrochemistry of cytochrome c. J. Phys. Chem. B 111, 1491–1495.

Nakao, K., Kurita, N., Fujita, M., 1994. Ab initio molecular-orbital calculation for C_{70} and seven isomers of C_{80}. Phys. Rev. B 49, 11415–11420.

Nasibulin, A.G., Pikhitsa, P.V., Jiang, H., Brown, D.P., Krasheninnikov, A.V., Anisimov, A.S., Queipo, P., Moisala, A., Gonzalez, D., Lientschnig, G., Hassanien, A., Shandakov, S.D., Lolli, G., Resasco, D.E., Choi, M., Tománek, D., Kauppinen, E.I., 2007. A novel hybrid carbon material. Nat. Nanotechnol. 2, 156–161.

Okada, S., Saito, S., Oshiyama, A., 2001. Energetics and electronic structures of encapsulated C_{60} in a carbon nanotube. Phys. Rev. Lett. 86, 3835–3838.

Popov, V.N., 2004. Curvature effects on the structural, electronic and optical properties of isolated single-walled carbon nanotubes within a symmetry-adapted non-orthogonal tight-binding model. New J. Phys. 6, 17.

Qian, D., Liu, W.K., Ruoff, R.S., 2001. Mechanics of C_{60} in nanotubes. J. Phys. Chem. B 105, 10753–10758.

Qian, D., Liu, W.K., Subramoney, S., Ruoff, R.S., 2003. Effect of the interlayer potential on mechanical deformation of multiwalled carbon nanotubes. J. Nanosci. Nanotechnol. 3, 185–191.

Rivera, J.L., McCabe, C., Cumming, P.T., 2005. The oscillatory damped behavior of incommensurate double-walled carbon nanotubes. Nanotechnology 16, 186–198.

Rubio, A., Corkill, J.L., Cohen, M.L., 1994. Theory of graphitic boron nitride nanotubes. Phys. Rev. B 49, 5081–5084.

Sha, J., Niu, J., Ma, X., Zhang, X., Yang, Q., Yang, D., 2002. Silicon nanotubes. Adv. Mater. 14, 1219–1221.

Shenderova, O.A., Zhirnov, V.V., Brenner, D.W., 2002. Carbon nanostructures. Crit. Rev. Solid State Mater. Sci. 27, 227–356.

Smith, B.W., Monthioux, M., Luzzi, D.E., 1998. Encapsulated C_{60} in carbon nanotubes. Nature 396, 323–324.

Thamwattana, N., Hill, J.M., 2008. Oscillation of nested fullerenes (carbon onions) in carbon nanotubes. J. Nanopart. Res. 10, 665–677.

Troche, K.S., Coluci, V.R., Braga, S.F., Chinellato, D.D., Sato, F., Legoas, S.B., Rurali, R., Galv ao, D.S., 2005. Prediction of ordered phases of encapsulated C_{60}, C_{70}, and C_{78} inside carbon nanotubes. Nano Lett. 5, 349–355.

Ueno, H., Osawa, S., Osawa, E., Takeuchi, K., 1998. Stone–Wales rearrangement pathways from the hinge-opened [2+2] C_{60} dimer to Ipr C_{120} fullerenes. Vibrational analysis of intermediates. Fuller. Sci. Technol. 6, 319–338.

Ugarte, D., 1992. Curling and closure of graphitic networks under electron-beam irradiation. Nature 359, 707–709.

Wang, N., Tang, Z.K., Li, G.D., Chen, J.S., 2000. Single-walled 4 Å carbon nanotube arrays. Nature 408, 50–51.

Weng-Sieh, Z., Cherrey, K., Chopra, N.G., Blase, X., Miyamoto, Y., Rubio, A., Cohen, M.L., Louie, S.G., Zettl, A., Gronsky, R., 1995. Synthesis of $B_xC_yN_z$ nanotubules. Phys. Rev. B 51, 11229–11232.

Yao, Z., Postma, H.W.C., Balents, L., Dekker, C., 1999. Carbon nanotube intarmolecular junctions. Nature 402, 273–276.

Zhao, Y., Lin, Y., Yakobson, B.I., 2003. Fullerene shape transformations via Stone-Wales bond rotations. Phys. Rev. B 68, 233403.

Zheng, Q., Liu, J.Z., Jiang, Q., 2002. Excess van der Waals interaction energy of a multiwalled carbon nanotube with an extruded core and the induced core oscillation. Phys. Rev. B 65, 245409.

Zhi, C., Bando, Y., Tang, C., Golberg, D., 2005. Immobilization of proteins on boron nitride nanotubes. J. Am. Chem. Soc. 127, 17144–17145.

Index

Note: Page numbers followed by *b* indicate boxes, *f* indicate figures and *t* indicate tables.

369